RECENT PROGRESS IN
FOURIER ANALYSIS

NORTH-HOLLAND MATHEMATICS STUDIES 111
Notas de Matemática (101)

Editor: Leopoldo Nachbin

Centro Brasileiro de Pesquisas Físicas
Rio de Janeiro
and University of Rochester

NORTH-HOLLAND – AMSTERDAM • NEW YORK • OXFORD

RECENT PROGRESS IN FOURIER ANALYSIS

Proceedings of the Seminar on Fourier Analysis held in El Escorial, Spain, June 30 - July 5, 1983

Edited by

I. PERAL and J.-L. RUBIO de FRANCIA

Universidad Autónoma de Madrid
Madrid
Spain

1985

NORTH-HOLLAND – AMSTERDAM • NEW YORK • OXFORD

ISBN: 0 444 87745 2

Publishers:

ELSEVIER SCIENCE PUBLISHERS B.V.
P.O. Box 1991
1000 BZ Amsterdam
The Netherlands

Sole distributors for the U.S.A. and Canada:

ELSEVIER SCIENCE PUBLISHING COMPANY, INC.
52 Vanderbilt Avenue
New York, N.Y. 10017
U.S.A.

Library of Congress Cataloging in Publication Data

Seminar on Fourier Analysis (1983 : Escorial)
Recent progress in Fourier analysis.

(North-Holland mathematics studies ; 111) (Notas de matemática ; 101)
English or French.
1. Fourier analysis--Congresses. I. Peral, Ireneo. II. Rubio de Francia, J.-L., 1949- . III. Title. IV. Series. V. Series: Notas de matemática (Rio de Janeiro, Brazil) ; no. 101.
QA1.N86 no. 101a 510 s [515'.2433] 85-4531
[QA403.5]
ISBN 0-444-87745-2 (Elsevier Science Pub.)

PRINTED IN THE NETHERLANDS

RECENT PROGRESS IN FOURIER ANALYSIS

The following contributions were presented at the Seminar on Fourier Analysis which was held in El Escorial from 30 June to 5 July 1983. This meeting was sponsored by the Asociación Matemática Española with financial support from the Comisión Asesora de Investigación Científica y Técnica (project 4192).

A decisive factor with respect to the organization was the financial help, together with the facilities, provided by the Vicerrectorado de Investigación of the Universidad Autónoma de Madrid.

The articles we present give a good idea of how work in the area has evolved and of the scientific character of the meeting. The friendly and cordial atmosphere meant that the organization, far from being a chore, became a pleasurable experience. For this we owe our sincerest thanks to all participants.

Special thanks must also go to the invited speakers for their magnificent collaboration, and to Caroline, without whose presence we hate to think what could have happened!

We should also like to express our gratitude to our colleagues in the División de Matemáticas in the Universidad Autónoma de Madrid, for their help in correcting proofs, and to Soledad, for typing the manuscript.

The Editors

CONTENTS

Recent Progress in Fourier Analysis
I. Peral and J.-L. Rubio de Francia (Editors)

FUNCTIONS OF L^p-BOUNDED PSEUDO-DIFFERENTIAL OPERATORS

Josefina Alvarez Alonso
Universidad de Buenos Aires

The aim of this paper is to construct a functional calculus over an algebra of L^p-bounded pseudo-differential operators acting on functions defined on a compact manifold without boundary.

The operators we consider here depend on amplitudes or symbols with a finite number of derivatives, without any hypothesis of homogeneity. The manifolds where the operators act are also of class C^M for a suitable M. In this way is it possible to control the number of derivatives of f that we need in order to give meaning to $f(A)$, when A is a self-adjoint operator in that algebra.

Indeed, this program was carried out in [1] and [2] when $p = 2$. In [1] an algebra of pseudo-differential operators acting on functions defined in $\mathbb{R}^n$ is constructed. The main tool to do that is the sharp L^2 estimates obtained by R. Coifman and Y. Meyer in [3]. Then, functions of those operators are defined by means of the H. Weyl formula (see [4], for example). Since it seems not to be possible to obtain directly a polynomial estimate for the exponential $\exp(-2\pi itA)$ in terms of t, a roundabout argument is employed by introducing an adapted version of the characteristic operators defined by A. P. Calderón in [5].

All this machinery is extended in [2] to non-infinitely differentiable compact manifolds without boundary.

In order to get the L^p version of these results the first thing to do is to obtain the analogous of the algebra constructed in [1]. The main point is to observe that amplitudes in a subclass of $S^o_{1,1}$ give rise to operators on which the classical theory of Calderón and Zygmund works (see [6]). Unfortunately as far as I know, it is an open question to get in the euclidean case a non trivial estimate for the exponential $\exp(-2\pi itA)$. However, when the operators act on functions defined in compact manifolds, a suitable estimate can be obtained and so, a non-infinitely differentiable functional calculus runs.

Given $0 \leq \delta < 1$, $k = 1,2,\ldots,$ let

$$N = \begin{cases} k/1-\delta & \text{if this is an integer} \\ [k/1-\delta] + 1 & \text{if not} \end{cases}$$

We will consider operators K acting on S in the following way

$$Kf = \sum_{j=0}^{N-1} \int e^{-2\pi i x\xi} \, p_j(x,\xi)\hat{f}(\xi)d\xi + Rf$$

where

i) The function p_j belongs to the class S^j; that is to say, p_j is a continuous function defined on $\mathbb{R}^n \times \mathbb{R}^n$; it has continuous derivatives in the variable ξ up to the order $n+N+2-j$ and each function $D_\xi^\gamma \, p_j$ has continuous derivatives in x,ξ up to the order $2[n/2]+N+k+2-j$, satisfying

$$\sup_{\substack{x,\xi \in \mathbb{R}^n \\ \alpha,\, \beta,\, \gamma}} \frac{|D_s^\alpha D_\xi^\beta D_\xi^\gamma \, p_j(x,\xi)|}{(1+|\xi|)^{-j(1-\delta)+|\alpha|\delta-|\beta+\gamma|}} < \infty$$

ii) For $1 < p_0 \leq 2$ fixed, R is a linear and continuous operator from L^p into itself for $p_0 \leq p \leq p_0'$. Moreover, R and the adjoint R^* are continuous from L^p into L_k^p, where L_k^p denotes the Sobolev space of order k and p_0' is the conjugate exponent of p_0.

Let M_k be the class of the operators K.

Now, let X be a differentiable compact manifold of dimension n and class C^M, $M = 2[n/2]+n+2N+k+5$, without boundary; X has a measure μ which in terms of any local coordinate system $x = (x_1,\ldots,x_n)$ can be express as $G(x)dx_1 \ldots dx_n$, where $G > 0$ is a function of class C^{M-1}.

We will introduce the following notation.

Let U_1, U_2 be open bounded subsets of X or $\mathbb{R}^n$ let $\phi : U_1 \to U_2$ be a diffeomorphism of class C^M; if f is a function defined on the ambient space of U_2, $\phi^*(f)$ will denote the function defined on the ambient space of U_1 which coincides with $f\circ\phi$ on U_1 and vanishes outside U_1. On the other hand, if A is an operator acting on functions defined on the ambient space of U_2, by $\phi^*(A)$ we denote the operator acting on functions defined on the ambient space of U_1 as

$$\phi^*(A)(f) = \phi^*[A(\phi^{-1*}(f))]$$

Now, we are ready to define classes of operators on X.

Given $1 < p_0 \leq 2$, R belongs to $\mathcal{R}_k(X)$ if R is a linear continuous operator from $L^p(X)$ into itself for $p_0 \leq p \leq p_0'$ and R, R^* map continuously $L^p(X)$ into $L^p_k(X)$ for $p_0 \leq p \leq p_0'$.

$\mathcal{R}_k(X)$ is a self-adjoint Banach algebra with the norm

$$|R|_{\mathcal{R}_k} = |R|_{L^{p_0},L^{p_o}_k} + |R|_{L^{p_0'},L^{p_o'}_k} + |R^*|_{L^{p_0},L^{p_o}_k} + |R^*|_{L^{p_0'},L^{p_o'}_k}$$

Now, given $1 < p_0 \leq 2$, $\mathcal{M}_k(X)$ is the class of linear continuous operators A from $L^p(X)$ into itself for $p_0 \leq p \leq p_0'$, which satisfy the following two conditions

i) Given $\phi_1,\phi_2 \in C_0^M(X)$ with disjoint supports, the operator $\phi_1 A \phi_2$ belongs to $\mathcal{R}_k(X)$. Here ϕ_1, ϕ_2 stand for the operators of multiplication by the function ϕ_1, ϕ_2, respectively.

ii) Let $U \subset X$ be an open subset and let $\phi : U \to U_1$ be a diffeomorphism of class C^M extendable to a neighborhood of $\overline{U}$, where $U_1 \subset \mathbb{R}^n$. There exists an operator $A_1 \in \mathcal{M}_k$ such that if $\phi_1,\phi_2 \in C_0^M(U)$,

$$\phi_1 A \phi_2 = \phi_1 \phi^*(A_1)\phi_2$$

$\mathcal{M}_k(X)$ is a self-adjoint algebra and $\mathcal{R}_k(X)$ is a two-sided ideal of $\mathcal{M}_k(X)$; moreover, operators in $\mathcal{M}_k(X)$ are continuous from $L^p_m(X)$ into itself for $p_0 \leq p \leq p_0'$, $0 \leq m \leq k$.

It is possible to endow $\mathcal{M}_k(X)$ with a complete norm. In order to avoid technical details, we will not precise the definition. With this norm $\mathcal{R}_k(X)$ is continuously included in $\mathcal{M}_k(X)$ and $\mathcal{M}_k(X)$ is continuously included in $L(L^p_m(X))$, the space of linear and continuous operators from $L^p_m(X)$ into itself, for $p_0 \leq p \leq p_0'$, $0 \leq m \leq k$.

THEOREM 1. *Let* p_0, k *and* n *be such that* $1/p_0 - k/n \leq 1/2$.

Given a self-adjoint operator $A \in M_k(X)$ *and a function* f *in the Sobolev space* L_s^2, *where* $s > 2\mu + 5/2$, $\mu = 2[n/2]+n+k+ + N(N+3)/2+4$, *the Bochner integral*

$$\int_{-\infty}^{\infty} e^{-\pi itA} \hat{f}(t)dt$$

belongs to $R_k(X)$ *and coincides with* $f(A)$ *calculated by means of the spectral formula in* $L(L^2(X))$.

Remarks:

a) It is possible to impose on f additional conditions under which the operator $f(A)$ belongs to $R_k(X)$.

b) When $p_0 = 2$ the above theorem remains true with $s > \mu + 3/2$.

c) The Weyl's formula also allows to define functions of a tuple of non-commuting self-adjoint operators.

We will include here the proof of the theorem 1 in a particular but significant case.

Suppose that $\delta = 0$, $k = 1$; it follows that $N = 1$. It is clear that theorem 1 can be deduced from a suitable estimate for $|\exp(-2\pi itA)|_{M_1(X)}$ in terms of $t \in \mathbb{R}$.

In order to get this estimate, some notations and results will be needed. We fix in X coordinate neighborhoods U_j, diffeomorphisms $\phi_j : U_j \to \phi_j(U_j)$ of class C^M, where $M = 2[n/2]+n+8$, functions $\theta_j \in C_0^M(U_j)$, $\theta_j \geq 0$ and a finite partition of unity $\{\eta_j\}$ of class C^M, such that $\text{supp}(\theta_j) \subset U_i$ whenever $\text{supp}(\theta_j) \cap \text{supp}(\theta_i) \neq \emptyset$; $\theta_i = 1$ in a neighborhood of $\text{supp}(\eta_j)$ if $j = i$ or if $\text{supp}(\eta_j) \cap \text{supp}(\eta_i) \neq \emptyset$.

Now, we define an space of symbols for operators in $M_1(X)$. More exactly, for each j we consider the restriction to $\phi_j(U_j)$ of a function $p^{(j)} \in S^0$. We define a norm of such a restriction as

$$|p^{(j)}| = \sup \frac{|D_x^\alpha D_\xi^\beta D_\xi^\gamma\, p^{(j)}(x,\xi)|}{(1 + |\xi|)^{-|\gamma+\beta|}}$$

where the supremum is taken over $x \in \phi_j(U_j)$, $\xi \in R^n$, $|\xi| \leq n+3$,

$|\alpha+\beta| \leq 2[n/2]+4$, j.

We note $\mathbb{N}_1(X)$ this space. With the pointwise multiplication

$$(p^{(j)}).(q^{(j)}) = (p^{(j)}q^{(j)})$$

as a product, $\mathbb{N}_1(X)$ becomes a commutative Banach algebra.

LEMMA. *Let* $H = (p^{(j)})$ *be an element in* $\mathbb{N}_1(X)$; *we suppose that each* $p^{(j)}$ *is a real function. Then, if* $t \in \mathbb{R}$,

$$\|\exp(-2\pi i t H)\| \leq C|(1 + \|H\|)(1 + |t|)|^{\mu}$$

where $C = C(X) > 0$, $\mu = 2[n/2]+n+7$.

Proof

Since $\mathbb{N}_1(X)$ is a Banach algebra, the exponential $\exp(-2\pi i t H)$ is well defined; moreover it is equal to $\exp(-2\pi i t p^{(j)})_j$. According to the norm that the space $\mathbb{N}_1(X)$ has, the conclusion follows.

Now, we will introduce the space $\mathbb{M}_1(X)$ in the following way An element K of $\mathbb{M}_1(X)$ is an operator R in $R_1(X)$ and a vector $(p^{(j)})$ in $\mathbb{N}_1(X)$ subject to the condition that if $U_i \cap U_j \neq \emptyset$ and $\phi_{ij} = \phi_j \circ \phi_i^{-1}$, then

$$(p^{(i)}) = \phi_{ij}^{*}(p^{(j)}) \quad \text{in} \quad \phi_i(U_i \cap U_j).$$

Such an element K will be denoted as $\{(p^{(j)}),R\}$.

We define a norm in $\mathbb{M}_1(X)$ as follows

$$\|K\|_{\mathbb{M}_1} = \|(p^{(j)})\| + \|R\|_{R_1}$$

Given $K \in \mathbb{M}_1(X)$ we define an operator $\Lambda(K)$ in the following way

$$\Lambda(K) = \sum_j \eta_j \phi_j^{*}(A_j)\theta_j + R$$

where

$$(A_j f)(x) = \begin{cases} \int e^{-2\pi i x\xi}\, p^{(j)}(x,\xi)\hat{f}(\xi)d\xi & \text{if} \quad x \in \phi_j(U_j) \\ 0 & \text{if not} \end{cases}$$

It can be proved that $\Lambda(K)$ belongs to $M_1(X)$. Moreover the linear map

$$\begin{array}{ccc} \mathbb{M}_1(X) & \xrightarrow{\Lambda} & M_1(X) \\ K & \longrightarrow & \Lambda(K) \end{array}$$

is into and continuous. Furthermore, if $A \in M_1(X)$ is self-adjoint, $A = \Lambda(K)$ for some $K = \{(p^{(j)}),R\}$, with $p^{(j)}$ real for all j.

It is possible to define a product in $\mathbb{M}_1(X)$ in such a way that $\mathbb{M}_1(X)$ becomes a Banach algebra and the map Λ above is a continuous homomorphism of algebras.

Finally, let us consider the maps

$$\begin{array}{ccccc} \mathbb{M}_1(X) & \xrightarrow{\Omega} & \mathbb{N}_1(X) & \xrightarrow{\Omega_1} & \mathbb{M}_1(X) \\ \{(p^{(j)}),R\} & \longrightarrow & (p^{(j)}) & \longrightarrow & \{(p^{(j)}),0\} \end{array}$$

Ω is a continuous homomorphism of algebras and the linear map Ω_1 is a right continuous inverse of Ω.

THEOREM 2. *Suppose that* $1/p_0 - 1/n \leq 1/2$.

Let $H = \{(p^{(j)}),R\}$ *be an element of* $\mathbb{M}_1(X)$ *such that* $\Lambda(H)$ *is a self-adjoint operator and the functions* $p^{(j)}$ *are real for all* j. *Then, if* $t \in \mathbb{R}$,

$$|\exp(-2\pi itH)|_{\mathbb{M}_1} \leq C[(1 + |H|_{\mathbb{M}_1})(1 + |t|)]^{2\mu+2}$$

where $C = C(X) > 0$, $\mu = 2[n/2]+n+7$.

Proof

According to the notations above, we set $A = \Lambda(H)$, $K = \Omega(H) = (p^{(j)}) \in \mathbb{N}_1(X)$.

We assert that

$$e^{-2\pi itH} - \Omega_1(e^{-2\pi itK})$$

it an element of the form $\{(0),R(t)\}$.

In fact, since Ω is a continuous homomorphism of algebras and Ω_1 is a right inverse of Ω, we have

$$\Omega[e^{-2\pi itH} - \Omega_1(e^{-2\pi itK})] = e^{-2\pi itK} - \Omega\Omega_1(e^{-2\pi itK}) = 0.$$

On the other hand, since Ω_1 is a continuous map, according to the lemma it suffices to estimate the norm of $\{(0),R(t)\}$ in $IM_1(X)$, which coincides with the norm of $R(t)$ in $\mathcal{R}_1(X)$.

We have

$$R(t) = \Lambda\left[e^{-2\pi itH} - \Omega_1(e^{-2\pi itK})\right] =$$

$$= e^{-2\pi itA} - \Lambda\Omega_1(e^{-2\pi itK}).$$

If we denote with $\dot{R}(t)$ the derivative of $R(t)$ with respect to t, we get

$$\dot{R}(t) = e^{-2\pi itA}(-2\pi iA) - \Lambda\Omega_1(e^{-2\pi itK}(-2\pi iK)) =$$

$$= \left[e^{-2\pi itA} - \Lambda\Omega_1(e^{-2\pi itK})\right] (-2\pi iA) + \Lambda\Omega_1(e^{-2\pi itK})(-2\pi iA) -$$

$$- \Lambda\Omega_1(e^{-2\pi itK}(-2\pi iK)) = R(t)(-2\pi iA) + B_1(t) \tag{1}$$

Since

$$B_1(t) = \Lambda\left[\Omega_1(e^{-2\pi itK})(-2\pi iH) - \Omega_1(e^{-2\pi itK}(-2\pi iK))\right]$$

and

$$\Omega\left[\Omega_1(e^{-2\pi itK})(-2\pi iH) - \Omega_1(e^{-2\pi itK}(-2\pi iK))\right] = 0,$$

we deduce that $B_1(t)$ belongs to $\mathcal{R}_1(X)$ for each t.

Thus,

$$|B_1(t)|_{\mathcal{R}_1} = |\Omega_1(e^{-2\pi itK})(-2\pi iH) - \Omega_1(e^{-2\pi itK}(-2\pi iK))|_{IM_1}$$

$$\leq C[(1 + |H|_{IM_1})(1 + |t|)]^{\mu}$$

where $C = C(X) > 0$.

Since $R(0) = 0$, from (1) it follows that

$$R(t) = \int_0^t B_1(s)e^{-2\pi i(t-s)A}\, ds \tag{2}$$

But we can also write

$$\dot{R}(t) = (-2\pi iA)e^{-2\pi itA} - \Lambda\Omega_1(e^{-2\pi itK}(-2\pi iK))$$

or

$$\dot{R}(t) = (-2\pi iA)R(t) + B_2(t)$$

where $B_2(t) \in R_1(X)$ for each t and

$$|B_2(t)|_1 \leq C[(1 + |H|_{M_1})(1 + |t|)]^{\mu}, \quad c = C(X) > 0$$

So,

$$R(t) = \int_0^t e^{-2\pi i(t-s)A} B_2(s)ds \tag{3}$$

or

$$R^*(t) = \int_0^t B_2^*(s)e^{2\pi i(t-s)A} ds \tag{4}$$

where $*$ denotes the adjoint.

Now, suppose we show that

$$|e^{-2\pi i(t-s)A}|_{L^{p_0},L^p} \leq C[(1 + |H|_{M_1})(1 + |t-s|)]^{\mu+1} \tag{5}$$

We will get the same estimate for

$$|e^{-2\pi i(t-s)A}|_{L^{p'_0},L^{p'_0}}$$

Thus, according to (2) and (4), we can deduce that

$$|R(t)|_{R_1} \leq C[(1 + |H|_{M_1})(1 + |t|)]^{2\mu+2}$$

So, it remains to prove (5).

From the definition of the operator $R(t)$, it is clear that it suffices to obtain the estimate

$$|R(t)|_{L^{p_0},L^{p_0}} \leq C[(1 + |H|_{M_1})(1 + |t|)]^{\mu+1}$$

But according to the hypothesis $1/p_0 - 1/n \leq 1/2$, the Sobolev immersion theorem provides the continuous inclusion of $L_1^{p_0}(X)$ into $L^2(X)$; moreover, since $p_0 \leq 2$, we also have a continuous injection from $L^2(X)$ into $L^{p_0}(X)$. Thus, the desired estimate follows from (3).

This completes the proof of the theorem 2.

References

[1] J. Alvarez Alonso, A.P. Calderón: "Functional calculi for pseudo-differential operators, I". Proceedings of the Seminar on Fourier Analysis held in El Escorial, (1979), pp. 1-61.

[2] ________________, "Functional calculi for pseudo-differential operators, II". Proceedings of the MIT Congress in honour of I. Segal, (1979). Studies in Appl. Math., vol 8, (1983), pp. 27-72.

[3] R. Coifman, Y. Meyer: "Au delà des opératéurs pseudo-différentiels". Asterisque n° 57, (1978).

[4] M.E. Taylor: "Functions of several self-adjoint operators". Proc. Amer. Math. Soc. 19, (1968), 91-98.

[5] A.P. Calderon: "Algebras of singular integral operators", Proc. of Symp. in Pure Math., 10, (1965), 18-55.

[6] J. Alvarez Alonso: "An algebra of L^p-bounded pseudo-differential operators". Journal of Math. Analysis and Appl. 94, (1983), 268-282.

Recent Progress in Fourier Analysis
I. Peral and J.-L. Rubio de Francia (Editors)

ON PROBLEMS RELATED TO THEOREMS A and B WITH ESTIMATES

E. AMAR
Université de Bordeaux

Introduction

We are interested in theorems of type Cartan A and B, but with estimates on the growth of the functions.

Type A problems

a) The caracterisation of the zero set of a holomorphic function in a given class. G. Henkin and H. Skoda gave, independently, a complete answer for this problem in the case of the Nevanlinna class of the unit ball $\mathbf{B}$ of $\mathbb{C}^n$ [12], [18]; N. Varopoulos studied the case of the Hardy classes H^p of the unit ball [19].

All of them used the P. Lelong's method leading to solve a $\partial\bar\partial$ equation with estimates.

b) Another example of type A problem is the corona problem: let $f_1,\dots,f_k$ be bounded holomorphic functions in the unit ball of $\mathbb{C}^n$ such that:

$$|f_1| +\dots+ |f_k| \geq \delta > 0 \quad \text{in} \quad \mathbf{B};$$

are there $g_1,\dots,g_k$ bounded holomorphic functions in $\mathbf{B}$ with: $f_1g_1 +\dots+ f_kg_k = 1$?

This problem was solved in 1962 by L. Carleson [9] in the case $n = 1$; L. Hormander [14] showed that the problem can be replaced by the problem of solving a $\bar\partial$ equation with bounded solution, and in 1979 T. Wolff gave a very simple proof of the Corona in one variable. We will show in §1, in generalizing the proof of Wolff to the unit ball of $\mathbb{C}^n$, that in fact it is a $\partial\bar\partial$ equation we are led to solve. Unfortunately we have not the complete answer but the following:

Theorem [1]. *Let* $f = (f_1,\dots,f_k)$, k *functions in the Hardy class*

$H^p(B)$ $\ell < p < +\infty$, *with* $\ell = \inf(k,n+1)$, B *the unit ball of* $\mathbb{C}^n$, *such that*: $|f|^2 = \sum_{i=1}^{k} |f_i|^2 \geq \delta > 0$ *in* B. *Then there are* k *functions in* $H^q(B)$ *where* $q = \frac{p}{\ell}$ *such that*: $\Sigma f_i g_i = 1$.

Type B problems

Let Ω be a convex domain in $\mathbb{C}^n$, bounded with smooth boundary and let:

$$X = \{x \in \Omega \ / \ u(z) = 0\}$$

where u is holomorphic in Ω.

The question is: if f is holomorphic on X, is there a holomorphic F in Ω such that:

$$F|_X = f \ ?$$

Without estimates on F the answer is yes by Cartan's theorem B.

a) In the case $\Omega = \mathbf{B}$, the unit ball of $\mathbb{C}^n$, we are able to obtain estimates with the following hypothesis on u : $u \in A^\infty(\overline{\Omega})$ i.e. u is smooth up to the boundary and: $\partial u \neq 0$ on $X \cap \partial\Omega$ i.e. X is a manifold near $X \cap \partial\Omega$. Then we obtain:

Theorem [2]. *Let* f *be holomorphic and bounded on* X; *then there is a bounded holomorphic* F *in* $\mathbf{B}$ *such that:*

$$F|_X = f.$$

In §2 we show how this problem leads to the equation:

$$\overline{\partial} H = f \, \overline{\partial} \left[\frac{1}{u}\right].$$

b) When Ω is not strictly convex, we cannot apply this method because we have no longer Carleson measures nor "good" solving kernels. In this case we use a very simple proposition on the Hilbert transform and the $\overline{\partial}$-cohomology with C^∞ data [4] to solve the extension problem [3]:

Theorem. *Let* f *be holomorphic and bounded on* X, X *being transverse to* $\partial\Omega$ *and of dimension one, then there is a bounded holomorphic* F *in* Ω *such that:*

$$F|_X = f.$$

In §3 we will see why the fact of X being of complex dimension one is related to the Hilbert transform.

§1. Generalization of the proof of T. Wolff

Let: $\mathbb{B} = \{z = (z_1,\ldots,z_n) \in \mathbb{C}^n \,/\, |z|^2 = |z_1|^2 + \ldots + |z_n|^2 < 1\}$ be the unit ball of $\mathbb{C}^n$, and, for simplicity, let $k = 2$ in the data of the Corona problem:

(1.1) $\quad f_1, f_2 \in H^\infty(\mathbb{B}), \quad |f_1| + |f_2| \geq \delta > 0$ in $\mathbb{B}$.

Let:

(1.2) $\quad \tilde{g}_i = \dfrac{\bar{f}_i}{|f|^2}, \quad i = 1,2$ and $|f|^2 = |f_1|^2 + |f_2|^2$

then, of course:

(1.3) $\quad \tilde{g}_i \in C^\infty(\mathbb{B})$ and $f_1\,\tilde{g}_1 + f_2\,\tilde{g}_2 = 1$

Now suppose we find a h in $C^\infty(\mathbb{B})$ s.t.:

(1.4) $\quad \bar{\partial} h = \omega \quad$ with $\quad \omega = \dfrac{\bar{f}_2\, \overline{\partial f_1} - \bar{f}_1\, \overline{\partial f_2}}{|f|^4}$

where:

$$\bar{\partial} f_i = \frac{\partial f_i}{\partial \bar{z}_i}\, d\bar{z}_1 + \ldots + \frac{\partial f_i}{\partial \bar{z}_n}\, d\bar{z}_n$$

(ω is easily seen to be C^∞ in $\mathbb{B}$ and $\bar{\partial}$ closed), then with:

(1.5) $\quad g_1 = \tilde{g}_1 - f_2 h$ and $g_2 = \tilde{g}_2 + f_1 h.$

We get:

(1.6) $\quad g_1\, f_1 + g_2\, f_2 = 1$ and $\bar{\delta} g_i = 0, \quad i = 1,2,$

by straightforward computations, as was shown by L. Hörmander [14].

So the point is to solve (1.4) with good estimates on h.

If $a \in C^\infty(\mathbb{B})$, then:

$$\partial a = \sum_i \frac{\partial a}{\partial z_i}\, dz_i \in C^\infty_{(1,0)}(\mathbb{B})$$

$$\bar{\partial} a = \sum_i \frac{\partial a}{\partial \bar{z}_i}\, d\bar{z}_i \in C^\infty_{(0,1)}(\mathbb{B})$$

and: $d = \partial + \bar{\partial}$; $\partial^2 = \bar{\partial}^2 = d^2 = 0$.

Let us now start with:

(1.4) $\quad \bar{\partial} h = \omega$

where:

$$\omega \in C^\infty_{(0,1)}(\mathbf{B}) \quad \text{and} \quad \bar\partial\omega = 0$$

$C^\infty_{(p,q)}(\mathbf{B})$ meaning, as usual, (p,q) forms with $C^\infty(\mathbf{B})$ coefficients.

Now let:

(1.7) $\gamma = \partial\omega \in C^\infty_{(1,1)}(\mathbf{B}) \Rightarrow d\gamma = 0$

So this implies:

(1.8) $\exists S \in C^\infty(\mathbf{B})$ s.t.: $\partial\bar\partial S = \gamma$

Let us calculate:

(1.9) $d(\bar\partial S - \omega) = \partial\bar\partial S - \partial\omega = 0$

So there is $v \in C^\infty(\mathbf{B})$ s.t.:

(1.10) $dv = \bar\partial S - \omega$

but then we get, from

$$dv = \partial v + \bar\partial v :$$

(1.11) $\partial v = \bar\partial(S-v) - \omega \Rightarrow \partial v = 0$ and $\bar\partial(S-v) = \omega$

by degree consideration. So:

(1.12) $v = \bar F$ with F holomorphic in $\mathbf{B}$ and

(1.13) $h = S - \bar F$ satisfies $\bar\partial h = \omega$

The reason why we have to complicate things so much will appear now; but we need first some definitions

(1.14) $r > 0$, $\zeta \in \partial\mathbf{B}$, let $B(\zeta,r) = \{z \in \mathbf{B} / |1 - z.\bar\zeta| < r\}$

$$b(\zeta,r) = \{z \in \partial\mathbf{B} / |1 - z.\bar\zeta| < r\}$$

be the pseudo-ball of Koranyi, where:

$$z.\bar\zeta = \sum_{i=1}^{n} z_i.\bar\zeta_i \ .$$

Let $V^1(\mathbf{B})$ the space of Carleson measures on $\mathbf{B}$ i.e.:

(1.15) $\mu \in V^1(\mathbf{B}) \Leftrightarrow \exists C > 0$ s.t. $\forall\zeta \in \partial\mathbf{B}$, $\forall r > 0$

$$|\mu|(B(\zeta,r)) \le C|b(\zeta,r)|$$

where $|b|$ is the Lebesgue measure of b on $\partial\mathbf{B}$.

Let $V^o(\mathbf{B})$ the space of bounded measures on $\mathbf{B}$ and, for $\alpha \in [0,1]$ and $p \in [1,\infty]$ let us put:

(1.16) $\quad W^{\alpha,p}(\mathbf{B}) = [V^o(\mathbf{B}), V^1(\mathbf{B})]_{\alpha,p}$.

These interpolating spaces were introduced and characterized in a joint work with A. Bonami [6]; it seems that they are intimately connected with the tent spaces of R. Coifman, Y. Meyer and E. Stein.

Now with $\rho = |z|^2-1$ and $W^\alpha = W^{\alpha,p}$ if $p = \frac{1}{1-\alpha}$, let us define:

(1.17) $\quad W^\alpha_{(0,1)}(\mathbf{B}) = \{\omega \in C^\infty_{(0,1)}(\mathbf{B})/$the components of ω and $\dfrac{\omega \wedge \bar\partial \rho}{\sqrt{-\rho}}$ are in $W^\alpha(\mathbf{B})\}$

(1.18) $\quad W^\alpha_{(1,1)}(\mathbf{B}) = \{\gamma \in C^\infty_{(1,1)}(\mathbf{B})/$the components of $\rho\gamma$, $\sqrt{-\rho}\ \gamma\wedge\partial\rho$, $\sqrt{-\rho}\ \gamma\wedge\bar\partial\rho$, $\gamma\wedge\partial\rho\wedge\bar\partial\rho$ are in $W^\alpha\}$

Now we have the estimates on the solutions (1.8) and (1.12):

Theorem [1]. 1) *If* $\gamma = \partial\omega \in W^\alpha_{(1,1)}(\mathbf{B})$, $0 \le \alpha < 1$, *then there is a* S *in* $C^\infty(\mathbf{B}) \cap L^p(\partial\mathbf{B})$ *s.t.* $\partial\bar\partial S = \gamma$.

2) *Moreover, if* $-\rho|\omega|^2 \in W^\alpha(\mathbf{B})$ *then* $F \in H^p(\mathbf{B})$ *(the Hardy class of* $\mathbf{B}$*) and we get a* $L^p(\partial\mathbf{B})$ *solution* h *to* $\bar\partial h = \omega$, *with* $\alpha = 1 - \frac{1}{p}$.

If $\alpha = 1$, we get a B.M.O.$(\partial\mathbf{B})$ solution h.

Sketch of the proof [1]:

(1) H. Skoda [18] proved (1) for $\alpha = 0$ and N. Varopoulos [19] for $\alpha = 1$ using a linear operator so we get (1) by interpolation.

(2) for $\alpha < 1$ we use Stein's estimates on the function g and for $\alpha = 1$ we use the factorization theorem of H^1 functions in $\mathbf{B}$ proved by Coifman, Rochberg and Weiss [7] exactly as T. Wolff did for $n = 1$.

Now the reason why we use such a way is clear: if f_i are in $H^\infty(\mathbf{B})$, ω is not in $W^1_{(0,1)}(\mathbf{B})$ but $\gamma = \partial\omega$ and $-\rho|\omega|^2$ are in $W^1_{(1,1)}(\mathbf{B})$ and in $V^1(\mathbf{B})$ respectively. In that way we get a B.M.O. $(\partial\mathbf{B})$ solution and if $n = 1$ this implies that there is a $L^\infty(\partial\mathbf{D})$ solution [5] and gives the Wolff's proof of the Corona theorem.

§2. A lemma on extension of type B

Let Ω be a pseudo convex set in $\mathbb{C}^n$, bounded with smooth boundary and let $H(\Omega)$ the set of holomorphic functions in Ω.

Let:

(2.1) $X = \{z \in \Omega / u(z) = 0\}$ where $u \in H(\Omega)$.

If f is holomorphic on X i.e.: there is a neighbourhood V of X in (the open set) Ω such that f extends holomorphically on V, then does it exist a holomorphic function F on Ω s.t. $F|_X = f$?

The answer is yes by the well known theorem B of Cartan but now we want estimates. The following lemma will give an answer.

Lemma [2]. *Let* H *be a distribution in* Ω *such that:* $\bar\partial H = f\bar\partial\left[\frac{1}{u}\right]$, *in the distribution sense, and let* $S = uH$, *then:*

α) S *is holomorphic in* Ω

β) $S|_X = f$.

Proof. Let $\left[\frac{1}{u}\right]$ be any distribution in Ω s.t. $u\left[\frac{1}{u}\right] = 1$; we known by L. Schwartz [16] that such a distribution exists; because f is C^∞ in a neighbourhood of X, then $f\bar\partial\left[\frac{1}{u}\right]$ is a well defined distribution in Ω clearly supported by X. So the equation:

(2.2) $\bar\partial H = f\bar\partial\left[\frac{1}{u}\right]$

is well posed.

Now let χ be a C^∞ function with support in V and such that:

(2.3) $\chi \equiv 1$ on a neighbourhood of X.

since f is holomorphic in V, we have

(2.4) $f\chi$ is C^∞ in Ω

and:

(2.5) $\omega = f\dfrac{\bar\partial\chi}{u}$

is a $C^\infty(\Omega)$, (0,1) form closed in Ω because $\bar\partial\chi \equiv 0$ in a neighbourhood of the zero set of u.

Now let G be any solution of:

(2.6) $\quad \bar{\partial} G = \omega$

then G is C^∞ in Ω because of the hypoellipticity of $\bar{\partial}$ on $(0,1)$ forms.

Moreover we have:

(2.7) $\quad S = f\chi - uG$

is holomorphic in Ω and $S|_X = f$, because $uG = 0$ on X, G being C^∞ in Ω, and $\chi = 1$ on X and $\bar{\partial} S = 0$ in Ω.

Let H be a distribution satisfying (2.2), then:

(2.8) $\quad G = f\chi \, |\frac{1}{u}| - H$

is a distribution in Ω such that:

(2.9) $\quad \bar{\partial} G = f\bar{\partial}\chi \left[\frac{1}{u}\right] + f\chi\bar{\partial}\left[\frac{1}{u}\right] - \bar{\partial} H = f\,\frac{\bar{\partial}\chi}{u} = \omega$

so, in fact, G is C^∞ in Ω and

(2.10) $\quad S = f\chi - uG = f\chi - u\,\{f\chi \left[\frac{1}{u}\right] - H\} = uH$

proves the lemma.

Let me give 2 applications of this lemma 2 ;

a) <u>$n = 1$</u>: $\Omega = \mathbf{D}$, the unit disk in $\mathbf{C}$, and

$$X = \{z_k,\ k \in \mathbf{N}\}$$

a Carleson sequence in $\mathbf{D}$ i.e.:

(2.11) $\quad \inf_k \prod_{m \neq k} \left|\frac{z_m - z_k}{1 - \bar{z}_k z_m}\right| > 0$

and $f = \{\lambda_k,\ k \in \mathbf{N}\} \in \ell^\infty(\mathbf{N})$, and $u = B$, the Blaschke product associated to X, then we get in a nice way the theorem of Carleson [8]:

<u>Theorem</u>. <u>*There is a*</u> F <u>*in*</u> $H^\infty(\mathbf{D})$ s.t. $F|_X = f$.

Because $f\bar{\partial}\left[\frac{1}{B}\right]$ is a Carleson measure in $\mathbf{D}$.

In the same way we get the Shapiro and Shields theorem concerning interpolating sequences in $H^p(\mathbf{D})$. [17].

b) $n > 1$, Ω is a strictly pseudo-convex domain (the unit ball for instance), $u \in A^\infty(\bar{\Omega}) \cap C^\infty(\bar{\Omega})$ such that: $\partial u \neq 0$ on $X \cap \partial\Omega$; then we get, using Skoda's kernel and Carleson measures again:

Theorem 2. *Let* f *be a bounded holomorphic function on* $X \cap \Omega$, *then there is a* F *holomorphic and bounded in* Ω *such that:* $F_{|X} = f$.

The extension here is linear. The case of H^p spaces is also studied [2] using Carleson measures of type α.

Remark. There is no hypothesis on the way the manifold X reaches the boundary.

This theorem was also proved by Henkin and Leiterer [13] by different methods.

§3. The non strictly pseudo-convex case

Now suppose Ω is a (lineally) convex set in $\mathbb{C}^n$, bounded with smooth boundary. Because Ω is not necessarily strictly pseudo-convex there are no longer "good" kernels nor Carleson measures. So we cannot solve (2.2) by the method of §2.

Now suppose Ω is defined by:

(3.1) $\Omega = \{z \in \mathbb{C}^n, \quad \rho(z) < 0\}$

where ρ is $C^\infty(\mathbb{C}^n)$, $\partial\rho \neq 0$ on $\partial\Omega$ and suppose that:

(3.2) $X = \{x \in \bar{\Omega} / u_1(z) = \ldots = u_{n-1}(z) = 0\}$

is a 1 dimensional complex variety s.t.:

(3.3) $\partial\rho \wedge \partial u_1 \wedge \ldots \wedge \partial u_{n-1} \neq 0$ on $X \cap \partial\Omega$

i.e. X is transverse to $\partial\Omega$ and is a manifold near $\partial\Omega$.

Then we still have [3]:

Theorem. *Let* f *be a bounded holomorphic function on* $X \cap \Omega$; *then there is a* F *holomorphic and bounded such that:* $F_{|X} = f$.

The way of proving this theorem is completely classical:

a) to find a bounded local extension

b) to patch together these extensions in a bounded way.

For finding a bounded local extension we use the Cauchy integral formula and this is the point where the dimension of X has to be one: we n eed a reproducing kernel for holomorphic functions holomorphic in both variables as Cauchy kernel is.

Generally the Cauchy kernel sends bounded fonctions to BMO ones, but here we still get a bounded function, because of the following simple theorem [3].

Theorem. *Let* $\{a_\lambda\}$ *be a family of* C^∞ *diffeomorphisms of a neighbourhood of a simply connected domain* D *of* $\mathbb{C}$ *into* $\mathbb{C}$, *depending* C^∞ *on the parameter* λ *in* $\mathbb{R}^n$. *Then we have: if* f *is in* $L^1(\partial D)$ *then:*

$$C_{\Gamma_\lambda}(f \circ a_\lambda^{-1}) \circ a_\lambda - C_\Gamma f \in C^\infty(\partial D \times \mathbb{R}^n).$$

Where C_Γ is the radial limit of the Cauchy integral of f in D, with $\Gamma = \partial D$, and $\Gamma_\lambda = a_\lambda(\Gamma)$. Because boundary values of C_Γ and the Hilbert transform are closely related, this theorem is a simple consequence of a similar proposition on the Hilbert transform.

Now in part (b) we are given a covering $\{\mathcal{U}_i\}$ of $\overline{\Omega}$ and a family of bounded extensions f_i in $\mathcal{U}_i \cap \Omega$. So we consider:

$$(3.4) \qquad g_{ij} = f_i - f_j \quad \text{in} \quad \mathcal{U}_i \cap \mathcal{U}_j \cap \Omega.$$

This is a bounded cocycle with value in the sheaf of holomorphic germs which are zero on X. Now we don't know if this bounded cohomology is zero! But using again the preceeding theorem we get in fact that the g_{ij} are not only bounded but smooth up to $\partial\Omega$, so we can apply the fact that the C^∞ $\bar\partial$-cohomology is zero [4] using essentially the Kohn's estimates [15] for smoothly bounded pseudo-convex domains.

References

[1] Amar, E., Generalisation d'un theorème de Wolff à la boule de $\mathbb{C}^n$. Prepublications Anal. Harm. Orsay (472) 1980.

[2] Amar, E., Extension de fonctions holomorphes et courants. Bull. des Sc. Math. 2e série, 107, 1983 p. 25-48.

[3] Amar, E., Extension de fonctions holomorphes et intégrales singulieres. Prepub. d'Analyse, Bordeaux, (n° 8302), 1983.

[4] Amar, E., $\bar\partial$ cohomology C^∞ et applications. Prepub. Anal. Harm. Orsay (400) 1980. a paraître au: Journal London Math. Soc.

[5] Amar, E., Representation des fonctions de B.M.O. et solutions de l'équation $\bar\partial_b$. Math. Ann. 239, p. 21-33, 1979.

[6] Amar, E. et Bonami A., Mesures de Carleson d'ordre α et solutions au bord de l'équation $\bar{\partial}$. Bull. Soc. Math. France 107, p. 23-48, 1979.

[7] Coifman, R., Rochberg R., Weiss, G., Factorization theorems for Hardy spaces in several variables. Ann. Math. 103 p. 611-635 (1976).

[8] Carleson, L., An interpolation problem for bounded analytic functions. Amer. J. Math. 80, 1958.

[9] Carleson, L., The Corona Theorem. Proc. 15 th Scandinavian Congress, Oslo 1968.

[10] Cumenge, A., Extension dans les classes de Hardy de fonctions holomorphes. A paraître Ann. de l'Institut Fourier.

[11] Henkin, G.M., Continuation of bounded holomorphic functions... Math. of the U.S.S.R. Izvestija t.6 p. 536-563, 1972.

[12] Henkin, G.M., Solutions with estimates of the H. Lewy and Poincaré-Lelong equations. Dokl. Akad. Nauk SSR 225, 771-774, 1975.

[13] Henkin, G.M., Leiterer, J., Theory of functions on strictly pseudo-convex sets with non-smooth boundary. Akademie der Wissenschaften der D.D.R. Institut fur Mathematik, Berlin, 1981.

[14] Hormander, L., Generators for some rings of Analytic functions. Bull. Amer. Math. Soc. 73, 1967, p. 943-949.

[15] Kohn, J.J., Global regularity for $\bar{\partial}$ on weakly pseudo-convex manifolds. T.A.M.S. 181 (1973) p. 273-292.

[16] Schwartz, L., Theorie des distributions, Nouvelle édition, Paris, Hermann, 1966.

[17] Shapiro, H., Shields, A.L., On some interpolations problems for analytic functions, Amer. J. Math. 83 (1961) 513-522.

[18] Skoda, H., Valeurs au bord pour les solutions de l'opérateur d",... Bull. Soc. Math. France t. 104, 1976, p. 225-299.

[19] Varopoulos, N., Zeros of H^p-functions in several complex variables. Preprint Orsay.

Recent Progress in Fourier Analysis
I. Peral and J.-L. Rubio de Francia (Editors)

THE DUAL OF THE BERGMAN SPACE A^1 IN THE TUBE OVER THE SPHERICAL CONE

DAVID BEKOLLE
Univ. Bretagne Occidentale

Introduction

Let Ω be the tube $\Omega = \mathbb{R}^{n+1} + i\Gamma$ over the spherical cone Γ in $\mathbb{R}^{n+1}$: $\Gamma = \{y = (y_0,y_1,y_2,\ldots,y_n) : y_0y_1 - y^2_2 - \ldots - y^2_n > 0, y_0 > 0\}$.

Let V denote the Lebesgue measure in Ω and $H(\Omega)$ the space of holomorphic functions in Ω. The Bergman space $A^p(\Omega)$, $0 < p < +\infty$, is defined by $A^p(\Omega) = H(\Omega) \cap L^p(dV)$.

In what follows, we give an affirmative answer to the following conjecture of R. Coifman and R. Rochberg [6]: the dual of the Bergman space $A^1(\Omega)$ can be realized as the Bergman projection of $L^\infty(\Omega)$ and coincides with the Bloch space of holomorphic functions in Ω. This conjecture is known to be true for $n = 0$ (Ω is the upper half-plane) and $n = 1$ (Ω is the product of two upper-half planes). For a proof of this fact, see [6].

From now on, we shall suppose $n \geq 2$; in this case, Ω is no more biholomorphic to a product of Siegel domains of lower dimension (e.g. half-planes) : this follows from the classification of symmetric Siegel domains by E. Cartan [5]. In the first part of the following, we deal with the case $n = 2$. We first recall some facts and state a useful estimate about the Bergman kernel of Ω (§1); next, we show that the dual of $A^1(\Omega)$ coincides with the Bloch space of Ω (§2) and in the last paragraph, we indicate how to realize the dual of $A^1(\Omega)$ as the Bergman projection of $L^\infty(\Omega)$.

The second part deals with the case $n > 2$: there, we only point some striking differences between this case and the preceding case $n = 2$.

In a last shorter part, we indicate an application of our results.

PART I. THE CASE n = 2

Here, $\Omega = \mathbb{R}^3 + i\Gamma$, where

$$\Gamma = \{(y_0, y_1, y_2) \in \mathbb{R}^3 : y_0\, y_1 - y_2^2 > 0,\ y_0 > 0\}.$$

§1. Results about the Bergman kernel of Ω.

The Bergman kernel $B(\zeta, z)$ of Ω is given by the following proposition, proved by S.G. Gindikin [8]:

Proposition I.1.1. *The Bergman kernel* $B(\zeta,z)$ *of* Ω *has the following two expressions:*

$$(1)\quad B(\zeta,z) = C[(\zeta_0 - \bar{z}_0)(\zeta_1 - \bar{z}_1) - (\zeta_2 - \bar{z}_2)^2]^{-3}$$

$$(2)\quad\quad = C' \int_\Gamma (\lambda_0 \lambda_1 - \lambda_2^2)^{3/2} \exp\ (i < \lambda,\ \zeta - \bar{z}>)\ d\lambda,$$

where $<\lambda,\omega> = (\lambda_0\, \omega_0)/2 + (\lambda_1\, \omega_1)/2 + \lambda_2\, \omega_2$, $\lambda \in \Gamma$, $\omega \in \Omega$.

We shall use the following estimate on $B(\zeta,z)$:

Lemma I.1.1. *The Bergman Kernel* $B(\zeta,z)$ *of* Ω *is in* $L^p(dV(z))$ *if and only if* $p > 7/6$; *in that case*

$$\int_\Omega |B(\zeta,z)|^p\, dV(z) = C_p\, B(\zeta,\zeta)^{p-1}.$$

For the proof of Lemma I.1.1., we use two results of [8]; the first one is a Plancherel formula:

Lemma I.1.2. *For* $f \in H(\Omega)$, *the following two properties are equivalent:*

(i) $f \in A^2(\Omega)$;

(ii) f *can be written in the form*

$$f(z) = \int_\Gamma \tilde{f}(\lambda)\ \exp\ (i <\lambda, z>)\ d\lambda,\quad z \in \Omega,\ \textit{with}$$

$$I(\tilde{f}) = \int_\Gamma |\tilde{f}(\lambda)|^2\, (\lambda_0\, \lambda_1 - \lambda_2^2)^{-3/2}\ d\lambda < +\infty.$$

In that case $I(\tilde{f}) \sim \|f\|_{A^2}$.

The second result is the following:

Lemma I.1.3. *Let* ρ_0, ρ_1 *be real numbers such that* $\rho_0 > 1/2$ *and* $\rho_1 > 0$. *Then, for any* z *in* Ω, *one has the equality*

$$z_o^{\rho_1-\rho_o}(z_o z_1 - z_2^2)^{-\rho_1} = C \int_\Gamma \lambda_1^{\rho_1-\rho_o}(\lambda_o\lambda_1-\lambda_2^2)^{3/2} \exp(i<\lambda,z>)d\lambda.$$

Proof of lemma I.1.1. First, write $B(\zeta,z)^{p/2}$ as a Fourier integral, using lemma I.1.3:

$B(\zeta,z)^{p/2} = C_p \int_\Gamma (\lambda_o\lambda_1-\lambda_2^2)^{3(p-1)/2} \exp(i<\lambda,\zeta-\bar{z}>)d\lambda$; then use Plancherel formula (lemma I.1.2) to obtain

$$\int_\Omega |B(\zeta,z)|^p \, dV(z) = C_p \int_\Gamma (\lambda_o\lambda_1-\lambda_2^2)^{3(p-3/2)} e^{(-2<\lambda,t>)} d\lambda$$

$$t = \mathrm{Im}\,\zeta.$$

Now, in order to estimate the right hand side of this equality, first integrate with respect to λ_o, using the change of variable $u = \lambda_o\lambda_1 - \lambda_2^2$: the related integral converges if and only if $p > 7/6$.

Remark. Hence, in lemma I.1.1, the condition $p > 7/6$ is necessary: this contradicts the statement given by R. Coifman and R. Rochberg in [6] with the condition $p > 1$.

§2. The dual of A^1 and Bloch functions

To define the Bloch space of Ω, we introduce the Riemann-Liouville differential operator $\Box$ of Ω (the wave operator, "the box"):

$$\Box_z = 4\frac{\partial^2}{\partial z_o \partial z_1} - \frac{\partial^2}{\partial z_2^2}$$

A function $g \in H(\Omega)$ will be called a Bloch function if

$$\|g\|_* = \sup_{z\in\Omega} \{|\Box g(z)| \, B^{-1/3}(z,z)\} < +\infty.$$

Let N denote the subspace of holomorphic functions in Ω whose box is zero; the Bloch space $\mathcal{B}$ of Ω is then defined to be the quotient space of Bloch functions by N.

The Bloch space $\mathcal{B}$ has the following property:

Lemma I.2.1. $\mathcal{B}$ *is a Banach space with the norm induce by* $\|\ \|_*$.

The main result in this paragrah is the following:

Theorem I.2. *The dual of* $A^1(\Omega)$ *coincides with the Bloch space* $\mathcal{B}$ *of* Ω.

Proof. Let us first prove that the dual $(A^1)^*(\Omega)$ of $A^1(\Omega)$ coincides with a subspace of $\mathcal{B}$. Let L be an element of $(A^1)^*(\Omega)$; by the Hahn-Banach theorem, there exists a bounded function ℓ in Ω such that for any $f \in A^1(\Omega)$: $L(f) = \int_\Omega \ell \bar{f}\, dV$.

By lemma I.1.1, we can define a function $G \in H(\Omega)$ by

(3) $G(\zeta) = C \int_\Omega B^{4/3}(\zeta,z)\, \ell(z)\, dV(z)$ and G satisfies

(4) $\sup_{\zeta\in\Omega} \{|G(\zeta)|\, B^{-1/3}(\zeta,\zeta)\} < +\infty$.

Let us associate to this function G an element L' of $(A^1)^*(\Omega)$ by

(5) $L'(f) = \int_\Omega G(\zeta)\, \bar{f}(\zeta)\, B^{-1/3}(\zeta,\zeta)\, dV(\zeta), \quad f \in A^1(\Omega)$.

We claim that $L = L'$ in $(A^1)^*(\Omega)$. For the proof of this claim, use (3), (4), (5) and the Fubini theorem to get

$$L'(f) = C \int_\Omega \left(\int_\Omega B^{4/3}(\zeta,z)\, \bar{f}(\zeta)\, B^{-1/3}(\zeta,\zeta)\, dV(\zeta)\right) \ell(z)\, dV(z)$$

and conclude by the reproducing formula:

$$\int_\Omega B^{4/3}(\zeta,z)\, \bar{f}(\zeta)\, B^{-1/3}(\zeta,\zeta)\, dV(\zeta) = C^{-1}\, \bar{f}(z), \quad f \in A^1(\Omega).$$

Let us next prove that the linear functional L can be represented by a Bloch function g : with G given by (3), g is defined by the following lemma [10]:

Lemma I.2.2. *For any* $G \in H(\Omega)$, *there exists a function* $g \in H(\Omega)$ *such that* $\Box g = G$.

The converse, $\mathcal{B} \subset (A^1)^*(\Omega)$, follows from the fact that any Bloch function g defines an element L of $(A^1)^*(\Omega)$ by

$$L(f) = \int_\Omega \Box g(z)\, \bar{f}(z)\, B^{-1/3}(z,z)\, dV(z), \quad f \in A^1(\Omega).$$

§3. The Bergman projection of $L^\infty(\Omega)$

We intend to define the Bergman projection of a bounded function ℓ in Ω; first of all, realize that the expression

$$P\ell(\zeta) = \int_\Omega B(\zeta,z)\, \ell(z)\, dV(z)$$

does not make sense since by lemma I.1.1, $B(\zeta,z)$ is not in $L^1(dV(z))$.

Like in the case $n = 0$, to give sense to $P\ell$, we use the same trick that defines the Hilbert transform of a bounded function: we substract from the Bergman kernel $B(\zeta,z)$ a kernel $B_o(\zeta,z)$ satisfying the following properties:

1°) with respect to ζ, $B_o(\zeta,z)$ is a holomorphic function orthogonal to A^1;

2°) $\int_\Omega |(B-B_o)(\zeta,z)|\, dV(z) < +\infty, \quad \zeta \in \Omega.$

For the construction of $B_o(\zeta,z)$, we first look for a sufficient condition of orthogonality to A^1; our condition is "esentially" $\Box_\zeta B_o(\zeta,z) \equiv 0$.

Next, recall the expression (2) of $B(\zeta,z)$:

$$(2) \quad B(\zeta,z) = C \int_\Gamma (\lambda_o \lambda_1 - \lambda_2^2)^{3/2} \exp\,(i\langle\lambda, \zeta-\bar{z}\rangle)\, d\lambda.$$

Notice that $B(\zeta,z)$ is not in $L^1(dV(z))$ because of a bad behavior when z tends to infinity; hence, we shall construct $B_o(\zeta,z)$ "close enough" to $B(\zeta,z)$ when z tends to infinity. The idea is to replace the exponential function in the right hand side of (2) by an exponential solution of $\Box_\zeta = 0$, "close enough" to the initial exponential function when z tends to infinity; such a solution will be of the form

$$\exp\,\left(\frac{\mu_2^2}{\mu_1} \frac{\zeta_o}{2} + \frac{\mu_1 \zeta_1}{2} + \mu_2 \zeta_2\right)$$

In fact, integrate the right hand side of (2) with respect to λ_o, using the change of variable $u = \lambda_o \lambda_1 - \lambda_2^2$; one obtains:

$$(6) \quad B(\zeta,z) = C(\zeta_o - \bar{z}_o)^{-5/2} \int_{(\lambda_1,\lambda_2) \in \mathbb{R}^+ \times \mathbb{R}} \lambda_1^{3/2} \times$$
$$\exp\left[i\left(\frac{\lambda_2^2}{\lambda_1} \frac{\zeta_o - \bar{z}_o}{2} + \lambda_1 \frac{\zeta_1 - \bar{z}_1}{2} + \lambda_2 (\zeta_2 - \bar{z}_2)\right)\right] d\lambda_1\, d\lambda_2.$$

Observe now that the exponential function in the right hand side of (6) is a solution of $\Box_\zeta = 0$.

By (6), the first term of $B_o(\zeta,z)$ will be taken to be $\frac{(\zeta_o-\bar{z}_o)^{5/2}}{(i-\bar{z}_o)^{5/2}} B(\zeta,z)$ and we shall add some terms independent of ζ_1 and ζ_2 (hence of box zero) in order to obtain the estimate

$$\int_\Omega |(B-B_o)(\zeta,z)|\, dV(z) < +\infty.$$

More precisely, we prove the following main lemma:

Main Lemma. *Define the kernel* $B_o(\zeta,z)$ *by*

$$(B-B_o)(\zeta,z) = \frac{(i-\bar{z}_o)^{5/2}-(\zeta_o-\bar{z}_o)^{5/2}}{(i-\bar{z}_o)^{5/2}}\,[B(\zeta,z)-B((\zeta_o,i,o),z)].$$

Then,

1°) *with respect to* ζ, $B_o(\zeta,z)$ *is a holomorphic function orthogonal to* A^1;

2°) *with respect to* z, $(B-B_o)(\zeta,z) \in L^1(dV(z))$.

The proof of 2°) essentially uses lemmas I.1.1, I.1.2 and I.1.3; for details, see [3].

Lemma I.2.1 and the main lemma yield the following theorem:

Theorem I.3.1. *For any* $\ell \in L(\Omega)$, *one defines a Bloch function in* Ω *(the Bergman projection* $P\ell$ *of* ℓ*) by*

$$(7) \qquad P\ell(\zeta) = \int_\Omega (B-B_o)(\zeta,z)\,\ell(z)\,dV(z), \quad \zeta \in \Omega.$$

Furthermore, the Bergman projection of $L^\infty(\Omega)$ *is a subspace of the dual of* $A^1(\Omega)$.

Let us prove now that the dual of $A^1(\Omega)$ coincides with the Bergman projection of $L^\infty(\Omega)$; it suffices by theorem I.2. to show that any Bloch function g in Ω can be written $g = P\ell + h$, where $\ell \in L^\infty(\Omega)$ and $h \in N$.

Since we take $\ell(z) = C\,B^{-1/3}(z,z)\,\Box g(z)$, we must prove that

$$g(\zeta) - C\int_\Omega (B-B_o)(\zeta,z)\quad g(z)\,B^{-1/3}(z,z)\,dV(z) \text{ is in } N;$$

if we set $G = \Box g$, this is equivalent to the following proposition:

Proposition I.3.1. *For any* $G \in H(\Omega)$ *satisfying*

$$(\ast) \qquad \sup_{z\in\Omega} \{B^{-1/3}(z,z)\ |G(z)|\} < +\infty,$$

we have the reproducing formula:

$$(8) \quad G(\zeta) = C\int_\Omega B^{4/3}(\zeta,z)\,G(z)\,B^{-1/3}(z,z)\,dV(z), \quad \zeta \in \Omega.$$

Remarks. 1°) This result is trivial for bounded domains, but seems to be new for unbounded domains: with respect to the measure $B^{-1/3}(z,z)\, dV(z)$, the related Bergman kernel $B^{4/3}(\zeta,z)$ does not only reproduce any $f \in A^p$, $1 \leq p < +\infty$, but also any $f \in H(\Omega)$ satisfying the uniform estimate (*). Let us point however that in the unbounded case, the presence of a weighted measure is necessary to give sense to the right hand side of (8).

2°) For the case $n = 0$, where Ω is the upper half-plane Π^+ in $\mathbb{C}$ and the Riemann-Liouville operator is $\frac{d}{dz}$, the analogous result yields the following corollary:

Corollary. *Let* $G \in H(\Pi^+)$ *satisfy*

$$\sup_{z = x+iy \in \Pi^+} \{y\, |G(z)|\} < +\infty.$$

We have the following identity

$$\int_i^\zeta G(z)\, dz = C \int_{\Pi^+} G(z) y \left[(\zeta-\bar{z})^{-2} - (i-\bar{z})^{-2}\right] dV(z) + \text{constant}$$

For the proof of proposition I.3.1, carry the problem from Ω onto a bounded circular representation D of Ω: the equivalence of Ω and such a bounded domain D was proved by A. Koranyi and J.A. Wolff [9]: The result follows by taking advantage of the boundedness of D; for details, see [3].

A summary of part I is given by the following theorem:

Theorem I.3.2. *Let* Ω *be the tube over the spherical cone in* $\mathbb{R}^3$: *The dual of* $A^1(\Omega)$ *coincides with the Bloch space* $\mathcal{B}$ *of* Ω *and the Bergman projection* P, *taken in the sense of (7), is a bounded operator from* $L^\infty(\Omega)$ *onto* $\mathcal{B}$.

PART II. THE CASE $n > 2$

Let now Ω denote the tube $\Omega = \mathbb{R}^{n+1} + i\Gamma$, $n > 2$, where Γ is the spherical cone in $\mathbb{R}^{n+1}$. Again in this case, in order to define the Bergman projection of $L^\infty(\Omega)$, we look for a kernel $B_0(\zeta,z)$ satisfying the following properties:

1°) with respecto to ζ, $B_0(\zeta,z)$ is a holomorphic function orthogonal to A^1;

2°) with respect to z, $(B-B_0)(\zeta,z)$ is in $L^1(dV(z))$.

The wave operator $\square$ in Ω is

$$\square_z = 4 \frac{\partial^2}{\partial z_0 \partial z_1} - \frac{\partial^2}{\partial z_2^2} - \ldots - \frac{\partial^2}{\partial z_n^2}$$

Now, since we wish to differentiate $P\ell$, $\ell \in L^\infty(\Omega)$, by differentiating the integrand of its expression

$$P\ell(\zeta) = \int_\Omega (B-B_0)(\zeta,z)\, \ell(z)\, dV(z), \quad \zeta \in \Omega,$$

our sufficient condition of orthogonality to A^1_ζ will no more be $\square_\zeta B_0(\zeta,z) \equiv 0$ because we cannot give sense to the expression

$$P\ell(\zeta) = c_n \int_\Omega B^{1+\frac{1}{n+1}}(\zeta,z)\, \ell(z)\, dV(z), \quad \zeta \in \Omega;$$

the reason is that $B^{1+\frac{1}{n+1}}(\zeta,z)$ is not in $L^1(dV(z))$. In fact, in this case, $B(\zeta,z)$ is in $L^p(dV(z))$ if and only if $p > \frac{3n+1}{2(n+1)}$, and to generalize the results of part I, the sufficient condition we take is $\square^{(m)} B_0(\zeta,z) \equiv 0$, $m \in \mathbb{N}$, $m > \frac{n-1}{2}$.

Let us next define the Bloch space $\mathcal{B}$ of Ω. Let m denote the smallest integer greater than $\frac{n-1}{2}$. A function $g \in H(\Omega)$ is a Bloch function if

$$\sup_{z\in\Omega} \{|\square^{(m)} g(z)|\; B^{-\frac{m}{n+1}}(z,z)\} < +\infty.$$

Let $\mathcal{N}$ denote the space of holomorphic solutions in Ω of the equation $\square^{(m)} = 0$; then, the Bloch space $\mathcal{B}$ of Ω will be the quotient space of Bloch functions by $\mathcal{N}$.

Our result is the following:

Theorem II. *Let Ω be the tube over the spherical cone in $\mathbb{R}^{n+1}$, $n > 2$. The dual of $A^1(\Omega)$ coincides with the Bloch space $\mathcal{B}$ of Ω and the Bergman projection P of Ω is a bounded operator from $L^\infty(\Omega)$ onto $\mathcal{B}$.*

Remarks

1°) In the same way, we may associate a Bloch space $\mathcal{B}_\alpha$ to any $\alpha \in \mathbb{N}^*$; then $\mathcal{B} = \mathcal{B}_m$ and it is easy to prove that $\mathcal{B}_\alpha \subset \mathcal{B}_\beta$ if $\alpha < \beta$.

In the other hand, in the classical cases of the unit disc, the upper half-plane and the Cayley transform of the unit ball (for this last case, see [1]), it is well known that all Bloch spaces $\mathcal{B}_\alpha$, $\alpha \in \mathbb{N}^*$, are equal (to the dual of A^1). Nevertheless, in the present case, our methods only yield the equality $\mathcal{B}_\alpha = \mathcal{B}_\beta$ when α and β are both greater than $\frac{n-1}{2}$.

2°) As a consequence of theorem II, we obtain that the conjecture of R. Coifman and R. Rochberg is also true for a cartesian product of upper half-planes, Cayley transforms of unit balls and tubes over spherical cones.

3°) Finally, let us mention that the equality between the dual of A^1 and the Bloch space can "easily" be extended, with the same proof as theorem I.2, to any symmetric Siegel domain of type II.

PART III. AN APPLICATION

The above kernel $B_o(\zeta, z)$ can be used to extend to the wave operator $\Box$ in the tube Ω over the spherical cone in $\mathbb{R}^{n+1}$ some well-known results of Hardy and Littlewood about the operator $\frac{d}{dz}$ in the unit disc of the complex plane (cf. chapter 5 of [7]). Our result is the following for $n = 2$:

Theorem III. *Let Ω be the tube over the spherical cone in $\mathbb{R}^3$. For any $p \in \left]0, +\infty\right]$, there exists a linear operator T_p defined in the Bergman space A^p such that $\Box T_p = Id_{A^p}$ and satisfying the following properties:*

1°) *if* $0 < p < \frac{12}{7}$, T_p *is a bounded operator from* A^p *to* $A^{\frac{3p}{3-p}}$

2°) *if* $\frac{12}{7} \leq p < 3$, T_p *is a bounded operator from* A^p *to the dual of* $A^{\frac{3p}{4p-3}}$;

3°) *if* $p = 3$, T_3 *is bounded from* A^3 *to the Bloch space* $\mathcal{B}$;

4°) *if* $3 < p < +\infty$, T_p *is bounded from* A^p *to the Lipschitz space* $\Lambda_{1-\frac{3}{p}}$.

A detailled discussion of this last result is presented in [4].

References

[1] Bekolle, D. Le dual de la classe de Bergman A^1 dans le transformé de Cayley de la boule unité de $\mathbb{C}^n$. Comptes-Rendus de l'Académie des Sciences de Paris, tome 296, série I, 377-380 (1983).

[2] Bekolle, D. Le dual de la classe de Bergman A^1 dans le complexifié du cône sphérique. C.R. Acad. Sc. Paris, tome 296, série I, 581-583 (1983).

[3] Bekolle, D. Le dual de l'espace des fonctions holomorphes intégrables dans des domaines de Siegel (to appear).

[4] Bekolle, D. Solutions avec estimations de l'équation des ondes (to appear in Prépublications de l'Université d'Orsay, Paris Sud).

[5] Cartan, E. Oeuvres completes.

[6] Coifman R. and Rochberg, R. Representation theorems for holomorphic and harmonic functions in L^p. Astérisque 77, 11-66, Soc. Math. France (1980).

[7] Duren, P. Theory of H^p spaces. Academic Press, New-York, (1971).

[8] Gindikin, S.G. Analysis in homogeneous domains. Russian Math. Surveys 19(4), 1-89 (1964).

[9] Koranyi, A. and Wolff, J.A. The realization of Hermitian symmetric spaces as generalized half-planes. Annals of Math. 81, 265-288 (1965).

[10] Treves, F. Linear partial differential equations with constant coefficients. Mathematics and its applications, Vol. 6, Gordon and Breach, (1966).

Recent Progress in Fourier Analysis
I. Peral and J.-L. Rubio de Francia (Editors)

BOUNDARY VALUE PROBLEMS FOR THE LAPLACE EQUATION IN LIPSCHITZIAN DOMAINS

A. P. Calderón*

University of Chicago
and
Instituto Argentino de Matemáticas

INTRODUCTION

Let D be a bounded domain in $\mathbf{R}^n$ whose boundary ∂D is locally the graph of a Lipschitzian function. We shall consider the problem of finding solutions G of the Laplace equation in D taking prescribed values g on ∂D (Dirichlet problem), or with $(\nabla G.v) = g$ where v is a prescribed continuous unit vector valued function on ∂D (oblique derivative problem). The precise sense in which these conditions are to be satisfied is described below. Dalhberg [3] has shown that the Dirichlet problem is always solvable uniquely if f is square integrable with respect to the surface area $d\sigma$ of ∂D. We shall prove that this also the case if $f \in L^p(d\sigma)$ for all p, $p_0 < p \leq 2$ where p_0, $1 \leq p_0 < 2$, depends on the domain D. We shall also prove that the oblique derivative problem is solvable with finitely many linear conditions imposed on g if the normal component of v has a positive lower bound. The case in which v coincides with the normal to ∂D (Neumann problem) is not covered by our results. For the solution of this problem in the case p =2 see [4]. As a consequence of our results on the Dirichlet problem we shall also show that harmonic measure for D belongs to every $L^q(d\sigma)$, $q < \frac{p_0}{p_0-1}$, where p_0 is the same as above.

* This research was partly suported by NSF Grant MCS 8203319.

I. LIPSCHITZIAN DOMAINS

In this section we shall discuss some properties of Lipschitzian domains which we will use later. We shall say that a bounded subdomain D of $\mathbf{R}^n$ is Lipschitzian if every point of its boundary ∂D has a cylindrical neighborhood 0 with the following properties:

i) There exists a cartesian coordinate system $(x_1,\dots,x_n)$ in $\mathbf{R}^n$, with origin at the point, in which 0 is defined by $x_1^2+\dots+x_{n-1}^2 < \delta^2$, $-a < x_n < a$,

ii) $0 \cap \partial D$ is given by $x_n = \Phi(x_1,\dots,x_{n-1})$, $x_1^2+\dots+x_{n-1}^2 < \delta^2$, where Φ is a Lipschitzian function such that $-a+\delta < \Phi < a-\delta$,

iii) $0 \cap D$ is given by $-a < x_n < \Phi(x_1,\dots,x_{n-1})$, $x_1^2+\dots+x_{n-1}^2 < \delta^2$.

For the purpose of proving uniqueness of solution of the Dirichlet problem, we shall construct a family of domains D_t, $D_t \subset D$, approximating D in a certain manner.

If 0 is a neighborhood of a point of ∂D as above, we let $0'$ be the smaller neighborhood defined by $x_1^2+\dots+x_{n-1}^2 < \delta^2/4$ and $-a+\delta/2 < x_n < a-\delta/2$. If x denotes the point of coordinates $x_1,\dots,x_n$, or a vector with these components, we denote by $\bar{x}$ the point of coordinates $(x_1,\dots,x_{n-1},0)$ and write $x = (\bar{x},x_n)$.

Now we cover ∂D with finitely many neighborhoods $0_j'$. If $d_j(y)$ denotes the distance of the complement 0_j^c of 0_j from y, u_j denotes the vector of components $(0,0,\dots,1)$ in the coordinate system associated with 0_j, and $n(y)$ denotes the outer unit vector normal to ∂D at y, the scalar product $n(y) \circ u_j$ has a positive lower bound in $0_j \cap \partial D$ and $u(y) = (\Sigma\, d_j(y)u_j)|\Sigma d_j(y)u_j|^{-1}$ is a Lipschitzian unit vector valued function on ∂D such that $n(y) \circ u(y)$ has a positive lower bound. Thus, there exists a Lipschitzian vector valued function $u(y)$, $|u(y)| = 1$ on ∂D, such that $u(y) \circ n(y) \geq \varepsilon > 0$. The definition and argument that follow will apply to any such function $u(y)$. Let now D_t be defined by

$$D_t = D \setminus \{x \mid x = y - su(y);\ y \in \partial D,\ 0 \leq s \leq t\}.$$

We will show that for sufficiently small t, D_t is a Lipschtzian domain and that

$$\partial D_t = \{x | x = y - tu(y),\ y \in \partial D\}.$$

First we observe that if t is sufficiently small then

$$D_t \cap O' = D \cap O' \setminus \{x | x = y - su(y);\ y \in \partial D \cap O,\ 0 \leq s \leq t\}.$$

Let $y \in \partial D \cap O$ and $x = y - su(y) \in D \cap O'$.

Then we have $y_n = \Phi(\bar{y})$ and

$$(1) \qquad \begin{aligned} x_n &= \Phi(\bar{y}) - su_n(y) \\ \bar{x} &= \bar{y} - s\bar{u}(y) \end{aligned}$$

Clearly, the mapping taking $\bar{y}$ to $\bar{x}$ above is Lipschitzian and has a Lipschitzian inverse for small s. Thus we have $x_n = \Phi_s(\bar{x})$, where $\Phi_s(\bar{x})$ is Lipschitzian and well defined for $|\bar{x}| < \delta/2$ and s small. If we show that $\Phi_s(\bar{x})$ is a strictly decreasing function of s for s small, it will follow that

$$D_t \cap O' = \{x | x_n < \Phi_t(\bar{x}),\ |\bar{x}| < \delta/2\},$$

and from this, which will hold for each O_j' of the covering above, we infer that D_t is a Lipschitzian domain for small t and that $\partial D_t = \{x | x = y - tu(y),\ y \in \partial D\}$. Let $\bar{x}_0 = \bar{y}_0 - s_0\bar{u}(y_0)$, $y_0 \in \partial D \cap O$, and consider, for given small s, the map T_s, $T_s(\bar{y}) = \bar{y} - s\bar{u}(y) = \bar{x}$.

As was pointed out above, T_s is Lipschitzian and has a Lipschitzian inverse $T_s^{-1}(\bar{x})$. If we set $\bar{x} = \bar{x}_0$, then $\bar{y} = T_s^{-1}(\bar{x}_0)$ becomes a Lipschitzian function of s. In fact, we have

$$\begin{aligned} &T_{s_1}^{-1}(\bar{x}_0) - T_{s_2}^{-1}(\bar{x}_0) = \bar{y}_1 - \bar{y}_2 = \\ &= s_1\bar{u}[\bar{y}_1,\Phi(\bar{y}_1)] - s_2\bar{u}[\bar{y}_2,\Phi(\bar{y}_2)] \\ &= s_1\{\bar{u}[\bar{y}_1,\Phi(\bar{y}_1)] - \bar{u}[\bar{y}_2,\Phi(\bar{y}_2)]\} + (s_1-s_2)\bar{u}[\bar{y}_2,\Phi(\bar{y}_2)] \end{aligned}$$

From the Lipschitzian character of $\bar{u}(y)$, we infer that

$$|\bar{y}_1-\bar{y}_2| \leq s_1M|\bar{y}_1-\bar{y}_2| + N|s_1-s_2|$$

which clearly implies that $|\bar{y}_1-\bar{y}_2| \leq c|s_1-s_2|$ if s_1 is small. Furthermore

$$\frac{d}{ds} T_s^{-1}(\bar{x}_0) = \frac{d\bar{y}}{ds} = \bar{u}(y) + s\,\frac{d\bar{u}(y)}{ds}$$

which shows that

$$\frac{d}{ds} T_s^{-1}(\bar{x}_0) = \bar{u}(y) + O(s)$$

Now, if we set $\bar{y} = T_s^{-1}(x_0)$ in (1), we obtain

$$x_n = \Phi_s(\bar{x}_0) = \Phi(\bar{y}) - su_n(y), \quad y = (\bar{y}, \Phi(\bar{y}))$$

and differentiating with respect to s we obtain

$$\frac{d}{ds}\Phi_s(\bar{x}_0) = \nabla\Phi(\bar{y}) \circ \frac{d\bar{y}}{ds} - u_n(y) + O(s)$$

$$= \nabla\Phi(\bar{y}) \circ \bar{u}(y) - u_n(y) + O(s)$$

But

$$\nabla\Phi(\bar{y}) \circ \bar{u}(y) - u_n(y) = -[n(y) \circ u(y)](1 + |\nabla\Phi|^2)^{1/2}$$

where $n(y)$ is the outer normal to ∂D at the point $y = (\bar{y}, \Phi(\bar{y}))$. Thus, since $n(y) \circ u(y)$ has a positive lower bound, the preceding expression has a negative upper bound and

$$\frac{d}{ds}\Phi_s(\bar{x}_0) < 0$$

if s is sufficiently small, uniformly for $|\bar{x}_0| < \delta/2$, which proves our assertion.

There are two features of these domains D_t that we will use. One is that, whenever $n(y)$ is defined, $y \in \partial D$, the point $x = y - su(y)$, $0 < s \leq \varepsilon$ is contained in a non-tangential domain $\Gamma_\delta(y)$, $\Gamma_\varepsilon(y) = \{x | x \in D;\ d(x, \partial D) > \varepsilon|x-y|\}$, where $d(x, \partial D)$ denotes the distance from x to ∂D, provided s is sufficiently small. Thus, as $s \to 0$, x approaches y non-tangentially to ∂D. This is an immediate consequence of the fact that $u(y) \circ n(y) \geq \varepsilon > 0$. The other is this: if we denote by $n_s(x)$, $x \in \partial D_s$ the outer unit vector normal to ∂D_s at x, then whenever $n_s(y - su(y))$ and $n(y)$ are defined (which is the case for almost all (y,s)) we have

$$|n_s(y - su(y)) - n(y)| < Cs.$$

Locally, this follows from (1) by observing that the vectors

$$\left(\frac{\partial \bar{x}}{\partial y_j}, \frac{\partial x_n}{\partial y_j}\right) \qquad j = 1,2,\dots,n-1$$

which span the hyperplane tangent to ∂D_s at $y - su(y)$, differ from the vector of components

$$\left(\delta_{ij}, \frac{\partial \Phi}{\partial y_j}\right)$$

which span the hyperplane tangent to ∂D, at y, by less than Cs in norm. Globally our assertion follows by covering ∂D with the sets O_j'.

Finally, we introduce the following notation: if F is a function defined in D, $m_\varepsilon(F)(y)$, $y \in \partial D$, is the function $\sup_{x \in \Gamma_\varepsilon(y)} |F(x)|$.

2. POTENTIALS OF DISTRIBUTIONS ON D.

Consider the Newtonian potential of a mass distribution on ∂D.

$$F(x) = \frac{1}{(n-2)\omega_n} \int_{\partial D} \frac{f(y)\,d\sigma}{|x-y|^{n-2}}, \quad x \in D, \quad n > 2$$

(2)

$$F(x) = \frac{1}{2\pi} \int_{\partial D} \log \frac{1}{|x-y|} f(y)\,d\sigma, \quad x \in D, \quad n = 2$$

where ω_n is the surface area of the unit sphere in $\mathbb{R}^n$, f is a function on ∂D whose p-th power, $1 < p < \infty$, is integrable with respect to the surface area $d\sigma$ of ∂D, i.e. it belongs to $L^p(d\sigma)$. Then $F(x)$ is harmonic in D and its gradient is given by

$$(\nabla F)(x) = \frac{-1}{\omega_n} \int \frac{(x-y)}{|x-y|^n} f(y)\,d\sigma.$$

This gradient has a limit as x approaches a point z in ∂D non-tangentially to ∂D, for almost all z, given by

$$(3) \qquad \lim_{x \to z} (\nabla F)(x) = \frac{1}{2} f(z) n(z) - \lim_{\varepsilon \to 0} \frac{1}{\omega_n} \int_{|y-z|>\varepsilon} \frac{(z-y)}{|z-y|^n} f(y)\,d\sigma,$$

where, again, $n(z)$ denote the outer unit normal vector to ∂D at z, (see [4], section 1) and the expression on the right represents an operator taking functions on ∂D into vector valued functions on ∂D, which is bounded with respect to the norm of $L^p(d\sigma)$.
Furthermore, $m_\varepsilon(|\nabla F|)(z)$, $z \in \partial D$ belongs to $L^p(d\sigma)$ if f does

(see section 1 for the definition of m_ε). Given a bounded vector valued function $v(y)$ on ∂D, consider also the function

$$(4) \qquad G(x) = \frac{-1}{\omega_n} \int_{\partial D} \frac{(x-y) \circ v(y)}{|x-y|^n} f(y)\, d\sigma, \qquad x \in D$$

where f is again a function on ∂D in $L^p(d\sigma)$. This function is also harmonic in D and has a limit as $x \to z \in \partial D$ non-tangentially, for almost all z, given by

$$(5) \qquad \lim_{x \to z} G(x) = \frac{1}{2}\,[v(z) \circ n(z)]\, f(z) - $$

$$- \frac{1}{\omega_n} \lim_{\varepsilon \to 0} \int_{|y-z|>\varepsilon} \frac{(z-y) \circ v(y)}{|z-y|^n} f(y)\, d\sigma$$

and the expression on the right represents a bounded operator in $L^p(d\sigma)$, $1 < p < \infty$. Here also $m_\varepsilon(G)(z)$, $z \in \partial D$, belongs to $L^p(d\sigma)$ if f does.

3. THE DIRICHLET PROBLEM

We seek solutions of the problem of finding harmonic functions $G(x)$ in D such that

$$\lim_{\substack{x \to z \\ z \in \partial D}} G(x) = g(z), \quad \text{a.e.}$$

where $g(z)$ is a given function on ∂D belonging to $L^p(d\sigma)$, and x approaches z non-tangentially.

We shall show that there exists a p_0, $1 < p_0 < 2$, depending on D, such that this problem has a unique solution with the property that

$$(6) \qquad \int_{\partial D} |G[z - tu(z)] - g(z)|^p\, d\sigma \to 0$$

as $t \to 0$ and where $u(y)$ is a Lipschitzian vector valued function on ∂D, $|u(y)| = 1$, with the property that $u(y) \circ n(y) \geq \varepsilon > 0$, $y \in \partial D$, provided that $p_0 < p \leq 2$. This p_0 can be estimated in terms of the local oscillation of $n(y)$:

$$\lim_{\varepsilon \to 0} \sup_{|y-z|<\varepsilon} |n(y) - n(z)|,$$

and can be shown to tend to 1 as the local oscillation of $n(y)$ tends to zero. This, however, we will not do in this paper.

The solution to the Dirichlet problem will have the additional property that, for any $x_0 \in D$, $G(x_0)$ will depend continuously on g, that is, there exists a function $h_{x_0}(y)$ on ∂D in $L^q(d\sigma)$, $q = \frac{p}{p-1}$ such that

$$G(x_0) = \int_{\partial D} g(z) h_{x_0}(z)\, d\sigma$$

and, thus, it will follow that the harmonic measure associated with a point x_0 in D belongs to every $L^q(d\sigma)$, $q < \frac{p_0}{p_0-1}$. That for certain domains this cannot hold for arbitrarily large q can easily bee seen by considering the image of the domain $\{z \mid |z-1| < 1\}$ in the complex plane under the map $z \to w = z^a$, $1 < a < 2$.

In this case, the harmonic measure associated with $w = 1$ is $2\pi . a^{-1} |w|^{\frac{1-a}{a}} d\sigma$ and this does not belong to $L^q(d\sigma)$ for $q \geq a/(a-1)$.

Let us turn now to the proof of our statements. Consider (5) with $v(y) = u(y)$ and denote by Af its right hand side. We shall prove that A is a Fredholm operator of index zero in $L^p(d\sigma)$ for $p_0 < p < \frac{p_0}{p_0-1}$, $1 < p_0 < 2$. For this purpose, consider the function $\phi(y) = [u(y) \circ n(u)]^{-1/2}$. Since $u(y) \circ n(y)$ is bounded and has a positive lower bound, our assertion about A will hold if and only if it holds for the operator

$$(\phi A \phi)(f) = \frac{1}{2} f(z) - \frac{1}{\omega_n} \lim_{\varepsilon \to 0} \int_{|y-z|>\varepsilon} \frac{\phi(z)[(z-y) \circ u(y)]\phi(y)}{|z-y|^n} f(y)\, d\sigma$$

$$= \frac{1}{2} f + Bf$$

Now, consider the operator $\frac{I}{2} + \lambda B$, where $0 \leq \lambda \leq 1$ and I is the identity, and

$$(7) \qquad \begin{aligned} (\tfrac{I}{2} + \lambda B)(\tfrac{I}{2} + \lambda B^*) &= \tfrac{I}{4} + \tfrac{\lambda}{2} (B + B^*) + \lambda^2 BB^* \\ (\tfrac{I}{2} + \lambda B^*)(\tfrac{I}{2} + \lambda B) &= \tfrac{I}{4} + \tfrac{\lambda}{2} (B + B^*) + \lambda^2 B^*B \end{aligned}$$

Since $u(y)$ is Lipschitzian, the operator

$$B + B^* = \frac{1}{\omega_n} \lim_{\varepsilon \to 0} \int \frac{\phi(z)[(z-y) \circ (u(y)-u(z))]\phi(y)}{|y-z|^n} f(y)\, d\sigma.$$

is compact in every $L^p(d\sigma)$, $1 < p < \infty$, and maps $L^p(d\sigma)$ continuously into $L^q(d\sigma)$, $\frac{1}{q} \geq \frac{1}{p} - \frac{1}{n-1}$, $1 < p < n-1$. On the other hand the operators

$$\frac{I}{4} + \lambda^2 BB^*, \quad \frac{I}{4} + \lambda^2 B^*B$$

as operators in $L^2(d\sigma)$, are positive and have $\frac{1}{4}$ and $\frac{1}{4} + \lambda^2 |B|^2_{L^2}$ as lower and upper bounds respectively. Thus if we set

$$D_1 = -\frac{1}{2} |B|^2_{L^2} I + BB^*$$

$$D_2 = -\frac{1}{2} |B|^2_{L^2} I + B^*B$$

these operators have norms in $L^2(d\sigma)$ not exceeding $\frac{1}{2} |B|^2_{L^2}$, and

$$\frac{I}{4} + \lambda^2 BB^* = (\frac{1}{4} + \frac{1}{2} \lambda^2 |B|_{L^2}) I + \lambda^2 D_1$$

(8)

$$\frac{I}{4} + \lambda^2 B^*B = (\frac{1}{4} + \frac{1}{2} \lambda^2 |B|_{L^2}) I + \lambda^2 D_2.$$

But the operators B and B^* are continuous in every $L^p(d\sigma)$, $1 < p < \infty$, (see [2]) and, consequently, so are the operators D_1 and D_2 and their norms as operators in $L^p(d\sigma)$ are logarithmically convex functions of $\frac{1}{p}$ and there is a p_0, $1 < p_0 < 2$, such that

$$|D_1|_{L^p} < \frac{1}{4} + \frac{1}{2} |B|^2_{L^2}$$

$$|D_2|_{L^p} < \frac{1}{4} + \frac{1}{2} |B|^2_{L^2}$$

for $p_0 < p < \frac{p_0}{p_0 - 1}$. Thus, for $p_0 < p < \frac{p_0}{p_0 - 1}$ and $0 \leq \lambda \leq 1$ we have

$$|\lambda^2 D_1|_{L^p} < \frac{1}{4} + \frac{1}{2} \lambda^2 |B|^2_{L^2}$$

$$|\lambda^2 D_2|_{L^p} < \frac{1}{4} + \frac{1}{2} \lambda^2 |B|^2_{L^2}$$

and the operators in (8) are invertible.

Consequently, since $B + B^*$ is compact, the operators in (7) are Fredholm operators, and, as is readily verified, this implies that both $\frac{1}{2} I + \lambda B$ and $\frac{1}{2} I + \lambda B^*$ have finite dimensional nullspaces and their ranges contain a closed subspace of finite codimension, and therefore, they themselves are also closed and of finite codimension. Thus they are both Fredholm operators for $0 \leq \lambda \leq 1$ and their index is zero, since it is zero for $\lambda = 0$. Furthermore, their nullspaces consist of functions belonging to every $L^p(d\sigma)$, $p < \frac{p_0}{p_0-1}$, and in particular to $L^2(d\sigma)$. To see this suppose that f is in the nullspace of $\frac{1}{2} I + \lambda B$ as an operator in $L^p(d\sigma)$, i.e. suppose that $f \in L^p(d\sigma)$ and that $\frac{1}{2} f + \lambda Bf = 0$. Then from (7) it follows that

$$-\lambda(B + B^*)f = (\frac{I}{2} + 2\lambda^2 B^*B)f$$

and, since $B + B^*$ maps $L^p(d\sigma)$ into $L^q(d\sigma)$ for $\frac{1}{q} \geq \frac{1}{p} - \frac{1}{n-1}$ and the operator on the right above is invertible in this space provided that $p_0 < q < \frac{p_0}{p_0-1}$, we conclude that $f \in L^q(d\sigma)$. Iterating this argument, if necessary, we obtain the desired conclusion. The same argument applies to $\frac{1}{2} I + \lambda B^*$. Let us turn now to the Dirichlet problem. Suppose we are given a function g on ∂D in $L^p(d\sigma)$, $p_0 < p < \frac{p_0}{p_0-1}$. Since the space C of continuous functions on ∂D is dense in $L^p(d\sigma)$ and the range R_A of A as an operator on $L^p(d\sigma)$, is closed and of finite codimension, $R_A + C = L^p(d\sigma)$ and we can find a continuous function g_1 and a function g_2 in R_A such that $g = g_1 + g_2$.

Now the classical Dirichlet problem is solvable in D, i.e. there exists a continuous function G_1 in the closure of D, coinciding with g_1 on ∂D, and harmonic in D.

Furthermore, since $g_2 \in R_A$ there exists a function f in $L^p(d\sigma)$ such that $Af = g_2$ and, consequently, setting

$$G_2(x) = -\frac{1}{\omega_n} \int_{\partial D} \frac{(x-y).u(y)}{|x-y|^n} f(y) d\sigma$$

we find that $G(x) = G_1(x) + G_2(x)$ solves our problem.

That G also satisfies (6) is an immediate consequence of the continuity of G_1 and the properties of G_2 described in section 2.

In order to prove the uniqueness of solutions satisfying (6) we must show that if G is harmonic in D, $p_0 < p \leq 2$, and

$$\int_{\partial D} |G[x-tu(z)]|^p \, d\sigma \to 0$$

as $t \to 0$ then $G = 0$.

Let $D_t \subset D$ be the domain whose boundary ∂D_t is the set $\{x | x = z - tu(z),\ z \in \partial D\}$ and let A_t be the operator associated with D_t as A is associated with D. If we identify ∂D_t with ∂D by identifying the point $z-tu(z)$ with z, and denote by $n_t(z-tu(z))$ the normal to ∂D_t at $z-tu(z)$, then

$$A_t f = [n_t(z-tu(z)) \circ u(z)] f(z) +$$

$$+ \lim_{\varepsilon \to 0} \int_{|z-y|>\varepsilon} \frac{[(z-y)-t(u(z)-u(y))] \circ u(y)}{|z-y-t\ u(z)-u(y)||^n} f(y) \left(\frac{d\sigma_t}{d\sigma}\right) d\sigma$$

where $d\sigma_t$ represents the area element of ∂D_t.

Now, as was shown in section 1, $n_t(z-tu(z))$ converges uniformly to $n(z)$ as $t \to 0$, and as is readily verified, $(d\sigma_t)(z-tu(z))/(d\sigma)(z)$ tends uniformly to 1. Thus, it follows from the work of R. R. Coifman, G. David and Y. Meyer (see [2]) that A_t converges to A in the operator norm of every $L^p(d\sigma)$ $1 < p < \infty$. Now let f_i, $i=1,2,\ldots,k$ be a basis for the nullspace N_A of A and g_i, $i=1,2,\ldots,k$ be continuous functions on ∂D which are linearly independent modulo R_A. Let P be the operator defined by

$$Pf = \sum_i^k g_i \int_{\partial D} f f_i \, d\sigma$$

Since the codimension of R_A equals the dimension of N_A, the g_i span a linear complement of R_A, and since the elements of N_A belong to every $L^q(d\sigma)$, $q < \frac{p_0}{p_0 - 1}$, P is well defined and continuous in $L^p(d\sigma)$, $p > p_0$. Thus the operator $A + P$ is invertible in $L^p(d\sigma)$, and so is $A_t + P$ for t sufficiently small and its inverse has bounded norm.

Now let g_t be the restriction of G to ∂D_t.

By assumption we have $\|g_t\|_{L^p(d\sigma_t)} \to 0$, as $t \to 0$. Thus if we set $g_t = (A_t + P) f_t$, then also $\|f_t\|_{L^p(d\sigma_t)} \to 0$. Now consider the functions in D_t

$$G_1(x) = -\frac{1}{\omega_n} \int_{\partial D} \frac{(z-y+tu(y)) \circ u(y)}{|x-y+tu(y)|^n} f_t(y) \left(\frac{d\sigma_t}{d\sigma}\right) d\sigma, \quad x \in D_t$$

and $G_2(x)$, the continuous harmonic extension to D_t of the image of the function Pf_t on ∂D under the inverse of the map taking $z \in \partial D$ to $z - tu(z) \in \partial D_t$.

Clearly $G_2(x)$ exists since Pf_t is continuous and the classical Dirichlet problem is solvable in D_t. Since $\|f_t\|_{L^p(d\sigma_t)} \to 0$, or, which is the same, $\|f_t\|_{L^p(d\sigma)} \to 0$, it follows that Pf_t converges uniformly to zero as $t \to 0$, and thus $G_2(x)$ also tends to zero uniformly on every closed subset of D, as $t \to 0$. Clearly the same conclusion holds for $G_1(x)$.

Now $G_1(x) + G_2(x)$ represents in a sense a solution of the Dirichlet problem in D_t with boundary data g_t and therefore $G_1(x) + G_2(x)$ presumably coincides with $G(x)$ in D_t. That this is indeed the case follows from the fact that, since $A_t + P$ is also invertible in $L^2(d\sigma)$ for t sufficiently small, $f_t \in L^2(d\sigma_t)$ and therefore the non-tangential maximal function $m_\varepsilon(G_2)$ in D_t also belongs to $L^2(d\sigma_t)$, and thus $G_1 + G_2$ represents the unique solution to the Dirichlet problem in the sense of Dahlberg (see [3], and [4] corollary 3.3). Thus we have $G(z) = G_1(x) + G_2(x)$ in D_t, and since $G_1 + G_2 \to 0$ as $t \to 0$ on closed subsets of D we conclude that $G = 0$ as we wished to show.

The fact that $G(x)$ depends continuously on the boundary data g in the sense that for any $x_0 \in D$, $G(x_0)$ is a continuous linear functional of g as an element of $L^p(d\sigma)$ follows immediately from the above construction of G.

4. THE OBLIQUE DERIVATIVE PROBLEM

By dualizing the construction in the preceding section we can solve the following problem:

Given a continuous unit vector valued function $v(x)$ on ∂D with the property that $v(z) \circ n(z)$ has a positive lower bound, and a function $f(z) \in L^p(d\sigma)$ on ∂D, find a harmonic function $F(x)$ in D such that

$$m_\varepsilon(|\nabla F|) \in L^p(d\sigma)$$

and

$$\lim_{\substack{x \to z \\ z \in \partial D}} (\nabla F).v(z) = f(z), \quad \text{a.e.}$$

as x approaches z non-tangentially. For simplicity we will restricted ourselves to the case where the complement of D is connected and we will show that if $p_0 < p < \frac{p_0}{p_0-1}$, where p_0 is the same as in the preceding section, then the problem has solutions F provided that f satisfies finitely many linear conditions. Furthermore, these solutions have the property that

$$(9) \qquad \lim_{t \to 0} \int_{\partial D} |(\nabla F)(z-tu(z)) \circ v(z) - f(z)|^p d\sigma \to 0$$

as $t \to 0$, where $u(z)$ is the same as in section 1, and the dimension of the space of solutions of the homogeneous problem, i.e. with $f = 0$, satisfying (9) with $f = 0$ and $m_\varepsilon(\nabla F) \in L^p(d\sigma)$ is the same as the number of linearly independent conditions to be imposed on f for solvability. It is to be observed that the Neumann problem is not a special case of the preceding one since $n(z)$ is not continuous except in the case of C^1-domains. For the Neumann problem in the case $p = 2$ see [4].

Now let us proceed to the proof of the preceding assertions. Consider the operator $\tilde{A}$ obtained from the right hand side of (5) by changing the sign of the second term. If $v(z)$ is Lipschitzian, then the same argument used in section (3) to show that A is a Fredholm operator, shows that A is Fredholm in $L^p(d\sigma)$ for $p_0 < p < \frac{p_0}{p_0-1}$, and that the nullspace of $\tilde{A}^*$ consists of functions belonging to every $L^p(d\sigma)$, $p_0 < p < \frac{p_0}{p_0-1}$. If $v(z)$ is merely continuous, then $\tilde{A}$ and $\tilde{A}^*$ are still Fredholm operators in $L^p(d\sigma)$, $p_0 < p < \frac{p_0}{p_0-1}$. This follows from the fact that if a sequence $v_n(z)$ of Lipschitzian functions converges uniformly to v, then the corresponding operators $\tilde{A}_n$ converge to $\tilde{A}$ in the operator norm of $L^p(d\sigma)$, $p_0 < p < \frac{p_0}{p_0-1}$. Furthermore, the dimensions of the nullspaces of $\tilde{A}$ and $\tilde{A}^*$ coincide.

Suppose now that $f \in L^p(d\sigma)$ is in the range of $\tilde{A}^*$, which will be the case if f satisfies finitely many appropiate linear conditions. Let $f = \tilde{A}^*g$ and, as in (2), set

$$F(x) = \frac{1}{(n-2)\omega_n} \int_{\partial D} \frac{g(y)}{|x-y|^{n-2}} d\sigma \quad \text{if} \quad n > 2$$

(10)

$$F(x) = \frac{1}{2\pi} \int_{\partial D} \log \frac{1}{|x-y|} g(y) d\sigma \quad \text{if} \quad n = 2$$

Then, as was asserted in section 2,

$$\lim_{\substack{x \to z \\ z \in \partial D}} |(\nabla F)(x) \circ v(z)| = \tilde{A}^* g = f, \quad \text{a.e.}$$

and

$$m_\varepsilon(|\nabla F|) \in L^p(d\sigma)$$

and F is a solution of our problem. That F satisfies (9) is an immediate consequence of the above convergence and the fact that $m_\varepsilon(|\nabla F|)(z) \in L^p(d\sigma)$.

In order to show that the dimension of the space of solutions of our problem with $f = 0$ coincides with the number of conditions to be imposed on f, or, which is the same, with the codimension of the range of $\tilde{A}^*$, which equals the dimension of the nullspace of $\tilde{A}$, we shall prove that if

$$m_\varepsilon(|\nabla F|) \in L^p(d\sigma), \quad \lim_{\substack{x \to z \\ z \in \partial D}} |\nabla F . v(z)| = 0$$

then F is given by (10) with $g \in L^p(d\sigma)$ and $\tilde{A}^* g = 0$. Since, as we will show, the map $g \to F$ given by (10) is injective, the desired conclusion will follow.

Consider the operator on the right hand side of (3) acting on $L^p(d\sigma)$, $p_0 < p < \frac{p_0}{p_0 - 1}$,

$$Hg = \tfrac{1}{2} n(z) g(z) - \lim_{\varepsilon \to 0} \frac{1}{\omega_n} \int_{|z-y|>\varepsilon} \frac{(z-y)}{|z-y|^n} g(y) d\sigma,$$

taking the scalar function g on ∂D into the vector valued function Hg, which coincides with the limit values of ∇F on ∂D, where F is as in (10). If we compose H on the left with scalar multiplication by $v(z)$ we obtain the operator $\tilde{A}^*$. If $v(z)$ is Lipschitzian then the nullspace of $\tilde{A}^*$ is finite dimensional and consists of functions belonging to every $L^p(d\sigma)$, $p_0 < p < \frac{p_0}{p_0 - 1}$, and consequently, since the nullspace of H is contained in that of $\tilde{A}^*$, the same holds for H. Furthermore this nullspace is one-dimensional. To see this suppose that $Hg = 0$. Then $g \in L^2(d\sigma)$ and the space of such g is one-dimensional as follows from the results of G. Verchota (see [4], theorem 3.4). Next let us consider the range of H as an operator in $L^p(d\sigma)$, $p_0 < p < \frac{p_0}{p_0-1}$. If we denote by v^* the adjoint of the operator scalar multiplication by the vector valued function v, we have $H^*v^* = \tilde{A}$ and the range of H^* contains that of $\tilde{A}$ which is closed and of finite codimension. Thus the range of H^* is closed and so is that of H. Now suppose that $F(x)$ is harmonic in D and continuous with continuous derivatives in the closure of D. Then according to [4], 3.13 there exists a unique $g \in L^2(d\sigma)$ such that $F(x)$ is representable as in (1). Now this function g belong also to every $L^p(d\sigma)$, $p_0 < p < \frac{p_0}{p_0-1}$. To see this let $v(z)$ be a vector valued Lipschitzian function on ∂D, and consider $(\nabla F)\circ v(z)$ on ∂D. Then from the result [4] just quoted it follows that $(\nabla F)\circ v(z)$ is in the range of $\tilde{A}^*$ as an operator in $L^2(d\sigma)$ and thus, is orthogonal to the nullspace of $\tilde{A}$ as an operator on $L^2(d\sigma)$. But this nullspace coincides with the nullspace of $\tilde{A}$ as an operator in $L^q(d\sigma)$, $q = \frac{p}{p-1}$. Therefore $(\nabla F)\circ v(z)$ is also orthogonal to the nullspace of $\tilde{A}$ as an operator in $L^q(d\sigma)$ and therefore it belongs to the range of $\tilde{A}^*$ as an operator in $L^p(d\sigma)$. Thus there exists a g_1 in $L^p(d\sigma)$ such that $\tilde{A}^*g_1 = (\nabla F)\circ v(z)$ and consequently, if $p > 2$ (in the case $p \leq 2$ it is obvious that $g \in L^p(d\sigma)$), $g-g_1$ is in the nullspace of $\tilde{A}^*$ as an operator in $L^2(d\sigma)$. But this nullspace consists of functions belonging to every $L^r(d\sigma)$, $r < \frac{p_0}{p_0-1}$. Thus $g-g_1 \in L^p(d\sigma)$ and we find that $g \in L^p(d\sigma)$.

Let now F be harmonic in D and such that $m_\varepsilon(|\nabla|F) \in L^p(d\sigma)$, $(p_0 < p < \frac{p_0}{p_0-1})$. We are now ready to show that in this case F is still given by (10) with a uniquely determined function g in $L^p(d\sigma)$. After showing this the proof of our assertions will be complete.

For the sake of brevity we shall hence-forth denote the right hand side of (10) by $P(g)$. Consider the approximating domains D_t in section 1 and the corresponding operators H_t and P_t which are defined like H and P in the case of D. Since F is continuously differentiable in D there exists a function g_t on ∂D_t belonging to $L^p(d\sigma)$ such that $P_t(g_t) = F$ in D_t. Let also h_t denote the function on ∂D_t such that $P_t(h_t) = 1$ in D_t. This function also belongs to $L^p(d\sigma)$ of ∂D_t. If ϕ denotes the map of ∂D into ∂D_t which takes the point $z \in \partial D$ into the point $z-tu(z)$ (see section 2), and g is a function on D we set

$$\tilde{H}_t(g) = [H_t(g\circ\phi^{-1})]\circ\phi$$

and an application of the results in section 1 and in [2] shows that $\tilde{H}_t$ converges to H in the operator norm of $L^p(d\sigma)$ of ∂D. Since the nullspace of H is one-dimensional and its range is closed in $L^p(d\sigma)$, there exists a closed subspace M of $L^p(d\sigma)$ of codimension 1 such that $\|H(g)\|_p > \varepsilon\,\|g\|$, for $g \in M$. Since $\tilde{H}_t$ converges to H in norm we find that for $0 \leq t < t_0$ and $g \in M$ we will have $|\tilde{H}_t g|_p > \varepsilon/2|g|_p$. Let us set now $\tilde{g}_t = g_t\circ\phi$, $\tilde{h}_t = h_t\circ\phi$. Since $P_t(h_t) = 1$ in D_t we have $H_t h_t = 0$ and consequently, since the nullspace of H_t is one-dimensional, it is generated by h_t, and the nullspace of $\tilde{H}_t$ is generated by $\tilde{h}_t$. Now if $0 \leq t < t_0$, $\tilde{H}_t$ is invertible on M and consequently its nullspace is a linear complement of M. Thus there exists a number λ_t such that $g_t - \lambda_t\,\tilde{h}_t \in M$ and

$$\tilde{H}_t(\tilde{g}_t - \lambda_t\tilde{h}_t) = \tilde{H}_t(\tilde{g}_t) = (\nabla F)[z-tu(z)]$$

Since the norm of $(\nabla F)[z-tu(z)]$ in $L^p(d\sigma)$ of ∂D is bounded we find that the norm of $\tilde{g}_t - \lambda_t\tilde{h}_t$ is bounded and we can find a sequence $t_k \to 0$ such that for $t = t_k$ $\tilde{g}_t - \lambda_t\tilde{h}_t$ converges weakly to a limit g_1 in $L^p(d\sigma)$. This clearly implies that

$$P_t(g_t - \lambda_t h_t) = F(x) - \lambda_t$$

in D_t converges pointwise to a limit as $t = t_k$ tends to zero. This in turn implies that λ_t, $t = t_k$, converges to a limit λ and we find that

$$P(g_1) = \lim_{t = t_k \to 0} P_t(g_t - \lambda_t h_t) = F(x) - \lambda.$$

Now let h be such that $P(h) = 1$ in D. As we pointed out above $h \in L^p(d\sigma)$ and therefore $g = g_1 + \lambda h$ also belongs to $L^p(d\sigma)$, and $P(g) = F$ in D.

That g is unique follows from the fact that if $g \in L^p(d\sigma)$ and $P(g) = 0$ in D, then $Hg = 0$ and therefore $g \in L^2(d\sigma)$ and 3.13 in [4] yields $g = 0$.

References

[1] Fabes, E.B., Jodeit, M., Riviere, N.M., Potential Techniques for boundary value problems on C^1 Domains, Acta Math. 141 (1978), 165-186.

[2] Coifman, R.R., David, G., Meyer, Y., La Solution des conjectures de Calderón, Advances in Math., 48, nº 2 (1983), 144-148.

[3] Dahlberg, B.E.J., On the Poisson integral for Lipschitz and C^1 Domains, Studia Math. 66, (1979), 7-24.

[4] Verchota, G.C., Layer Potentials and Boundary Value Problems for Laplace's Equation on Lipschitz Domains, Dissertation, Dept. of Math. University of Minnesota, 1982.

Recent Progress in Fourier Analysis
I. Peral and J.-L. Rubio de Francia (Editors)

RADIAL FOURIER MULTIPLIERS AND ASSOCIATED MAXIMAL FUNCTIONS

Anthony Carbery*
California Institut of Technology

I. In this talk we intend to show how a certain square function introduced by E. M. Stein can be used to obtain "general" multiplier and maximal multiplier theorems for radial Fourier multipliers. The multiplier theorem extends the theorem of Carleson and Sjölin, [5] to radial multipliers of $\mathbb{R}^2$ which are not smooth away from a one dimensional "singularity" (as are the Bochner-Riesz multipliers $(1 - |\xi|^2)_+^\alpha$, $\alpha > 0$) and the maximal theorem generalizes the result of [1] concerning almost-everywhere convergence of Bochner-Riesz means on $\mathbb{R}^2$ to a wider class of functions, as well as providing a unified approach to certain other opertors associated to maximal and pointwise convergence problems, including Stein's spherical maximal function, [14] and the solution operator to the linearised Schrödinger equation $\Delta u = i\partial u/\partial t$, $u(x,0) = f$.

Let us begin with the square function. For $\alpha > 1/2$, let $\widehat{\theta^\alpha}(\xi) = |\xi|(|\xi| - 1)_+^{\alpha-1}$ and let $G^\alpha(f)(x) = (\int_0^\infty |\theta_t^\alpha * f(x)|^2\, dt/t)^{1/2}$ (where $\psi_t(x) = t^{-n}\psi(x/t)$). Stein introduced G^α in [12] where he used it to study the L^2 behaviour of the maximal Bochner-Riesz operator. In fact, the easy result about G^α on L^2 is that if $\alpha > 1/2$, then $\|G^\alpha(f)\|_2 \leq C_\alpha \|f\|_2$. Sunouchi [18] observed that if $\alpha > \frac{n+1}{2}$, then G^α behaves as a vector-valued Calderón-Zygmund singular integral operator, and so $\|G^\alpha(f)\|_p \leq C_{p,n,\alpha}\|f\|_p$ for $1 < p < \infty$. This seems to be all that was known about G^α until fairly recently. However we now have further knowledge of G^α's $L^p(\mathbb{R}^n)$ ranges of boundedness when $n = 1$ or 2.

* Research supported in part by NSF Grant MCS 820-3319.

Theorem 1. *Let* $n = 1$ *or* 2, $\alpha > 1/2$, *and* $1 < p < \infty$. *Then* $\|G^\alpha(f)\|_p \leq C_{p,n,\alpha}\|f\|_p \iff \frac{2n}{n-2\alpha} > p > \frac{2n}{n+2\alpha-1}$.

The case $\alpha > \frac{n+1}{2}$ of the theorem is Sunouchi's result, and the case $\alpha > 1/2$, $p = 2$ is Stein's. The case $\alpha > \frac{n}{2}$, $p \geq 2$ seems to be due to Córdoba [7] when $n = 1$, and is possibly new in general*. The case $\alpha > 1/2$, $2 \leq p \leq 4$, $n = 2$ is a consequence of Carbery, [1]. (This is the only "critical" case which does not at present generalise to all dimensions). The remaining positive cases follow by complex interpolation using Stein's theorem (see [17]) for analytically varying families of operators. (Unfortunately we have to complicate G^α a little by letting α take on complex values; but using the formula (1) below and arguing as in [17] we see that $\|G^\alpha(f)\|_p \leq C\|f\|_p \Rightarrow \|G^{\alpha+\beta}(f)\|_p \leq C(\Gamma(\mathrm{Re}\ \beta)/|\Gamma(\beta)|)\|f\|_p$ when $\mathrm{Re}\ \beta > 0$ and $\mathbb{R}$, which is good enough for us to apply Stein's theorem).

Than G^α cannot be bounded unless $p > 2n/n+2\alpha-1$ is a corollary of theorem 4 (see application 3 below), and the necessity of the condition $p < 2n/n-2\alpha$ follows from applying theorem 2 below to the Bochner-Riesz means. (These statements remain valid in all dimensions).

Before proceeding to state and prove our main theorems, we need a variant of the formula

$$(1 - |\xi|^2)_+^\alpha = C_{\alpha,\rho} \int_0^1 (1-t^2)^{\alpha-\rho-1}\, t^{2\rho+1} \left(\frac{1-|\xi|^2}{t^2}\right)_+^\rho dt \qquad (1)$$

(valid when $\mathrm{Re}\ \alpha > \rho + 1/2$) which appears in the works of Stein on the maximal Bochner-Riesz and spherical maximal operators, [12] and [14]. This variant is contained in the following well-known Riemann-Liouville formula:

Lemma. *For* $h \in L^2(\mathbb{R})$ *and* $\alpha \geq 0$ *let* $\{(d/dt)^\alpha\ h\}^\wedge(\nu) = (-i\nu)^\alpha h^\wedge(\nu)$. *Suppose that* $\mathrm{supp}\ h \subseteq (-\infty, a]$ *and that* $(d/dt)^\alpha\ h \in L^2(\mathbb{R})$ *for some* $\alpha > 1/2$. *Then* $\mathrm{supp}(\frac{d}{dt})^\alpha h \subseteq (-\infty, a]$ *and*

$$h(x) = c_\alpha \int_x^\infty (t-x)^{\alpha-1} \left(\frac{d}{dt}\right)^\alpha h(t)\,dt \quad \text{(a.e)}. \qquad (2)$$

---

* Details will appear elsewhere.

Proof. For $\varepsilon, \alpha > 0$, h in the hypothesis of the lemma and $f \in L^2(\mathbb{R})$, introduce $\{(d/dt)^\alpha_\varepsilon h\}^\sim(\nu) = (\varepsilon - i\nu)^\alpha \hat{h}(\nu)$ and $I^\alpha_\varepsilon f(x) = \int_x^\infty (t-x)^{\alpha-1} e^{-\varepsilon(t-x)} f(t)dt$. Notice that since $c_\alpha(\varepsilon - i\nu)^{-\alpha}$ is the Fourier transform of the L^1 function $(-x)_+^{\alpha-1} e^{\varepsilon x}$, $h = c_\alpha I^\alpha_\varepsilon (d/dt)^\alpha_\varepsilon h$ and moreover $(d/dt)^\alpha_\varepsilon h = C_\alpha (d/dt)_\varepsilon^{[\alpha]+1} I_\varepsilon^{1-\alpha+[\alpha]} h$. Thus $\operatorname{supp} (d/dt)^\alpha_\varepsilon h \subseteq (-\infty, a|$, and an application of the dominated convergence theorem shows that $(d/dt)^\alpha_\varepsilon h \to (d/dt)^\alpha h$, and that for each x, $e^{-\varepsilon(t-x)} \chi_{[x,\infty)}(t)(d/dt)^\alpha_\varepsilon \to \chi_{[x,\infty)}(t)(\frac{d}{dt})^\alpha h$ in L^2 as $\varepsilon \to 0$. Therefore, since the function $(t-x)^{\alpha-1} \chi_{[x,a]}(t)$ belongs to L^2 when $\alpha > 1/2$, we have for almost every x that

$$h(x) = c_\alpha I^\alpha_\varepsilon (d/dt)^\alpha_\varepsilon h(x)$$

$$= c_\alpha \int_x^\infty (t-x)^{\alpha-1} e^{-\varepsilon(t-x)} (d/dt)^\alpha_\varepsilon h(t)dt \underset{\varepsilon\to 0}{\longrightarrow} c_\alpha \int_x^\infty (t-x)^{\alpha-1} (d/dt)\ h(t)dt,$$

concluding the proof of the lemma.

II. We are now in a position to state the main inequality from which multiplier theorems may be deduced. Let ϕ be a fixed nonnegative smooth bump function supported in $[1,2]$. For $1 < q < \infty$ and $\alpha > 1/q$ let $R(q,\alpha) = \{m \in L^\infty(0,\infty) \mid \|m\|_{R(q,\alpha)} = \sup_{0<u<\infty} \|(d/dt)^\alpha \{\phi(t)\ m\ (ut)\}\|_{L^q(\mathbb{R})} < \infty\}$. Let $\hat{\psi}(\xi) = |\xi|\ \phi(\xi)$, and $g(f)(x) = (\int_0^\infty |\psi_t * f(x)|^2 dt/d)^{1/2}$.

Theorem 2. *Let* $\alpha > 1/2$ *and suppose that* $m \in R(2,\alpha)$. *Let* $Tf(\xi) = m(|\xi|)\ \hat{f}\ (\xi)$. *Then*

$$g(Tf)(x) \leq C_\alpha \|m\|_{R(2,\alpha)}\ G^\alpha(f)(x) \tag{3}$$

Corollary 3. ([2]). Let $n = 2, \alpha > 1/q$ and suppose that $m \in R(q,\alpha)$. Then $m(|\xi|)$ is an $L^p(\mathbb{R}^2)$ multiplier when $\left|\frac{1}{p} - \frac{1}{2}\right| < \frac{1}{2}(\alpha - (\frac{1}{q} - \frac{1}{2})_+)$.

Remarks 1. The most interesting special case of the corollary is the case $q = 2, \alpha > 1/2$. It then states that a radial function, which, when regarded as a function on $(0,\infty)$ belongs to $R(2,\alpha)$, $\alpha > 1/2$, is an $L^p(\mathbb{R}^2)$ multiplier for $4/3 \leq p \leq 4$. This extends the

Carleson-Sjölin theorem because $(1-t^2)_+^\lambda \in R(q,\alpha)$ when $\alpha < \lambda + 1/q$. This special case follows from theorem 2, the case $n = 2$, $p = 4$ of theorem 1 and the inequality $\|Tf\|_p \leq C_{p,n}\|g(Tf)\|_p$, $1 < p < \infty$, which is a standard result from singular integrals, [13].

2. The case $q = 2$, $\alpha > 1$ of the corollary is a version of Hörmander's multiplier theorem for radial functions, and follows from the case $n = 2$, $\alpha > 1$, $p \geq 2$ of theorem 1.

The remaining cases follow from embedding and interpolation properties of $R(q,\alpha)$ spaces. (For these properties and a comparison with WBV spaces, see [3]).

3. The corollary is best possible in the sense that index $1/2(\alpha - (\frac{1}{q} - \frac{1}{2})_+)$ appearing in its statement cannot be increased. This may be seen in the case $q \leq 2$ by examining the Bochner-Riesz multipliers, and in the case $q \geq 2$ smooth multipliers of the form $e^{i|\xi|^a}/|\xi|^{\alpha a}$, (large $|\xi|$), $0 < a < 1$ which belong to $R(q,\alpha)$ for $1 < q < \infty$. The precise range of p's for which these latter multipliers give bounded operators on L^p was determined in [10].

Proof of Theorem 2. Apply the lemma with $h(s) = \phi(su)\, m\,(s)$: $\phi(su)\, m\,(s) = C_\alpha \int_s^\infty (t-s)^{\alpha-1}(\frac{d}{dt})^\alpha [\phi(tu)\, m\,(t)]dt$, which is valid by the assumption on m. Therefore, $\hat{\psi}(\xi u)\, m\,(|\xi|)\hat{f}(\xi) =$
$= |\xi|u\phi(|\xi|u)\, m\,(\xi)\, \hat{f}(\xi)$

$$= C_\alpha\, |\xi|u\int_{|\xi|}^\infty (t-|\xi|)^{\alpha-1}(\frac{d}{dt})^\alpha[\phi(tu)m(t)]dt\ \hat{f}(\xi)$$

$$= C_\alpha\, u\int_0^{2/u} (\frac{1-|\xi|}{t})_+^{\alpha-1}\, \frac{|\xi|}{t}\ t^\alpha(\frac{d}{dt})^\alpha\ [\phi(tu)m(t)]dt\ \hat{f}(\xi),$$

since $\operatorname{supp}\,(d/dt)^\alpha\ [\phi(ut)m(f)] \subseteq (-\infty,2/u|$. Thus

$$\psi_u * Tf(x) = C_\alpha\, u\int_0^{2/u} \theta_{1/t}^\alpha * f(x)t^\alpha\ (\frac{d}{dt})^\alpha[\phi(tu)m(t)]dt,$$

so $|\psi_u * Tf(x)| \leq C_\alpha\ (u\int_0^{2/u} |\theta_{1/t}^\alpha * f(x)|^2\ dt)^{1/2}\ \|m\|_{R(2,\alpha)}$,

and $G(Tf)(x) \leq C_\alpha\|m\|_{R(2,\alpha)}\ (\int_0^\infty u(\int_0^{2/u} |\theta_{1/t}^\alpha * f(x)|^2 dt\ \frac{du}{u})^{1/2}$

$$\leq C_\alpha\ \|m\|_{R(2,\alpha)}\ G^\alpha(f)^*(x)$$

as required.

III. We turn now to the maximal operator associated to a radial Fourier multiplier $m(|\xi|)$.

If

$$\widehat{T_t f}(\xi) = m(t|\xi|)\hat{f}(\xi), \quad \text{let}$$

$$T_* f(x) = \sup_{0<t<\infty} |T_t f(x)|.$$

We seek "general" conditions on a multiplier m that allow us to assert an inequality of the form $\|T_* f\|_{L^p(\mathbb{R}^n)} \leq C_{p,n}\|m\|_X \|f\|_{L^p(\mathbb{R}^n)}$ for an appropriate Banach space X of bounded functions defined on $(0,\infty)$. Since $|Tf(x)| \leq T_* f(x)$, one is led to believe the appropriate dilation invariant spaces X to consider are embedded in the spaces $R(q,\alpha)$ - in fact we need potential spaces of which the $R(q,\alpha)$ are localized versions. For $\alpha > 1/2$, let L^2_α be the completion of the C^∞ functions of compact support in $(0,\infty)$ under the norm $\|m\|^s_{L^2_\alpha} = \int_0^\infty \left| s^{\alpha+1} (d/ds)^\alpha \left\{\frac{m(s)}{s}\right\}\right|^2 \frac{ds}{s}$. A simple computation shows that the Mellin transform of $s^{\alpha+1}(d/ds)^\alpha \left\{\frac{m(s)}{s}\right\}$ is essentially $(it + 1)^\alpha$ times the Mellin transform of m, and so the space L^2_α is just the usual Bessel potential space $L^2_\alpha(R)$, (as found for example in Stein [13] Ch. 5) under the change of variables $s \to \exp s$.

Theorem 4. *Suppose* $m \in L^2_\alpha$ *for some* $\alpha > 1/2$. *With* $T*$ *as above*

$$T_* f(x) \leq C_\alpha \|m\|_{L^2_\alpha} G^\alpha(f)(x).$$

Corollary 5. Let $n = 1$ or 2, $\alpha > 1/2$ and suppose that $m \in L^2_\alpha$. Then the maximal operator associated with m is bounded on $L^p(\mathbb{R}^n)$ when

$$\frac{2n}{n-2\alpha} > p > \frac{2n}{n+2\alpha-1}$$

Remarks 1. The corollary follows directly from Theorems 1 and 4. By considering the Bochner-Riesz means, we see that the index $2n/(n-2\alpha)$ cannot be raised; to see that $2n/(n+2\alpha-1)$ cannot be lowered, see application 3 below.

2. If $D^s f(\xi) = |\xi|^s \hat{f}(\xi)$ for $s \in \mathbb{R}$ we see immediately that

$$|T_t f(x)| \leq C_\alpha t^s \left\|\frac{m(.)}{|.|^s}\right\|_{L^2_\alpha} G^\alpha(D^s f)(x).$$

A corresponding simple maximal inequality (valid in all dimensions and not reflecting the full "Bochner-Riesz" characterístics of G^α) is

$$\left\| \sup_{t>0} \frac{|T_t f(x)|}{t^s} \right\|_{L^p(\mathbb{R}^n)} \leq C_\alpha \left\| \frac{m(.)}{|.|^s} \right\|_{L^2_\alpha} \| D^s f \|_{L^p(\mathbb{R}^n)},$$

where $\alpha > 1/2$ and $\frac{2n-2}{n-2\alpha} < p < \frac{2n}{n+2\alpha-1}$.

Proof of Theorem 4. We proceed as in the proof of Theorem 2. Let $m \in C_c^\infty(0,\infty)$ and apply the lemma:

$$\frac{m(u)}{u} = C_\alpha \int_u^\infty (v-u)^{\alpha-1} \left(\frac{d}{dv}\right)^\alpha \frac{m(v)}{v}\, dv.$$

Therefore $m(u) = C_\alpha \int_0^\infty (1 - \frac{u}{v})_+^{\alpha-1} \frac{u}{v} v^\alpha \left(\frac{d}{dv}\right)^\alpha \left[\frac{m(v)}{v}\right] dv.$

Hence $T_t f(x) = C_\alpha \int_0^\infty \theta^\alpha_{\frac{t}{v}} * f(x) v^\alpha \left(\frac{d}{dv}\right)^\alpha \left(\frac{m(v)}{v}\right) dv.$

Finally, applying the Cauchy-Schwarz inequality, we obtain

$$T_* f(x) \leq C_\alpha \|m\|_{L^2_\alpha} G^\alpha(f)(x).$$

Applications 1. Let $M(\xi) = (1 - |\xi|^2)_+^\alpha$, $\alpha > 0$, $\xi \in \mathbb{R}^2$. Then the corresponding maximal operator is bounded on $L^4(\mathbb{R}^2)$. [1]

2. Let ϕ be a C^∞ bump function supported on the annulus $\{\xi \in \mathbb{R}^2 : ||\xi| - 1| < \delta\}$. Then for small δ, the corresponding maximal operator T_* satisfies $\|T_*^\delta f\|_{L^4} \leq C_\varepsilon (1/\delta)^\varepsilon \|f\|_4$ for all $\varepsilon > 0$. In fact, a slightly finer analysis (see the lecture of Córdoba in these proceedings) gives a power of $\log(1/\delta)$ in place of $O((1/\delta)^\varepsilon)$ $\forall\varepsilon > 0$.

3. Let $m_\beta(\xi)$, $-n/2 < \beta \leq 0$, be a smooth multiplier vanishing near zero and of the form $C_\beta e^{i|\xi|}/|\xi|^{\frac{n+1}{2}+\beta}$ for $|\xi| \geq 1$, and let $T^\beta f(\xi) = m_\beta(\xi)\hat{f}(\xi)$. Modulo a term belonging to S, the kernels of the multiplier operators T^β are the characteristic function of the unit ball when $\beta = 0$ and the uniform surface measure on the unit sphere in $\mathbb{R}^n$ $(n \geq 2)$ when $\beta = -1$ respectively and so we are considering the Hardy-Littlewood maximal function and Stein's spherical maximal function [14] in these special cases. Combining remark 2 with the obvious L^∞ estimate when $\beta = -1$ yields that $\|T_*^\beta f\|_{L^p(\mathbb{R}^n)} \leq C_{\beta,n} \|f\|_{L^p}$ when $\max\{0, \frac{-(\beta+1)}{n-2}\} < \frac{1}{p} < \frac{n+\beta}{n}$ for $-n/2 < \beta \leq 0$. (When $n = 2$, read $0 < \frac{1}{p} < \frac{2+\beta}{2}$ for $-1 < \beta \leq 0$). When $-1 \leq \beta \leq 0$, it is known [16]

that the range of parameters $n/(n+\beta) < p$ cannot be improved - consequently G^{α} cannot be bounded on $L^p(\mathbb{R}^n)$ unless $p > \frac{2n}{n+2\alpha-1}$.

4. Let $m(\xi) = e^{i|\xi|^2}$. Then the linearised Schrödinger equation $\Delta u(x,t) = i\,\partial u / \partial t(x,t)$, $t > 0$, $u(x,0) = f(x)$, has a solution operator $u(x,t) = T_{\sqrt{t}}\, f(x)$. Conditions on f sufficient to imply that $u(x,t) \to f(x)$ a.e. as $t \to 0$ have been studied by Carleson, [4], Dahlberg and Kenig, [9] and Kenig and Ruiz [11]. The results of these papers show that if $f \in L^2_\alpha(\mathbb{R}^n)$, $\alpha \geq n/4$, then $\sup_{0<t<1} |T_t\, f(x)|$ is locally in $L^2(\mathbb{R}^n)$. ([11]). Here we show that $\sup_{0<t<1} |T_t\, f(x)|$ is in $L^2(\mathbb{R}^n)$ if $\alpha > 1$. (This result has also been obtained by M. Cowling [8] and E. M. Stein [15]). Since the proof uses the pointwise inequality of theorem 4, the result extends to the context of $L^p_\alpha(\mathbb{R}^n)$ spaces, $1 < p < \infty$. In fact if we subtract a term $\psi \in S$ from $e^{i|\xi|^2}$ agreeing with it on $|\xi| \leq 1/2$, the resulting multiplier $n(\xi) = e^{i|\xi|^2} - \hat{\psi}(\xi)$ satisfies

$$|S_t\, f(x)| \leq C_\alpha\, t^s \left\| \frac{m(.)}{|.|^s} \right\|_{L^2_\alpha} G^{\alpha}(D^s f)(x)$$

when $\alpha > 1/2$ by remark 2, (where $\hat{S}\, g(\xi) = n(\xi)\hat{g}(\xi)$). Consequently

$$|T_t\, f(x)| \leq |\psi_t * f(x)| + C_{s,\alpha} t^s \; G^{\alpha}(D^s f)(x)$$

when $s > 2\alpha$.

References

[1] A. Carbery, The Boundedness of the Maximal Bochner-Riesz operator on $L^4(R^2)$, Duke Math. J. 50 (1983), 409-416.

[2] A. Carbery, G. Gasper and W. Trebels, Radial Fourier Multipliers of $L^p(R^2)$, preprint.

[3] A. Carbery, G. Gasper and W. Trebels, Manuscript in preparation.

[4] L. Carleson, Some analytic problems related to Statistical Mechanics, in Lecture Notes in Math., 779 (1979), 5-45.

[5] L. Carleson and P. Sjolin, Oscillatory Integrals and a Multiplier problem for the disc, Studia Math. 44, (1972), 287-299.

[6] A. Córdoba, A Note on Bocher-Riesz Operators, Duke Math. J. 46 (1979), 505-511.

[7] A. Córdoba, Some Remarks on the Littlewood-Paley Theory, Supplemento ai Rendiconti del Circolo Mat. di Palermo, II, 1, (1981), 75-80.

[8] M. Cowling, Pointwise Behaviour of Solutions to Schrödinger equations, preprint.

[9] B. Dahlberg and C. Kenig, A Note on the Almost everywhere behaviour of Solutions to the Schrödinger equation, in Lecture Notes in Math, 908 (1982), 205-209.

[10] C. Fefferman and E.M. Stein, H^p spaces of several variables, Acta Math., 129 (1972), 137-193.

[11] C. Kenig and A. Ruiz, personal communication.

[12] E. M. Stein, Localisation and Summability of Multiple Fourier Series, Acta Math., 100 (1958), 92-147.

[13] E. M. Stein, Singular Integrals and Differentiability Properties of Functions, Princeton U. Press, Princeton N.J. (1970).

[14] E. M. Stein, Maximal Functions - Spherical means, P.N.A.S. U.S.A. 73, (1978), 2174-5.

[15] E. M. Stein, personal communication via C. Kenig.

[16] E. M. Stein and S. Wainger, Problems in Harmonic Analysis related to curvature, BAMS 84, 6, (1978), 1239-1295.

[17] E. M. Stein and G. Weiss, Fourier Analysis on Euclidean Spaces, Princeton U. Press, Princeton, N.J. (1971).

[18] G. Sunouchi, On the Littlewood-Paley Function g^* of Multiple Fourier Integrals and Hankel Multiplier Transformation, Tôhoku Math. J. 19, 4, (1967), 496-523.

Recent Progress in Fourier Analysis
I. Peral and J.-L. Rubio de Francia (Editors)

RESTRICTION LEMMAS, SPHERICAL SUMMATION, MAXIMAL FUNCTIONS, SQUARE FUNCTIONS AND ALL THAT

A. Córdoba
Universidad Autónoma de Madrid

The Fourier transform, defined on integrable functions f by the formula $\hat{f}(\xi) = \int_{R^n} e^{-2\pi i x.\xi} f(x)dx$, plays a crucial role in many branches of Mathematics. Here we shall emphasize certain properties of this transformation which are not too well known and whose understanding, to a reasonable level of depth, remains to be completed.

To begin, let us consider, once more, the interplay between the Fourier transform and the family of dilations of euclidean space: given a real number $\delta > 0$ and a function f let $f_\delta(x) = f(\delta x)$ for every $x \in R^n$, we then have that

$$\hat{f}_\delta(\xi) = \delta^{-n} f(\delta^{-1}.\xi).$$

Another basic fact for our discussion is that the Gaussian $e^{-\pi|x|^2}$ is equal to its own transform. Combining this result with the previous one we get the family of formulas:

$$\widehat{\delta^{-n/4} e^{-\pi|x|^2/\delta}}(\xi) = \delta^{n/4} e^{-\pi\delta|x|^2}$$

which give us a good enough formulation, for our later considerations, of the socalled uncertainty principle.

We shall present now the basic operators: Hilbert transform (partial Sum operators) and Egorof operator, defined by the formulas:

$$\widehat{Hf}(\xi) = i \text{ Sign } (\xi) \hat{f}(\xi)$$

$$\widehat{S_n f}(\xi) = \chi_{[-n,+n]}(\xi) \hat{f}(\xi)$$

$$Ef(x) = \int_{R^n} e^{i S(x,\xi)} \hat{f}(\xi)d\xi, \quad \text{where the phase } S(x,\xi)$$

is a real valued homogeneous symbol of degree one.

And their estimates:

a) $\|(\Sigma |S_k f_k|^r)^{1/r}\|_p \leq C_p \|(\Sigma |f_k|^r)^{1/r}\|_p$

$$1 < r < \infty, \quad 1 < p < \infty.$$

b) $\|\operatorname{Sup}_k |S_k f|\|_p \leq C_p \|f\|_p, \qquad 1 < p < \infty.$

c) $\int_{-\infty}^{+\infty} |Hf(x)|^r \omega(x)\,dx \leq C_{r,s} \int_{-\infty}^{+\infty} |f(x)|^r [(\omega^s)^*(x)]^{1/s}\,dx,$

$$1 < r,s < \infty, \quad \omega \in L^1_{loc}(\mathbb{R})$$

where $*$ denotes the Hardy-Littlewood maximal function.

d) $\|Ef\|_2 \leq C\|f\|_2$.

Results

[1] Let $\gamma : [0,1] \to \mathbb{R}^n$ be a smooth curve such that $\{\gamma'(t),\ldots,\gamma^{(n)}(t)\}$ spands $\mathbb{R}^n$, $\forall t \in [0,1]$, and let $d\sigma$ be a smooth measure over γ.

THEOREM 1. *The Fourier transform extends to a continuous map from* $L^p(\mathbb{R}^n)$ *to* $L^q(d\sigma)$:

$$\left(\int_\gamma |\hat{f}(\xi)|^q \, d\sigma\right)^{1/q} \leq C_{p,q}(\gamma, d\sigma) \|f\|_{L^p(\mathbb{R}^n)},$$

$$1 \leq p < \frac{n(n+2)}{n(n+2) - 2}, \quad \frac{1}{q} \geq \frac{n(n+1)}{2}\left[1 - \frac{1}{p}\right]$$

(references: [1]).

[2] Let Γ be a circular cone in $\mathbb{R}^3$ with measure $d\sigma = dr\,d\theta$.

THEOREM 2. $\left|\int_\Gamma |\hat{f}(\xi)|^q d\sigma\right|^{1/q} \leq C_p \|f\|_{L^p(\mathbb{R}^3)}$

$$1 \leq p < 4/3, \quad \frac{1}{q} = 3\left[1 - \frac{1}{p}\right].$$

(reference [2]).

[3] Let $E \subset \mathbb{R}^n$ be a convex set, with a smooth boundary M whose curvature K is strictly positive everywhere.

Let us denote by $G : M \to S^{n-1}$ the Gauss map and let us define

$$\alpha(\xi) = G^{-1}\left(\frac{\xi}{|\xi|}\right).$$

Given $f d\sigma$ a smooth measure over M we have the following asymtotic formula:

$$\widehat{fd\sigma}(\xi) = C_n \frac{e^{2\pi i\alpha(\xi).\xi} f(\alpha(\xi))}{|\xi|^{\frac{n-1}{2}} \sqrt{K(\alpha(\xi))}} + $$

$$+ \overline{C_n} \frac{e^{2\pi i\alpha(-\xi).\xi} f(\alpha(-\xi))}{|\xi|^{\frac{n-1}{2}} \sqrt{K(\alpha(-\xi))}} + O(|\xi|^{-\frac{n+1}{2}})$$

(Reference [3]).

[4] Spherical summation below the critical index

With $\alpha > 0$ let us consider the family of operators:

$$\widehat{S_R^\alpha f}(\xi) = (1 - \frac{|\xi|^2}{R^2})_+^\alpha \cdot \hat{f}(\xi)$$

i.e. $S_R^\alpha f = K_R^\alpha * f$, where

$$K_1^\alpha(x) = \Gamma(1+\alpha)\pi^{-\alpha} |x|^{-\frac{n}{2}-\alpha} J_{\frac{n}{2}+\alpha}(2\pi|x|)$$

and J_ν denotes the Bessel function of order ν.

THEOREM 3. *In* $\mathbb{R}^2$, $0 < \alpha < 1/2$

a) $\|(\Sigma |S_{R_j}^\alpha f_j|^2)^{1/2}\|_p \le C_p \|(\Sigma |f_j|^2)^{1/2}\|_p$,

$$4/(3+2\alpha) < p < 4/(1-2\alpha),$$

Uniformly on any sequence of $\{R_j\}$.

b) *With* $S_*^\alpha f(x) = \sup_R |S_R^\alpha f(x)|$ *we have*

$$\|S_*^\alpha f\|_p \le C_p \|f\|_p, \quad 2 \le p < 4/(1-2\alpha).$$

c) *If* $\{R_j\}$ *is lacunary then:*

$$\lim_{R_j\to\infty} S_{R_j}^\alpha f(x) = f(x). \text{ a.e. } x \quad \textit{for every}$$

$$f \in L^p(\mathbb{R}^2), \quad 4/(3+2\alpha) < p < 4/(1-2\alpha).$$

References: [4], [5], [6].

Auxiliary operators

[1] Maximal functions

Let $\gamma : [0,1] \to S^{n-1}$ be a smooth curve.

Given $N >> 1$, let us consider

$$\mathcal{B}_N = \{\text{cylinders of eccent.} = \frac{\text{height}}{\text{radius}} = N \text{ and direction in } \gamma\}$$

and the operator

$$Mf(x) = \sup_{x \in R \in \mathcal{B}_N} \frac{1}{|R|} \int_R |f(y)|dy$$

THEOREM 5. *There exists a finite constant* C_γ, *independent of* N, *such that*

$$\|Mf\|_2 \leq C_\gamma |\log N|^2 \|f\|_2.$$

References: [7], [8].

[2] Square functions

(A) In $\mathbb{R}^n$ let us consider a cubic lattice $\{Q_\nu\}$ and associated operators

$$P_\nu f = \chi_{Q_\nu} \cdot \hat{f}$$

$$Gf(x) = (\Sigma |P_\nu f|^2)^{1/2}.$$

THEOREM 5. *For every* $s > 1$, *there exists a finite constant* C_s *such that, for every* $\omega \in C_0(\mathbb{R}^n)$ *we have:*

$$\int_{\mathbb{R}^n} |Gf(x)|^2 \omega(x)dx \leq C_s \int_{\mathbb{R}^n} |f(x)|^2 [(\omega^s)^*(x)]^{1/s} dx$$

Corollary. $\|Gf\|_p \leq C_p \|f\|_p$, $2 \leq p < \infty$.

(here $*$ denote the Hardy-Littlewood maximal functions).

References: [8].

(B) Let us divide $\mathbb{R}^2$ into N equal angles (see figure) and consider the operators

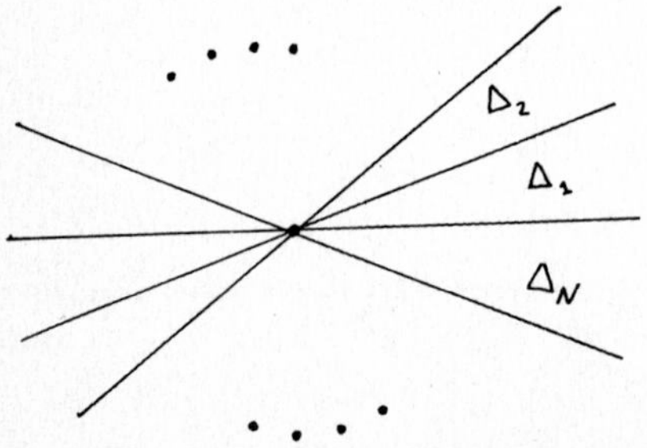

$$\widehat{P_j f}(\xi) = \chi_{\Delta_j}(\xi)\hat{f}(\xi)$$

$$Sf(x) = (\Sigma |P_j f|^2)^{1/2}$$

$$\Delta_j = \{\xi : \frac{2\pi j}{N} \leq \arg(\xi) \leq \frac{2\pi(j+1)}{N}\}$$

THEOREM 6. *There exist finite constants* C, a, *independent of* N, *such that*

$$\|Sf\|_4 \leq C\,|\log N|^a\,\|f\|_4$$

References: [8].

If we change now our division of $\mathbb{R}^2$ to the case of a sequence of lacunary angles $\{\Delta_j\}$ and we consider:

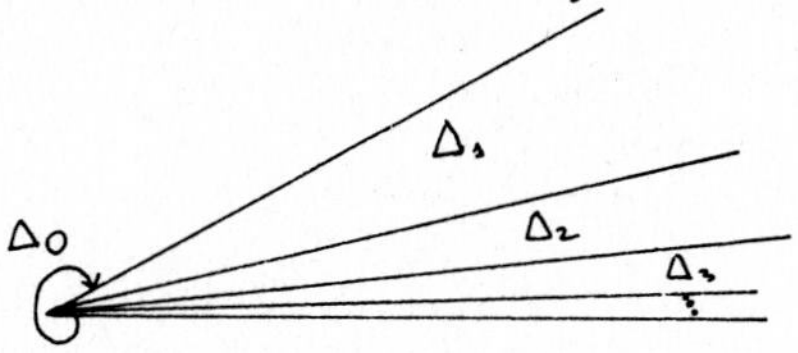

$$Sf = \left(\sum_{j=0}^{\infty} |P_j f|^2\right)^{1/2}$$

Then:

$$\|Sf\|_p \sim \|f\|_p, \quad 1 < p < \infty.$$

See references: [9], [10].

(C) Let $\phi \in C_0^\infty(\mathbb{R})$, $\mathrm{supp}(\phi) \subset (1-\delta, 1+\delta)$, $|\phi^{(n)}(x)| \leq \delta^{-n}$, $n = 0,1,2$, and define

$$\widehat{M_t f}(\xi) = \phi\left(\frac{|\xi|}{t}\right) \hat{f}(\xi), \quad \xi \in \mathbb{R}, \quad t > 0$$

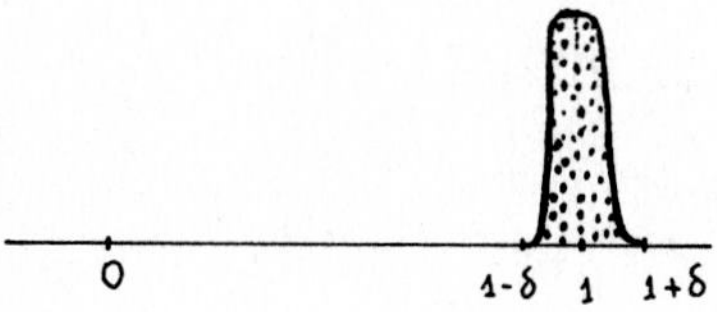

$$g(f)(x) = \left[\int_0^\infty |M_t f(x)|^2 \frac{dt}{t}\right]^{1/2}$$

THEOREM 7. *There exists a finite constant* C_0 *such that for every* s, $1 < s < \infty$, *we have the inequality:*

$$\int_{-\infty}^{+\infty} g(f)^2(x)\omega(x)\,dx \leq C\,\delta \int_{-\infty}^{+\infty} |f(x)|^2 A_s\omega(x)\,dx$$

(*where* $A_s\omega(x) = [(\omega^s)^*(x)]^{1/2}$ *as before*).

ONE PROOF

[A] Proof of part (b) of theorem 4.

First we decompose the multiplier $\left(1 - \frac{|\xi|^2}{R^2}\right)_+^\alpha$ and the operator S_R^α:

$$S_R^\alpha \sim \sum_{n=1}^{\infty} 2^{-n\alpha}\, T_R^{2^{-n}}$$

Where $\widehat{T_R^\delta f}(\xi) = \phi(\frac{|\xi|}{R})\hat{f}(\xi)$, ϕ as in

<u>THEOREM 7</u>.

$(1 - \frac{|\xi|^2}{R^2})_+^\alpha \sim$

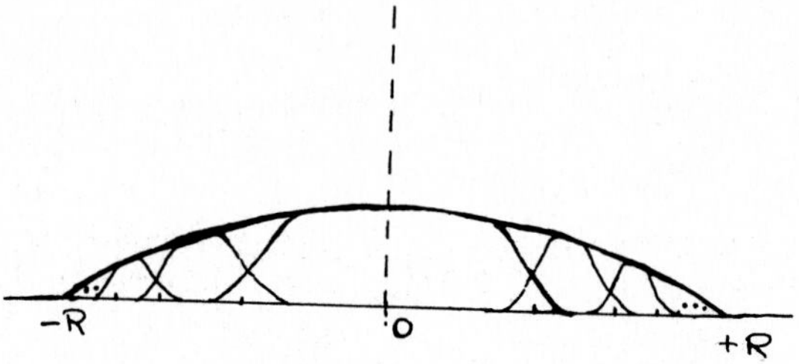

and

$$T_*^\delta f(x) = \sup_{R>0} |T_R^\delta f(x)|$$

<u>Claim</u>: For each p, $2 \leq p < \frac{4}{1-2\alpha}$, there exist finite constants $C = C(p)$ and $a = a(p)$ such that:

$$\|T_*^\delta f\|_p \leq C\,|\log\delta|^a\,\|f\|_p.$$

Observe that

$$(T_t^\delta f(x))^2 = \int_0^t \frac{d}{ds}(T_s^\delta f(x))^2 ds =$$

$$= 2\int_0^t T_s^\delta f(x)\,\frac{d}{ds}T_s^\delta f(x)ds \leq$$

$$\leq 2\left[\int_0^\infty |T_s^\delta f(x)|^2\frac{ds}{s}\right]^{1/2}\left[\int_0^\infty s\left|\frac{d}{ds}T_s^\delta f(x)\right|^2 ds\right]^{1/2}$$

Furthermore,

$$\frac{d}{dt}\widehat{T_t^\delta f}(\xi) = -\frac{|\xi|}{t^2}\phi'(\frac{|\xi|}{t})\hat{f}(\xi) = \delta^{-1}\frac{1}{t}\psi(\frac{|\xi|}{t})\hat{f}(\xi)$$

where $\psi(\frac{|\xi|}{t}) = -\delta\frac{|\xi|}{t}\phi'(\frac{|\xi|}{t})$ and ψ satisfies the same estimates that ϕ. Therefore

$$(T_t^\delta f(x))^2 \leq 2\,\delta^{-1}A_\phi f(x)\,A_\psi f(x)$$

where $A_\phi f(x) = \left[\int_0^\infty |T_t^\delta f(x)|^2\frac{dt}{t}\right]^{1/2}$ and analogously A_ψ.

It will be enough to show that:

$$\|A_\phi f\|_4 \leq C\,\delta^{1/2}(\log\delta)^a\,\|f\|_4.$$

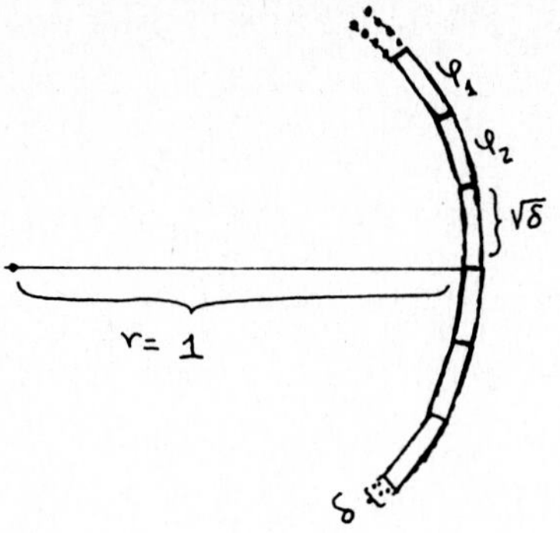

With $\delta > 0$ fixed, we decompose further $\phi = \sum_{j=1}^{\delta^{-1/2}} \phi_j$ and

$$T_t^\delta = \sum_j S_t^j.$$

See reference [7]

We have used the same geometric arguments as in references [4], [7].

$$\left\| \left[\int_0^\infty |T_t f(x)|^2 \frac{dt}{t}\right]^{1/2} \right\|_4 \leq$$

$$\leq C\,|\log \delta|\, \left\| \left[\sum_{j=1}^{\delta^{-1/2}} \int_0^\infty |S_t^j f(x)|^2 \frac{dt}{t}\right]^{1/2} \right\|_4$$

To finish we just compute:

$$\sum_j \int_0^\infty \int_{\mathbb{R}^2} |S_t^j f(x)|^2\, \omega(x)dx\, \frac{dt}{t}$$

With the notation of theorem 6. $N = \delta^{-1/2}$

$$S_t^j f = S_t^j P_j f$$

We may apply Theorem 7 to obtain:

$$I = \int_{\mathbb{R}^2} \left(\int_0^\infty |S_t^j f(x)|^2 \frac{dt}{t}\right) \omega(x)dx \leq$$

$$\leq C_s \delta \int_{\mathbb{R}^2} |P_j f(x)|^2 \left(M_j\, \omega^s(x)\right)^{1/s} dx$$

where M_j denotes the maximal function in the direction of the "rectangle" corresponding to the S_t^j kernels.

Therefore

$$I \leq C_s \delta \int_{\mathbb{R}^2} \sum_j |P_j f(x)|^2 \left[\operatorname{Sup}_j M_j\, \omega^s(x)\right]^{1/s} dx$$

$$\leq C_s \delta \left\| \left(\sum_j |P_j f(x)|^2\right)^{1/2} \right\|_4^2 \left\| (M\, \omega^s)^{1/s} \right\|_2$$

and we can invoke theorems 5 and 6 to finish the proof.

References

[1] E. Prestini, Thesis, University of Maryland.

[2] B. Barceló, Thesis, Universidad Autónoma de Madrid.

[3] C. Herz, Fourier transforms related to convex sets, Annals of Math. vol. 75, 1, 1962.

[4] A. Córdoba and B. López-Melero, Spherical Summation: A problem of E.M. Stein, Ann. Inst. Fourier, 1981.

[5] Igari, Decomposition theorem and lacunary convergence of Riesz-Bochner means of Fourier transforms of two variables. Tôhoku Math. Journal 33 (1981).

[6] A. Carbery. Thesis, U.C.L.A.

[7] A. Córdoba, The Kakeya maximal function and the spherical summation multipliers, Amer. J. of Math. 1977.

[8] __________, Geometric Fourier Analysis, Ann. Inst. Fourier, 1983.

[9] A. Córdoba and R. Fefferman, On the equivalence between maximal functions and Fourier multipliers on Fourier Analysis, Proc. Acad. Sci., USA, 1977.

[10] A. Nagel, E. Stein and S. Wainger, Differentiation on lacunary directions, Proc. Acad. Sci., USA, 1977.

Recent Progress in Fourier Analysis
I. Peral and J.-L. Rubio de Francia (Editors)

ENSEMBLES ALEATOIRES ET DIMENSIONS

Jean-Pierre Kahane
Université Paris-Sud, Orsay

PREMIERE LEÇON.- Mesures, capacités, dimensions

Mesures et Dimensions de Hausdorff. Lemme de Frostman

En 1919, Hausdorff a introduit les notions suivantes. E étant un espace métrique, h una application croissante de $\mathbb{R}^+$ sur $\mathbb{R}^+$ (donc h est continue et $h(0) = 0$), on associe à tout $\varepsilon > 0$ l'ensemble des recoubrements de E par des boules B_n de diamètre $\leq\varepsilon$:

$$E \subset \bigcup_n B_n, \qquad \text{diam } B_n \leq \varepsilon.$$

Soit $H(\varepsilon)$ la borne inférieure des sommes $\Sigma\, h(\text{diam } B_n)$ pour les recoubrements en question $(0 \leq H(\varepsilon) \leq \infty)$. Quand $\varepsilon \downarrow 0$, $H(\varepsilon) \uparrow H$ $(0 \leq H \leq \infty)$. Par définition, H est la mesure de Hausdorff de E par rapport à la fonction déterminante h, et on écrit

$$H = \text{mes}_h\, E.$$

C'est une mesure extérieure. En conclusion de son article, Hausdorff montre que, si h est concave, il existe un ensemble fermé linéaire tel que $0 < \text{mes}_h\, E < \infty$.

Quand $h(t) = t^\alpha$ $(\alpha > 0)$, on écrit $\text{mes}_\alpha E$ au lieu de $\text{mes}_h E$; c'est, par définition, la mesure de Hausdorff de E en dimension α. Comme fonction de α, $\text{mes}_\alpha E$ est une fonction décroissante, égale à ∞ ou 0 sauf peut-être en un point. On définit en tous cas

$$\alpha_0 = \{\sup \alpha \,|\, \text{mes}_\alpha E = \infty\} = \{\inf \alpha \,|\, \text{mes}_\alpha E = 0\}$$

et on appelle α_0 la dimension de Hausdorff de E:

$$\alpha_0 = \dim E.$$

Si E est plongé dans $\mathbb{R}^d$ euclidien, on a pour tout $\lambda > 0$

$$\text{mes}_\alpha(\lambda\, E) = \lambda^\alpha\, \text{mes}_\alpha E$$
$$\text{mes}_\alpha(E + x) = \text{mes}_\alpha E.$$

Lemme de Frostman (thèse, 1935, p. 89). Soit E un compact dans $\mathbb{R}^d$, et h une fonction déterminante telle que $h(2t) = O(h(t))$ $(t \to 0)$. Pour avoir $\text{mes}_h E > 0$, il faut et suffit que E porte une mesure positive μ (on écrit $\mu \in \Omega^+(E)$), non nulle, telle que $\mu(B) \leq h(\text{diam } B)$ pour toute boule B.

En application du lemme de Frostman, dans le cas où h est concave, on construit facilement un ensemble du type de Cantor $E(\xi_1,\xi_2,\ldots)$ tel que $\text{mes}_h E = 1$ (voir figure). C'est, en fait, la construction de Hausdorff.

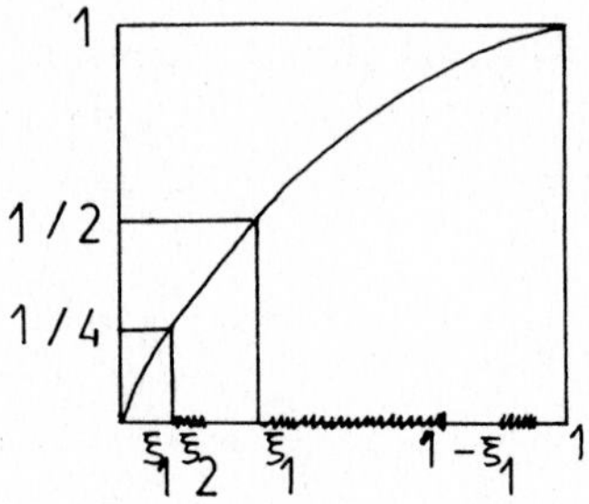

Capacités et dimensions capacitaires. Théorème de Frostman

Soit toujours E un compact dans $\mathbb{R}^d$. Soit $0 < \alpha < d$, et

$$\mathcal{n}(x) = |x|^{-\alpha}$$

(on écrit $|\ |$ pour la norme euclidienne). Ainsi

$$\hat{\mathcal{n}}(\xi) = c|\xi|^{\alpha-d}.$$

A toute mesure $\mu \in M^+(E)$ on associe son potentiel

$$p(x) = \mu * \mathcal{n}(x) = \int \frac{d\mu(y)}{|x-y|^\alpha}$$

et son intégrale d'énergie

$$I = \int p\, d\mu = \int \frac{d\mu(x)\,d\mu(y)}{|x-y|^\alpha} = \int |\hat{\mu}|^2\, \hat{\mathcal{n}}.$$

A toute distribution $T \in \mathcal{D}'(E)$ (portée par E) on associe aussi l'intégrale d'énergie

$$I(T) = \int |\hat{T}|^2\, \hat{\mathcal{n}}.$$

Les distributions d'énergie finie forment un espace de Hilbert, et les mesures positives d'énergie finie un cône convexe fermé dans cet espace de Hilbert (thèse Deny, 1950).

Il y a trois définitions naturelles de la capacité d'ordre α:

$$\text{Cap}_\alpha^{(0)} E = \sup \{\mu(E) \mid \sup_x p(x) \leq 1,\ \ \mu \in M^+(E)\}$$

$$\text{Cap}_\alpha^{(1)} E = \sup \{(\mu(E))^2 \mid I(\mu) \leq 1,\ \ \mu \in M^+(E)\}$$

$$\mathrm{Cap}_\alpha^{(2)} E = \sup\{|\langle T,1\rangle|^2 \mid I(T) \le 1, \quad T \in \mathcal{D}'(E)\}$$

et on vérifie facilement

$$\mathrm{Cap}_\alpha^{(0)} \le \mathrm{Cap}_\alpha^{(1)} \le \mathrm{Cap}_\alpha^{(2)}.$$

On voit aussi que, pour chacune de ces capacités,

$$\mathrm{Cap}_\alpha(\lambda E) = \lambda^\alpha \mathrm{Cap}_\alpha E$$

$$\mathrm{Cap}_\alpha(E + x) = \mathrm{Cap}_\alpha E.$$

La théorie du potentiel (principle du maximum, mesure d'équilibre) s'applique lorsque $d-2 \le \alpha < d$ (le cas $\alpha = d-2$ est celui du potentiel newtonien, $d \ge 3$); elle donne l'égalité

$$\mathrm{Cap}_\alpha^{(0)} = \mathrm{Cap}_\alpha^{(1)} = \mathrm{Cap}_\alpha^{(2)}.$$

Pour $\alpha < d-2$, ces égalités sont en défaut. Nous verrons cependant tout-à-l'heure que

$$\mathrm{Cap}_\alpha^{(2)} E > 0 \iff \mathrm{Cap}_\alpha^{(1)} E > 0 \iff \mathrm{Cap}_\alpha^{(0)} E > 0.$$

(La première équivalence m'a été révélée en 1977 par Hans Wallin, et Lars Hedberg m'en a donné une démonstration détaillée en 1978; la seconde résulte d'une observation de Peter Sjögren pendant le séminaire de l'Escorial).

On peut définir des dimensions capacitaires

$$\dim_0, \quad \dim_1, \quad \dim_2,$$

à partir de ces trois notions de capacité, comme on a défini la dimension de Hausdorff, à savoir

$$\dim_i E = \inf\{\alpha \mid \mathrm{Cap}_\alpha^{(i)} E = 0\}.$$

Les dimensions capacitaires $\dim_0$ et $\dim_1$ on été introduites par Polya et Szegö en 1932.

Théorème de Frostman (thèse, p. 90):

$$\dim = \dim_0 = \dim_1 \ (= \dim_2).$$

La dernière égalité résulte du théorème du folklore suédois, et naturellement elle n'était pas énoncée par Frostman. La démonstration de Frostman est donnée dans le cas $d = 2$, et elle repose sur la théorie du potentiel. On peut, heureusement, se passer de la théorie du potentiel, comme me l'a montré Jacques Peyrière.

Théorème. *Si* $0 < \alpha-\varepsilon < \alpha < d$,

$$\mathrm{Cap}_\alpha^{(1)} E > 0 \Rightarrow \mathrm{mes}_\alpha E > 0 \Longrightarrow \mathrm{Cap}_{\alpha-\varepsilon}^{(0)} E > 0$$

Preuve de la première implication. Supposons $\mathrm{Cap}_\alpha^{(1)} E > 0$. Soit $\mu \in M^+(E)$, $\mu \neq 0$, $I(\mu) < \infty$. Il existe une partie de E, soit F, telle que $\mu(F) > 0$ et $\sup_{x \in F} p(x) < \infty$. Soit B_n une suite de boules recouvrant F; on peut supposer qu'aucune n'est disjointe de F, et considérer $x_n \in B_n \cap F$; alors

$$\mu(B_n) \leq (\mathrm{diam}\, B_n)^\alpha \int_{B_n} \frac{d\mu(y)}{|x_n - y|^\alpha} \leq C(\mathrm{diam}\, B_n)^\alpha$$

donc
$$\mu(F) \leq C\, \Sigma(\mathrm{diam}\, B_n)^\alpha$$

donc $\mathrm{mes}_\alpha F > 0$, donc $\mathrm{mes}_\alpha E > 0$.

Preuve de la seconde implication. Supposons $\mathrm{mes}_\alpha E > 0$, et soit μ la mesure donnée par le lemme de Frostman. On a

$$\int \frac{d\mu(y)}{|x-y|^{\alpha-\varepsilon}} < C$$

indépendant de x en intégrant sur des couronnes $\{y ; 2^{-j-1} < |x-y| < 2^{-j}\}$.

Equivalence des trois notions de capacité

Nous allons maintenant démontrer le résultat annoncé.

THÉORÈME. $\mathrm{Cap}_\alpha^{(2)} E > 0 \Longrightarrow \mathrm{Cap}_\alpha^{(1)} E > 0 \Longrightarrow \mathrm{Cap}_\alpha^{(0)} E > 0$.

Preuve de la première implication (suivant Lars Hedberg). En fait, nous allons démontrer l'inégalité

$$\mathrm{Cap}_\alpha^{(2)} E \leq C(\alpha,d)\, \mathrm{Cap}_\alpha^{(1)} E,$$

$C(\alpha,d)$ étant une fonction positive de α et d que nous n'explici terons pas, sauf dans le cas $\alpha \geq d-2$, où $C(\alpha,d) = 1$ (théorème de Deny).

Pour toute fonction $p \in \mathcal{D}(R^d)$, posons

$$J(p) = \int |\hat{p}|^2 / \hat{n}$$

de sorte que, pour toute distribution T d'énergie finie,

$$|\langle T,p\rangle|^2 \leq I(T)\ J(p).$$

Le complété de $\mathcal{D}(\mathbb{R}^d)$ pour la norme hilbertienne $\sqrt{J(.)}$ est l'espace de Hilbert dual de l'espace de Hilbert des distributions d'énergie finie. Il est commode d'écrire

$$\mathrm{Cap}_\alpha^{(1)}E = (\inf\{I(\mu),\quad \mu \in M^+(E),\ \langle\mu,1\rangle = 1\})^{-1}$$

$$\mathrm{Cap}_\alpha^{(2)}E = (\inf\{I(T),\quad T \in \mathcal{D}'(E),\ \langle T,1\rangle = 1\})^{-1}$$

de sorte que par dualité on a:

<u>Lemme 1</u>.

$$\mathrm{Cap}_\alpha^{(1)}E = \inf\{J(p),\quad p \in \mathcal{D},\quad p \geq 1 \text{ sur } E\}$$

$$\mathrm{Cap}_\alpha^{(2)}E = \inf\{J(p),\quad p \in \mathcal{D},\quad p = 1 \text{ sur } E\}.$$

Posons maintenant, pour $p \in \mathcal{D}$,

$$\hat{T}_p = \hat{p}/\hat{n},\qquad p = T_p * n,\qquad J(p) = I(T_p).$$

On peut écrire

$$\mathrm{Cap}_\alpha^{(1)}E = \inf\{I(T_p),\ T_p * n \in \mathcal{D},\ T_p * n \geq 1 \text{ sur } E\}.$$

Le convexe

$$\{T_p | T_p * n \geq 1 \text{ sur } E,\quad T_p * n \in \mathcal{D}\}$$

admet un point frontière unique de norme minimum, soit T_o, dans l'espace de Hilbert des distributions d'énergie finie. Pour toute $\phi \in \mathcal{D}$, $\phi \geq 0$, on a

$$\langle T_o,\phi\rangle = \lim_{t\downarrow 0} \frac{1}{2t}\,(I(T_o + tT_\phi) - I(T_o)) \geq 0$$

donc T_o est une mesure positive portée par E, soit

$$T_o = \mu_o \in M^+(E)$$

$$\mathrm{Cap}_\alpha^{(1)}E = I(\mu_o).$$

Si les T_p appartiennent au convexe ci-dessus, et tendent vers μ_o dans l'espace des distributions d'énergie finie, leurs potentiels $T_p * n$ tendent vers $\mu_o * n$ dans l'espace $L^1(d\mu_o)$. Cela montre que $\mu_o * n \leq 1$ sur le support de $d\mu_o$.

Lemme 2. *Si* $\mu_o * n \leq 1$ *sur le support de* $d\mu_o$,

$$\mu_o * n \leq M(d) \quad \textit{sur} \quad \mathbb{R}^d$$

($M(d)$ *ne dépendant que de* d).

Preuve. Partageons $\mathbb{R}^d$ en M cônes convexes de sommet 0 tels que l'angle maximum de deux génératrices d'un même cône soit $\leq \frac{\pi}{3}$. Supposons $0 \notin$ Support μ_o. Soit Γ un cône de la famille considérée qui intersecte le support de μ_o, et soit x un point de $\Gamma \cap$ Support u_o, de norme minimum.
Pour tout autre point $y \in \Gamma \cap$ Support μ_o, on a

$$|y| \geq |y - x|$$

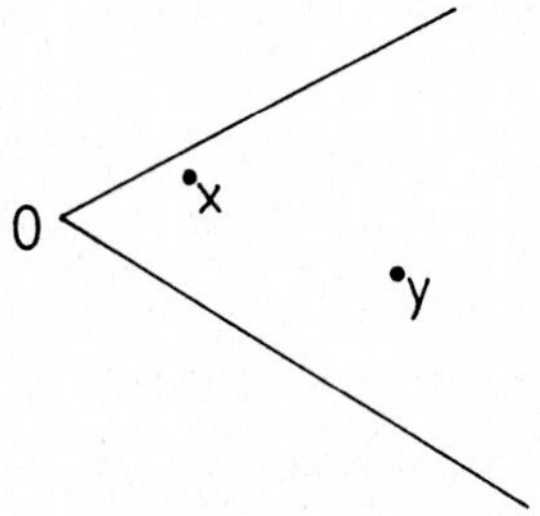

(voir figure, dans le plan contenant 0, x, y), donc

$$n(y) \leq n(y - x)$$

donc le potentiel en 0 de la mesure $\mu_o 1_\Gamma$ ne dépasse pas son potentiel en x, qui est majoré par 1. Donc le potentiel en 0 de μ_o ne dépasse pas M et, quitte à faire una translation,

$$\mu_o * n(x) \leq M$$

pour tout $x \notin$ Support μ_o. Cela démontre le lemme avec $M(d) = M$.

Lemme 3. *Pour* $M = M(d)$ *comme ci-dessus,*

$$\mathrm{Cap}_\alpha^{(1)} E = \inf\{J(p),\ p \in \mathcal{D},\ \ 0 \leq p \leq M+1,\ \ p \geq 1 \ \text{ sur } \ E\}$$

$$\mathrm{Cap}_\alpha^{(2)} E = \inf\{J(p),\ p \in \mathcal{D},\ \ p = 1 \ \text{ sur } \ E\}.$$

Preuve. Lemme 1, interprétation de μ_o, et lemme 2.

Désormais nous désignons par F une fonction de classe C^∞ nulle sur $]-\infty,0]$ et égale à 1 sur $[1,\infty[$.

Lemme 4. *Si* $p \in \mathcal{D}$ *et* $0 \leq p \leq M+1$,

$$J(F \circ p) \leq C\, J(p)$$

C *ne dépendant que de* d, α, M *et* F. *De plus, si* $d-2 \leq \alpha < d$, *on peut choisir* F *de façon que* C *soit arbitrairement proche de* 1.

Le théorème résulte de la comparaison des lemmes 3 et 4. Reste donc à prouver le lemme 4. On va poser $\beta = d - \alpha$, et considérer d'abord le cas $0 < \beta \leq 2$. Alors

$$\iint \frac{|p(x+y) - p(x)|^2}{|y|^{d+\beta}} dx\, dy = \iint \frac{|e^{i\xi.y} - 1|^2}{|y|^{d+\beta}} |\hat{p}(\xi)|^2 \, d\xi dy$$
$$= c \int |\hat{p}(\xi)|^2 |\xi|^\beta d\xi = c' J(p).$$

Il suit de là que

$$J(F \circ p) \leq \|F'\|_\infty J(p)$$

et comme $\|F'\|_\infty$ est arbitrairement proche de 1, on a la conclusion voulue (théorème de Deny) dans le cas $d-2 \leq \alpha < d$.

Dans le cas général $0 < \beta < d$, considérons l'espace P_β constitué par les fonctions p sommes de constantes et de fonctions indéfiniment dérivables à supports compact $(p - C_p \in \mathcal{D}(\mathbb{R}^d))$ telles que $J(p) < \infty$. On a J (Constante) $= 0$, donc $\sqrt{J(.)}$ est une semi-norme sur P_β. On va d'abord établir l'inégalité

$$(*) \qquad J(p^2) \leq C\|p\|_\infty^2 J(p)$$

$(p \in P_\beta,\ C = C(\beta, d))$.

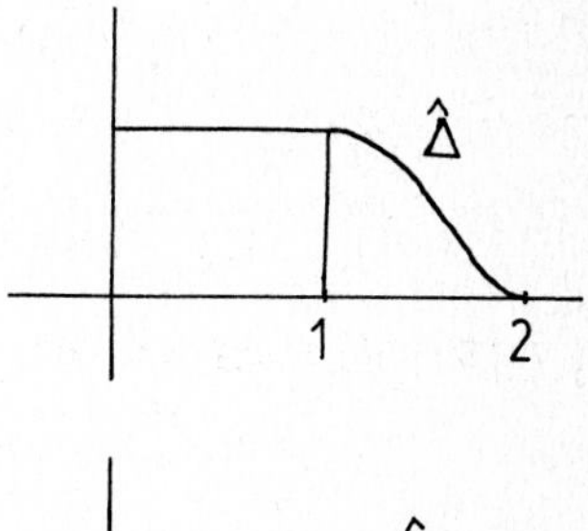

Pour cella, soit $\Delta(.)$ une fonction radiale telle que $\hat{\Delta}(\xi) = 1$ pour $|\xi| \leq 1$, $\hat{\Delta}(\xi) = 0$ pour $|\xi| \geq 2$ et $\hat{\Delta} \in \mathcal{D}(\mathbb{R}^d)$. Soit $\hat{\Delta}_j(\xi) = \hat{\Delta}(\xi 2^{-j})$

$$V_j = \Delta_j - \Delta_{j-1}$$

$$p_j = p * V_j \qquad (p \in P_\beta).$$

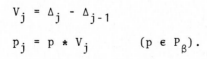

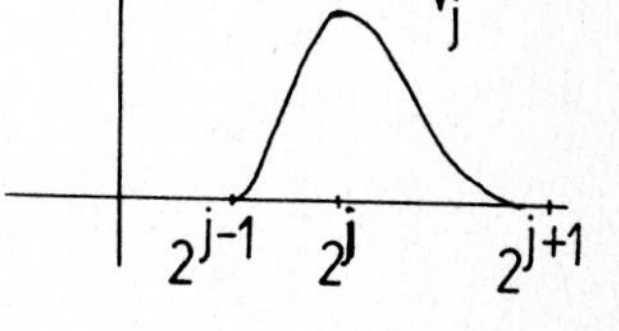

On a

$$p = C_p + \sum_{-\infty}^{+\infty} p_j$$

et la série est uniformément convergente. Ecrivons

$$p^2 = C_p^2 + \sum_{-\infty}^{+\infty} (q_k + r_k)$$

$$q_k = (2C_p + 2 \sum_{j \leq k-4} p_j) p_k$$

$$r_k = (2(p_{k-3} + p_{k-2} + p_{k-1}) + p_k)p_k.$$

Le spectre de q_k est contenu dans l'anneau

$$2^{k-2} \leq |\xi| \leq 2^{k+1} + 2^{k-2}$$

et le spectre de r_k dan le boule

$$|\xi| \leq 2^{k+2}.$$

Ecrivons, pour $\nu \in Z$

$$J_\nu(p) = \int_{2^{\nu-1} \leq |\xi| \leq 2^\nu} |\hat{p}(\xi)|^2 \, d\xi$$

et considérons $J_\nu(p^2)$. On a, compte tenu des spectres de q_k et r_k,

$$J_\nu(p^2) = \int |(\sum_{\nu-2 \leq k \leq \nu+1} q_k + \sum_{\nu-2 \leq k} r_k)\hat{\ }(\xi)|^2 \, d\xi$$

l'intégrale étant prise sur l'anneau $2^{\nu-1} \leq |\xi| \leq 2^\nu$, donc

$$J_\nu(p^2) \leq \int | \sum_{\nu-2 \leq k \leq \nu+1} q_k + \sum_{k \geq \nu-2} r_k|^2.$$

Or

$$|q_k| \leq C\|p\|_\infty |p_k|$$

donc

$$\int | \sum_{\nu-2 \leq k \leq \nu+1} q_k|^2 \leq C\|p\|_\infty^2 \int \sum_{\nu-2 \leq k \leq \nu+1} |p_k|^2.$$

D'autre part, par Littlewood-Paley

$$\begin{aligned}\int (\sum_{k \geq \nu-2} |p_k|^2)^2 &\leq C \int | \sum_{k \geq \nu-2} p_k|^4 \\ &\leq C\|p\|_\infty^2 \int | \sum_{k \geq \nu-2} p_k|^2 \\ &\leq C\|p\|_\infty^2 \int \sum_{k \geq \nu-2} |p_k|^2\end{aligned}$$

et, en remplaçant p_k par $p_{k-3} + p_{k-2} + p_{k-2} + p_{k-1} + p_k$ et en polarisant, on obtient

$$\int | \sum_{k \geq \nu-2} r_k|^2 \leq C\|p\|_\infty^2 \sum_{k \geq \nu-2} \int |p_k|^2$$

d'où finalement

$$\sum_{-\infty}^{+\infty} 2^{\beta\nu} J_\nu(p^2) \leq C \sum_{-\infty}^{+\infty} 2^{\beta\nu} J_\nu(p)$$

c'est-à-dire (*).

En polarisant (*), on obtient

$$J(pq) \leq C(\|p\|_\infty^2 + \|q\|_\infty^2)(J(p) + J(q))$$

et cela donne par induction

$$(**) \qquad J(p^m) \leq m^A \|p\|_\infty^{2(m-1)} J(p)$$

$(m = 1,2,\ldots;\ A = A(\beta,d))$.

Supposons maintenant $-B \leq p \leq B$. Si F est une fonction de classe C^∞ sur $[-B,B]$, on peut la prolonger para périodicité et écrire

$$F(x) = \sum_{m \in Z} \hat{F}_m e^{i\varepsilon mx} \qquad (\varepsilon < \frac{\pi}{B})$$

avec $\hat{F}_m = O(|m|^{-A-2})$. A partir de (**) on obtient

$$J(e^{i\varepsilon p}) \leq C\, J(p)$$

$(C = C(B,\beta,d))$, d'où

$$J(F \circ p) \leq C\, J(p)$$

$(C = C(B,\beta,d))$, ce qui achève la preuve du lemme 4.

Cela termine la preuve de la première implication du théorème.

Passons à la seconde partie, et démontrons

$$\mathrm{Cap}_\alpha^{(1)} E > 0 \implies \mathrm{Cap}_\alpha^{(0)} E > 0.$$

On opère comme dans la preuve du théorème de Frostman. Supposons $\mathrm{Cap}_\alpha^{(1)} E > 0$. Soit encore $\mu \in M^+(E)$, $\mu \neq 0$, $I(\mu) < \infty$. Il existe une partie de E, soit F, telle que $\mu(F) > 0$ et $\sup_{x \in F} p(x) < \infty$ $(p(x) = \mu * n(x))$. Soit μ_F la restriction de μ à F. On a $\mu_F \neq 0$ et $\mu_F * n(.)$ est borné sur le support de μ_F. D'après le lemme 2, $\mu_F * n(.)$ est borné partout, donc $\mathrm{Cap}_\alpha^{(0)} E > 0$.

Cela termine la démonstration du théorème.

Compléments sur mesure, capacités, ε-entropie

Désormais nous pouvons écrire $\mathrm{Cap}_\alpha E > 0$, sans préciser l'indice supérieur. En complément au théorème de Frostman, voici un résultat utile.

$I(\mu) < \infty$. Alors

$$\iint \frac{d\mu(x)\ d\mu(y)}{\varepsilon(x-y)|x-y|^2} < \infty$$

pour une fonction convenable $\varepsilon(.)$ positive, et tendant vers 0 au voisinage de 0. On peut supposer $\varepsilon(.)$ concave, et poser $h(t) = \varepsilon(t)t^\alpha$. En imitant la preuve du théorème de Frostman, on obtient $\mathrm{mes}_h\ F > 0$, d'où immédiatement $\mathrm{mes}_\alpha F = \infty$.

Nous utiliserons ce théorème plus tard sous la forme

$$\mathrm{mes}_\alpha\ E < \infty \implies \mathrm{Cap}_\alpha\ E = 0.$$

Etant donné E, compact dans $\mathbb{R}^d$, il est parfois suffisant, pour montrer $\mathrm{mes}_\alpha\ E < \infty$, d'utiliser des recouvrements par des boules égales. Nous utiliserons les notations suivantes:

$N_\varepsilon(E)$: nombre minimum de boules de rayon ε recoubrant E (l'ε-entropie de Kolmogorov est $\log N_\varepsilon(E)$)

$E_\varepsilon = E + B(0,\varepsilon)$ = ensemble des points dont la distance a E est inférieure à ε.

Il est à peu près évident que

$$\mathrm{mes}_d\ E_\varepsilon \asymp \varepsilon^d\ N_\varepsilon(E),$$

le signe $\asymp$ significant que le rapport des deux membres est compris entre deux nombres positifs qui ne dépendent que de d. Il es tout aussi évident que

$$\mathrm{mes}_h\ E_\varepsilon \le C \varliminf_{\varepsilon\to 0}\ (h(\varepsilon)\ N_\varepsilon(E)) \qquad (C = C(d)).$$

Pour des ensembles E construits de façon assez homogène, comme les ensembles de Cantor, on a souvent égalité. Mais on a aussi souvent inégalité stricte; par exemple, pour un dénombrable E ayant un seul point d'accumulation, le second membre peut être infini.

Désignons par S une suite strictement positive tendant vers 0. On se servira du fait suivant, très simple.

THEOREME. *Supposons que,* E *et* F *étant deux compacts de* $\mathbb{R}^d$, *on ait*

$$\text{mes}_d\, E_\varepsilon = O(\varepsilon^{d-\alpha}) \qquad (\varepsilon \in S)$$

$$\text{mes}_d\, F_\varepsilon = O(\varepsilon^{d-\beta}) \qquad (\varepsilon \in S).$$

Alors, pour presque tout $x \in \mathbb{R}^d$, *il existe une sous-suite infinie* $S(x)$ *de* S *telle que*

$$\text{mes}_d((E \cap (F+x))_\varepsilon) = O(\varepsilon^{2d-\alpha-\beta}) \qquad (x \in S(x)).$$

Preuve.

$$(E \cap (F+x))_\varepsilon \subset E_\varepsilon \cap (F_\varepsilon + x)$$

et

$$\int \text{mes}_d(E_\varepsilon \cap (F_\varepsilon + x))dx = \text{mes}_d\, E_\varepsilon . \text{mes}_d\, F_\varepsilon .$$

La conclusion résulte du lemme de Fatou.

Dans l'énoncé de ce théorème il est important que ce soit la même suite S qui serve à jauger E_ε et F_ε. Voici un exemple montrant qu'on peut très bien avoir

$$\underline{\lim}_{\varepsilon \to 0} (\varepsilon^{\alpha-d} \text{mes}_d\, E_\varepsilon) < \infty$$

$$\underline{\lim}_{\varepsilon \to 0} (\varepsilon^{\alpha-d} \text{mes}_d\, F_\varepsilon) < \infty$$

et, pour tout x appartenant à un ouvert non vide et tout $\alpha' < \alpha$

$$\underline{\lim}_{\varepsilon \to 0} (\varepsilon^{\alpha'-d}\ \text{mes}_d(E \cap (F+x))_\varepsilon) > 0,$$

avec $0 < \alpha < d$. Limitons nous au cas $d = 1$. Identifions $T = R/Z$ et l'intervalle $[0,1[$. Choisissons une suite d'entiers n_j très rapidement croissante, au sens

$$\log n_j = o(\log n_{j+1})$$

Posons, les ϕ_j étant des réels arbitrairement donnés, et $A > 0$,

$$D_j = \{x \in T;\ |\sin 2\pi n_j(x-\phi_j)| \leq n_j^{-A}\}$$

$$D = \bigcap_{j=1}^{\infty} D_j$$

$$E = \bigcap_{j=1}^{\infty} D_{2j}$$

$$F = \bigcap_{j=1}^{\infty} D_{2j-1}$$

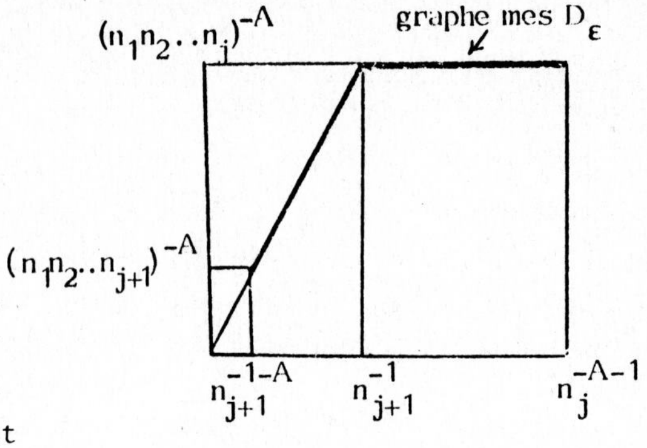

Le comportement de $\text{mes}_1\ D_\varepsilon$ est indiqué sur la figure (le palier horizontal représente en réalité une pente bornée). En choisissant $\alpha = \frac{1}{A+1}$ on a

$$\varliminf_{\varepsilon \to 0} (\varepsilon^{\varepsilon - 1}\ \text{mes}_1\ E_\varepsilon) = \lim_{\varepsilon = \varepsilon_{2j}} = 0$$

$$\varliminf_{\varepsilon \to 0} (\varepsilon^{\alpha - 1}\ \text{mes}_1\ F_\varepsilon) = \lim_{\varepsilon = \varepsilon_{2j-1}} = 0$$

$$\varliminf_{\varepsilon \to 0} (\varepsilon^{\alpha' - 1}\ \text{mes}_1\ D_\varepsilon) = \lim_{\varepsilon = \varepsilon_j} = \infty$$

et la même chose a lieu en remplaçant D par $E \cap (F+x)$ quel que soit x.

Compléments sur la dimension

Terminons en donnant une série d'énoncés intéressants, mais dont nous n'aurons pas à nous servir.

Si E et F son deux compacts dans $\mathbb{R}^d$, que peut-on dire de $E \times F$?.

THEOREME (Marstrand).

$$\dim E \times F \geq \dim E + \dim F.$$

On peut avoir inégalité stricte.

Preuve. L'inégalité résulte du lemme de Frostman. Pour un exemple d'inégalité stricte, possons,

$$D_j = \{x \in T\ ;\ |\sin 2\pi n_j x| \leq |\sin \pi n_j n_{j+1}^{-1}|\},$$

de sorte que D_j est une réunion d'intervalles de longueurs $2n_{j+1}^{-1}$, la distance de deux intervalles consécutifs étant inférieure a n_j^{-1}. Posons

$$E = \bigcap_{j \text{ pair}} D_j , \qquad F = \bigcap_{j \text{ impair}} D_j .$$

Comme $D_j + D_{j+1} = T$ quel que soit j, on a $E + F = T$. D'autre part, étant donné une fonction convexe $h(.)$ quelconque, on peut aisément, par induction, choisir les n_j de façon que $\text{mes}_h\, E = \text{mes}_h\, F = 0$. En particulier, on peut avoir $\dim E = \dim F = 0$, et $\dim(E+F) = 1$, donc $\dim E \times F \geq 1$.

Si E et F sont deux compacts dans $\mathbb{R}^d$, de dimension égale à d, peut-on affirmer que $\dim(E \cap (F+x)) > 0$ pour un ensemble convenable (non vide) de valeurs de x? La réponse est négative. On peut, quelle que soit la fonction $h(t) = o(t^d)$ $(t \to 0)$, construire un compact K dans $\mathbb{R}^d$ tel que $\text{mes}_h\, K = \infty$ et que K soit un ensemble indépendant sur les rationnels. En particulier, on peut avoir $\dim K = d$. En partageant K en deux parties disjointes E et F de dimension d, on a l'exemple voulu: quel que soit x, $E \cap (F = x)$ contient au plus un point.

Peut-on alors affirmer que $\dim(E \cap (AF + x)) > 0$ pour un ensemble convenable de transformations $A \in GL(d)$ et de $x \in \mathbb{R}^d$? La réponse est positive.

THEOREME. *Soit* G *un sous-groupe fermé de* $GL(d,\mathbb{R})$, *transitif sur* $\mathbb{R}^d \setminus \{0\}$, *et soit* τ *sa mesure de Haar. Si*

$$\dim E + \dim F - d = \delta > 0,$$

on a pour τ-*presque tout* $A \in G$

$$\dim(E \cap (AF + x)) \geq \delta$$

pour un ensemble de $x \in \mathbb{R}^d$ *de mesure de Lebesgue positive*

Pour la preuve, voir Séminaire d'Analyse Harmonique, Orsay 1983.

Considérons maintenant un compact $E \subset \mathbb{R}^n$, et une application f de $\mathbb{R}^n$ dans $\mathbb{R}^d$. Supposons $f \in \text{Lip}\, \alpha$ $(0 < \alpha < 1)$. Que peut-on dire de $f(E)$? Que peut-on dire des ensembles de niveau $f^{-1}(x)$? du graphe de f?

THEOREME. *Si* $f \in \text{Lip}\, \alpha(\mathbb{R}^n, \mathbb{R}^d)$,

$$\dim f(E) \leq \frac{1}{\alpha} \dim E$$

$$\dim \text{graphe } f|_E \leq d + \frac{1}{\alpha} \dim E$$

et, pour presque tout $x \in R^d$,

$$\dim f^{-1}(x) \leq \sup(0, n-d\alpha).$$

Les preuves sont faciles (voir par exemple SRSF[1], p. 142).

(1) Ici comme dans la suite SRSF désigne mon livre Some random series of functions, Heath 1968.

DEUXIEME LEÇON. Quelques processus stochastiques

Le processus de Wiener

Commençons par le processus de Wiener.

Soit $(\Omega,\mathcal{A},P)$ un espace de probabilité sans atome et complet -par exemple l'intervalle $[0,1]$ de la droite réelle, muni de la tribu de Lebesgue et de la mesure de Lebesgue-. Soit H un sous-espace fermé de $L^2(\Omega)$ réel; de dimension infinie, dont les éléments sont des variables aléatoires (v.a.) gaussiennes centrées; ainsi $X \in H$ signifie

$$E(e^{i\xi X}) = e^{-\frac{1}{2}\tau^2\xi^2}, \qquad \tau = \|X\|_2 = (E(X^2))^{1/2}$$

$$P(X < \lambda\tau) = \frac{1}{\sqrt{2\pi}} \int_{-\infty}^{\lambda} e^{-\frac{u^2}{2}} du.$$

L'orthogonalité dans H équivaut a l'indépendance.

Soit maintenant I un intervalle réel (éventuellement, la droite entiere, ou une demi-droite) muni de la tribu et de la mesure de Lebesgue. Considérons une isométrie linéaire

$$W : L^2(I) \longrightarrow H.$$

Supposons $0 \in I$, et posons, pour $t \in I$,

$$W(t) = W\, 1_{[0,t]}.$$

C'est une famille de variables aléatoires indexée par I (= un processus), à valeurs dans H. On a

$$(*) \qquad \|W(t') - W(t)\|_2^2 = |t - t'|, \qquad W(0) = 0$$

et les accroissements de $W(.)$ sur des intervalles disjoints sont indépendants. Inversement, tout processus gaussien, à accroissement indépendants, normalisé selon la formule (*), s'obtient de cette façon.

L'interprétation géométrique de (*) est que le point $W(t)$ décrit dans H une hélice (c'est-à-dire que la distance de deux points ne dépend que de la distance des paramètres), et que trois points quelconques sur cette hélice forment un triangle rectangle.

On appelle version du processus toute fonction $W(t,\omega)$ $(t \in I,\ \omega \in \Omega)$ telle que les v.a. $W(t,.)$ vérifient (*). On s'intéresse alors aux propriétés presque sûres des fonctions $W(.,\omega)$. Dans la théorie du mouvement brownien, on montre qu'il existe des versions telles que la fonction aléatoire $W(.,\omega)$ soit presque sûrement continue (Wiener) et nulle part différentiable (Paley Wiener Zygmund). C'est ce que nous allons faire, rapidement.

Si (e_n) est une base de $L^2(I)$, son image $(X_n = W(e_n))$ est une suite de variables gaussiennes normalisées indépendantes, et

$$W(t) = \Sigma\, X_n \int_o^t e_n(s)ds$$

(intégrale convergente dans $L^2(\Omega)$, et aussi presque sûrement). Choisissons pour base la base de Haar (I étant l'intervalle $[0,1]$) $h_0, h_{00}, h_{10}, h_{11}, h_{20}, \ldots$ ($h_o = 1$, h_{jn} est portée par l'intervalle $I_{jn} = [n2^{-j}, (n+1)2^{-j}]$, égale a $2^{j/2}$ sur la partie gauche de I_{jn} et a $-2^{j/2}$ sur la partie droite). Désignons par t, $\Delta_{00}(t)$, $\Delta_{10}(t)$, $\Delta_{11}(t)$, $\Delta_{20}(t), \ldots$ les primitives de ces fonctions nulles en 0. Ainsi

$$W(t) = X_o t + \sum_{j=0}^{\infty} \sum_{n=0}^{2^j-1} X_{jn}\, \Delta_{jn}(t)$$

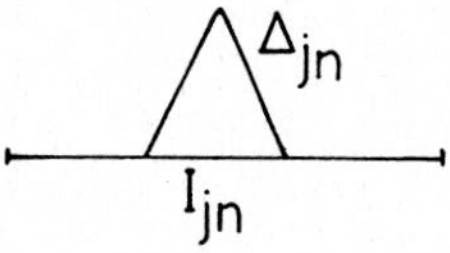

somme convergente dans $L^2(\Omega)$. Montrons que p.s. elle converge uniformément sur I. Pour chaque j,

$$P(\sup_n |X_{jn}| > \lambda \sqrt{\log j}) < \alpha^j$$

où $\alpha = \alpha(\lambda) < 1$ si $\lambda > \sqrt{\frac{2}{\log 2}}$. Comme $|\Delta_{jn}|_\infty = 2^{-j/2-1}$, on a p.s. à partir d'un certain rang $j_o = j_o(\omega)$

$$\Big|\sum_{n=0}^{2^j-1} X_{jn}\, \Delta_{jn}\Big|_\infty \leq \lambda \sqrt{\log j}\;\; 2^{-j/2-1}.$$

Cela montre non seulement la convergence uniforme, mais que le module de continuité $\omega(\delta)$ de la fonction $W(t)$ satisfait presque sûrement

$$\omega(\delta) = O(\sqrt{\delta \log \tfrac{1}{\delta}})$$

et même, en regardant un peu mieux,

$$\overline{\lim_{\delta \to 0}} \frac{\omega(\delta)}{\sqrt{2\delta \log \frac{1}{\delta}}} = 1 \text{ (Paul Lévy)}$$

Montrons maintenant que p.s. la fonction $W(t)$ n'est nulle part dérivable. Choisissons $\lambda > 0$ tel que

$$P(|X_{jn}| > \lambda) > \frac{1}{2}$$

(la probabilité écrite au premier membre ne dépend pas de (j,n)), et disons que l'intervalle I_{jn} est mauvais (pour un choix de ω) si $|X_{jn}| > \lambda$. Si l'on peint en noir, à chaque étape, les mauvais intervalles, les intervalles blancs de chaque étape forment une population qui se dédouble, mais meurt (noircit) avec une probabilité supérieure à $\frac{1}{2}$: cette population va donc mourir p. s.. Donc, p.s. l'intervalle I est recouvert une infinité de fois par des intervalles noirs I_{jn}, sur lesquels la variation de $X_{jn}\,\Delta_{jn}(t)$ dépasse $\lambda 2^{-j/2-1}$. On voit ainsi que, presque sûrement,

$$(*) \qquad \forall t \qquad \overline{\lim_{h\to 0}} \frac{|W(t+h) - W(t)|}{\sqrt{|h|}} > 0.$$

C'est une forme forte (Dvoretzky (1963)) d'un théorème de Paley, Wiener et Zygmund (1932) qui dit que p. s. pour tout $\varepsilon > 0$

$$\forall t \qquad \overline{\lim_{h\to 0}} \frac{|W(t+h) - W(t)|}{|h|^{1/2+\varepsilon}} = \infty.$$

En étudiant de près la démonstration de $(*)$, on voit que le résultat n'est pas améliorable: presque sûrement

$$\exists t \qquad W(t+h) - W(t) = O(\sqrt{|h|}) \qquad (h \to 0).$$

Ces "points lents", découverts il y a 10 ans, ont fait l'objet d'études récentes de Burgess Davis et d'Edwin Perkins.

A partir de maintenant, l'isométrie W qui nous a servi à définir la fonction $W(t)$ du mouvement brownien sera écrite comme "intégrale de Wiener"

$$W(f) = \int f(t)\, dW(t).$$

L'intégrale de droite a un sens presque sûr (par exemple au moyen d'une intégration par parties), lorsque $f \in L^2(\mathbb{R})$.

La fonction aléatoire $W(t)$ représente le mouvement brownien à valeurs dans $\mathbb{R}$. Le mouvement brownien a valeurs dans $\mathbb{R}^d$ est représenté par le d-uple

$$(W_1(t),\ldots,W_d(t)),$$

où les $W_j(t)$ sont indépendantes et représentent chacune le mouvement brownien linéaire. On peut encore le noter $W(t)$ et écrire

$$E(e^{i\xi.(W(t)-W(t'))}) = e^{-\frac{1}{2}|t-t'||\xi|^2} \qquad (\xi \in \mathbb{R}^d)$$

le point représentant le produit scalaire, et $|.|$ la norme euclidienne.

Les processus Gaussiens stationnaires, ou à accroissements stationnaires

A toute mesure positive $d\nu$, bornée ou non, on peut faire correspondre l'opérateur

$$W_\nu : L^2(d\nu) \longrightarrow H,$$

H représentant comme ci-dessus un espace de Hilbert de variables gaussiennes centrées réelles, ou plus généralement un espace de Hilbert de variables gaussiennes centrées a valeurs dans $\mathbb{R}^d$ (d entier ≥ 1 fixé).

Considérons d'abord le cas où $d\nu$ est une mesure bornée sur $\mathbb{R}^n$, et posons

$$X(t) = W_\nu(\xi \to e^{i\xi.t}) \qquad (t \in \mathbb{R}^n, \quad \xi \in \mathbb{R}^n).$$

Ainsi

$$(\dagger) \qquad E(e^{i\xi.(X(t)-X(t'))}) = e^{-\frac{1}{2}|\xi|^2\psi(t-t')}$$

où

$$(\ddagger) \qquad \psi(t) = \int_{\mathbb{R}^n} |1 - e^{i\xi.t}|^2 \, d\nu(\xi).$$

On dit que $X(t)$ est un processus gaussien stationnaire, à temps n-dimensionnel, à valeurs dans $\mathbb{R}^d$. Si maintenant $d\nu$ satisfait

$$\int_{|\xi|>1} d\nu(\xi) < \infty, \qquad \int_{|\xi|\leq 1} |\xi|^2 \, d\nu(\xi) < \infty$$

on peut définir $X(0) = 0$ et

$$X(t) = W_\nu(\xi \to e^{i\xi.t} - 1).$$

On a de nouveau $(\dagger)$ et $(\ddagger)$. On dit que $X(t)$ est un processus gaussien à accroissements stationnaires. Comme dans le cas de $W(t)$, on peut se représenter $X(t)$ comme parcourant une "hélice" dans H (c'est une courbe seulement quand $n = 1$). La "fonction d'hélice" (screw-

-function de Schoenberg) est la fonction $\psi(t)$ donnée par (‡) telle que

$$\|X(t) - X(t')\|_2^2 = \psi(t - t').$$

On dit aussi que ψ est une fonction de type négatif, ou définie-négative (définition de Beurling).

Si $d\nu(\xi)$ est le produit tensoriel de la mesure de probabilité équidistribuée sur S^{n-1}, et de la mesure radiale $\frac{dr}{r^{1+\gamma}}$, on obtient $\psi(t) = |t|^\gamma$ $(0 < \gamma < 2)$. Dans le cas $\gamma = 1$, on a la fonction du mouvement brownien avec temps dans $\mathbb{R}^n$ (définition de Paul Lévy).

Dans le cas $n = 1$, nous appellerons, comme Benoit Mandelbrot, "mouvements browniens fractionnaires" les processus correspondant a $\psi(t) = |t|^\gamma$. Voici une maniere intéressante de se les représenter. Posons pour un instant

$$X(t) = \int_{-\infty}^{t} (t-s)^{-\beta}\, dW(s) \;;$$

cette intégrale n'a pas de sens, parce que, quel que soit β,

$$\int_{-\infty}^{t} (t-s)^{-2\beta}\, ds = \infty.$$

Mais, si $-\frac{1}{2} < \beta < \frac{1}{2}$, $X(t) - X(t_o)$ a un sens, à savoir, pour $t_o < t$,

$$X(t) - X(t_o) = \int_{-\infty}^{t_o} ((t-s)^{-\beta} - (t_o - s)^{-\beta})\, dW(s) + \int_{t_o}^{t} (t-s)^{-\beta} dW(s)$$

et on trouve

$$\|X(t) - X(t_o)\|_2^2 = \int_{-\infty}^{t_o} ((t-s)^{-\beta} - (t_o - s)^{-\beta})^2 ds + \int_{t_o}^{t} (t-s)^{-2\beta} ds$$

$$= C|t - t_o|^\gamma, \quad \text{avec} \quad \gamma = 1 - 2\beta.$$

Soit H_γ l'hélice décrite par $X(t)$. Suivant que $\gamma = 1$, $\gamma > 1$, ou $\gamma < 1$, les triangles formés par trois points $X(t_o)$, $X(t_1)$, $X(t_2)$ $(t_o < t_1 < t_2)$ ont en leur sommet $X(t_1)$ un angle droit, obtus ou aigü (voir figure): pour $\gamma = 1$, le passé et le futur au temps t_1 sont indépendants; pour $\gamma > 1$, le passé "repousse" le futur, et pour $\gamma < 1$, le passé "attire" le futur.

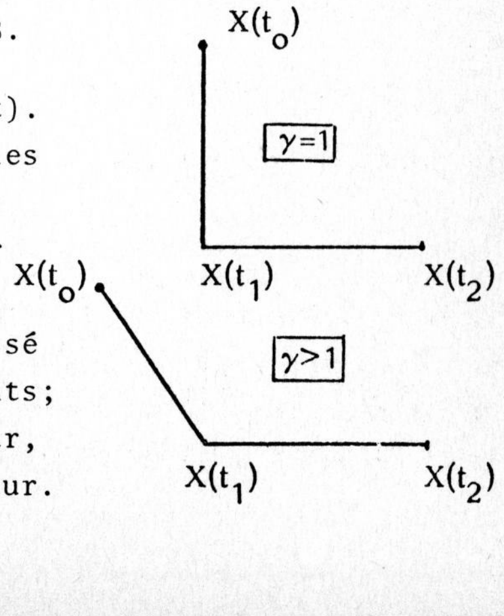

C'est d'ailleurs un exercice intéressant que de calculer la projection de $X(t)$ sur le passé de t_1 (c'est-à-dire le sous-espace de H engendré par les $X(s)$, $s \leq t_1$) sous la forme de l'intégrale de $X(s)$ par rapport à une mesure convenable, dépendant de t.

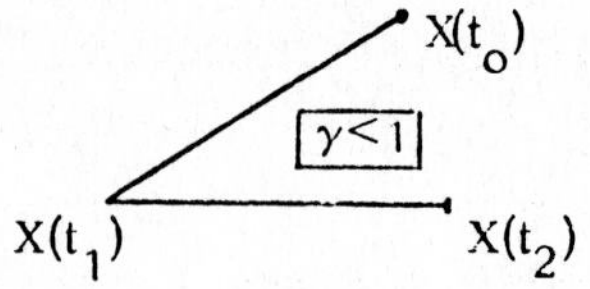

Les propriétés locales du mouvement brownien fractionnaire sont calquées sur celles du mouvement brownien. En particulier, le module de continuité vérifie p. s.

$$\omega(\delta) = O(\sqrt{\delta^\gamma \log \frac{1}{\delta}}),$$

et on a p. s.

$$\forall t \qquad \overline{\lim_{h\to 0}} \frac{|X(t+h) - X(t)|}{|h|^{\gamma/2}} > 0$$

et

$$\exists t \qquad X(t+h) - X(t) = O(|h|^{\gamma/2}).$$

Le processus de Lévy

Nous abandonnons maintenant l'espace H, et nous allons introduire d'autres processus à accroissements indépendants et stationnaires, en suivant Paul Lévy. Nous bornons au cas $n = 1$ (temps réel) et nous choisirons $\mathbb{R}^+$ comme ensemble des temps, t, toujours équipé de la mesure de Lebesgue dt.

Considérons le quadrant $\mathbb{R}^+ x \mathbb{R}^+$, muni d'une mesure

$$d\sigma(t,u) = dt \, d\nu(u)$$

ou $d\nu$ est une mesure positive sur $\mathbb{R}^+$ telle que

$$\int_0^1 u \, d\nu(u) = 1, \quad \int_1^\infty d\nu(u) < \infty$$

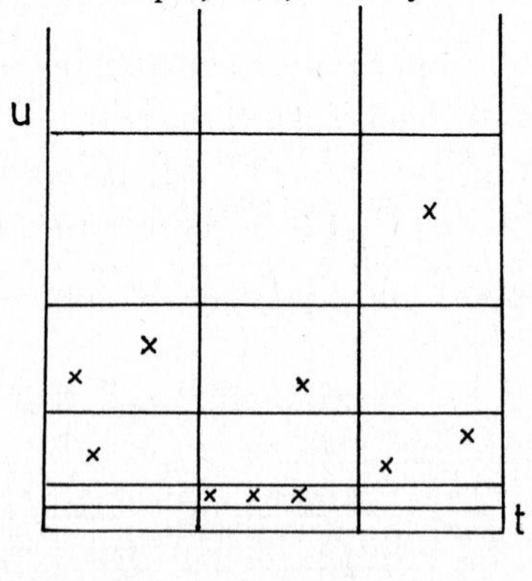

(par exemple $\frac{du}{u^{3/2}}$). On a représenté sur la figure un quadrillage correspondant (chaque "carré" a 1 pour σ-mesure). De façon indépendante pour les différents carrés, choisissons 1°) une variable aléatoire de Poisson p, de paramètre 1, 2°) p points au hasard, selon la mesure σ. Soit $d\mu = d\mu_\omega$ (mesure aléatoire) la somme des masses de Dirac au point de l'ensemble aléatoire que nous venons de construire. On vérifie facilement que, pour borélien $A \subset \mathbb{R}^+ \times \mathbb{R}^+$, $\mu_\omega(A)$ est une variable aléatoire de Poisson de parametre $\sigma(A)$, c'est-à-dire que

$$(\#) \qquad E(e^{i\xi\mu(A)}) = e^{-\sigma(A)(1-e^{i\xi})}.$$

On appellera μ la mesure de Poisson associée à la mesure positive σ. A partir de (#) on peut définir $\int f\, d\mu$ quand $f \geq 0$ et $f \in L^1(\sigma)$ (en commençant par le cas où f est une fonction étagée $\Sigma\, c_j\, 1_{A_j}$); c'est une variable aléatoire vérifiant

$$E(e^{i\xi\int f\, d\mu}) = e^{-\int (1-e^{i\xi f})d\sigma}.$$

Choisissons $f(t,u) = u\, 1_{[0,t_o]}$, et posons

$$X(t_o) = \int f\, d\mu = \int_o^{t_o}\int_o^\infty u\, d\mu(t,u).$$

Il est clair que la fonction aléatoire $X(t)$ est une fonction de sauts, dont les accroissements sont indépendants et stationnaires, et on a

$$E(e^{i\xi(X(t')-X(t))}) = e^{-(t'-t)\int_o^\infty (1-e^{i\xi u})d\nu(u)}$$

pour $t' > t$; la dernière intégrale a bien un sens, à cause des hypothèses faites sur $d\nu$. On pose (notation de Paul Lévy)

$$\psi(\xi) = \int_o^\infty (1 - e^{i\xi u})\, d\nu(u);$$

c'est encore une fonction de type négatif, mais qui n'est plus réelle. On obtient de cette façon - à l'addition près d'un terme linéaire - tous les processus croissants, à accroissements indépendants et stationnaires. Si on choisit

$$d\nu(u) = \frac{du}{u^{1+\gamma}}$$

on obtient

$$\psi(\xi) = \begin{cases} c\ \xi^\gamma & \text{pour } \xi > 0 \\ \overline{c}|\xi|^\gamma & \text{pour } \xi < 0. \end{cases}$$

$X(t)$ est le processus de Lévy croissant d'indice γ $(0 < \gamma < 1)$.

Considérons maintenant le demi-plan $\mathbb{R}^+ \times \mathbb{R}$, muni d'une mesure $d\sigma(t,u) = dt\, d\nu(u)$, ou $d\nu$ est une mesure positive localement bornée sur $\mathbb{R}\setminus\{0\}$, symmétrique, et telle que

$$\int_{-1}^{1} u^2\, d\nu(u) < \infty\ , \qquad \int_{|u|>1} d\nu(u) < \infty .$$

En s'inspirant de ce qui précède, on définit maintenant un processus $X(t)$ à acrroissements indépendants, continu à gauche, tel que

$$E(e^{i\xi(X(t')-X(t))}) = e^{-(t'-t)\psi(\xi)} \qquad (t' > t)$$

avec

$$\psi(\xi) = \int_{-\infty}^{\infty} (1-\cos \xi u)\, d\nu(u) = 2\int_{0}^{\infty} (1-\cos \xi u)\, d\nu(u).$$

Si on choisit

$$d\nu(u) = \frac{du}{|u|^{1+\gamma}}$$

on obtient

$$\psi(\xi) = c\, |\xi|^{\gamma} \qquad (c \text{ réel})$$

et $X(t)$ est le processus de Lévy symétrique d'indice γ $(0 < \gamma < 2)$.

Ces définitions se transcrivent aisément en considérant, au lieu de $\mathbb{R}^+$, un cône convexe Γ de sommet 0 dans $\mathbb{R}^d$ (qu'on supposera toujours d'intérieur non vide), et, au lieu de $\mathbb{R}$, $\mathbb{R}^d$ entier. Dans le premier cas, $d\nu$ sera une mesure dont le support engendre Γ, et intégrant $1 + |u|$ $(u \in \mathbb{R}^d)$. Dans le second cas, $d\nu$ sera une mesure symétrique, dont le support engendre tout $\mathbb{R}^d$, et intégrant $1 + |u|^2$. Les fonctions ψ correspondantes seront

$$\psi(\xi) = \int_{\Gamma} (1 - e^{i\xi . u})\, d\nu(u) \qquad (\xi \in \mathbb{R}^d)$$

et

$$\psi(\xi) = \int_{\mathbb{R}^d} (1 - \cos \xi . u)\, d\nu(u) \quad (\xi \in \mathbb{R}^d).$$

Lorsque d est le produit tensoriel d'une mesure portée par la sphère S^{d-1} et de la mesure radiale $\frac{dr}{r^{1+\gamma}}$, on obtient

$$\psi(\xi) = |\xi|^{\gamma}\, \psi(\xi') \qquad (\xi' = \frac{\xi}{|\xi|}).$$

Dans tous le cas, $\inf \operatorname{Re} \psi(\xi') > 0$: cela correspond au fait que la mesure $d\nu$ n'est pas contenu dans un hyperplan, c'est-à-dire que le processus est vraiment α-dimensionnel. Les valuers de γ permises sont $0 < \gamma < 1$ dans le cas du cône, et $0 < \gamma < 2$ dans le cas symétrique.

Remarques. Processus (d,γ) et processus (n,d,γ)

Dans la suite, quand nous parlerons d'un processus (d,γ), il s'agira d'un processus de Lévy stable d'exposant γ à valeurs dans $\mathbb{R}^d$ (si $0 < \gamma < 2$), ou du processus de Wiener à valeurs dans $\mathbb{R}^d$ (si $\gamma = 2$). Si $X(t)$ est un processus (d,γ), $\lambda^{-1/\gamma}X(\lambda t)$ est un processus qui lui est équivalent (c'est-à-dire que les deux processus ont même distribution): une dilatation du temps dans le rapport $\lambda > 0$ équivaut à une dilatation de l'espace dans le rapport $\lambda^{1/\gamma}$.

Lorsque nous parlerons d'un processus (n,d,γ), il s'agira d'un processus gaussien défini sur $\mathbb{R}^n$, à valeurs dans $\mathbb{R}^d$, à accroissements stationnaires et tels que

$$E(e^{i\xi.(X(t)-X(t'))}) = e^{-\frac{1}{2}|t-t'|^\gamma|\xi|^2}$$

ou, ce qui revient au même,

$$E(|X(t) - X(t')|^2) = \frac{1}{d}\,|t-t'|^\gamma.$$

Si $X(t)$ est un processus (n,d,γ), $\lambda^{-\gamma/2}\,X(\lambda t)$ est un processus qui lui est équivalent : une dilatation du temps dans le rapport $\lambda > 0$ équivaut à une dilatation de l'espace dans le rapport $\lambda^{\gamma/2}$.

Un théorème sur la distribution de processus arrêtés

Soit $X(t)$ un processus de Lévy ou de Wiener d-dimensionnel, d'exposant γ $(0 < \gamma \leq 2)$. Le "potentiel" correspondant est, selon la définition des probabilistes, la mesure

$$n = \int_0^\infty \mu_t\, dt,$$

où μ_t est la distribution de $X(t)$. Autrement dit,

$$\hat{n}(\xi) = \int_0^\infty E(e^{i\xi.X(t)})dt = \frac{1}{\psi(\xi)} \qquad (\xi \in \mathbb{R}^d).$$

L'hypothèse de d-dimensionalité signifie $\operatorname{Inf} \operatorname{Re} \psi(\xi') > 0$. Comme $\psi(\xi) = |\xi|^{\gamma}\psi(\xi')$, on a

$$\hat{n}(\xi) = |\xi|^{-\gamma}\,\hat{n}(\xi'), \quad \operatorname{Inf} \operatorname{Re} \hat{n}(\xi') > 0.$$

Le support de n est soit $\mathbb{R}^d$, soit le cône Γ engendré par la support de la mesure de Lévy $d\nu$. Dans la suite on ne distinguera pas la mesure n et sa densité, qui vérifie

$$n(x) = |x|^{\gamma-d}\, n(x') \quad \left(x \in \mathbb{R}^d,\ x' = \frac{x}{|x|}\right).$$

Le potentiel newtonien correspond à $\gamma = 2$, $n(x') = $ constante. A part ce cas, $n(x') = $ constante correspond aux processus de Lévy admettant la symétrie sphérique, c'est-à-dire que la mesure de Lévy $d\nu$ est le produit tensoriel d'une mesure équidistribuée sur S^{d-1} et de la mesure radiale $\frac{dr}{r^{1+\gamma}}$.

Désignons par τ_0 et τ deux "temps d'arrêt" pour $X(t)$, tels que $\tau_0 \le \tau$, et par μ_0 et μ les distributions de $X(\tau_0)$ et $X(\tau)$.

<u>THEOREME</u>. $\mu_0 * n \ge \mu * n$ *(le potentiel de* μ_0 *majore le potentiel de* μ*).*

Preuve. Ecrivons $\overset{F}{\le}$ pour l'inégalité usuelle entre transformées de Fourier:

$$\hat{a}(\xi) \overset{F}{\le} \hat{b}(\xi) \iff a(x) \le b(x).$$

Alors, visiblement,

$$E\int_{\tau_0}^{\infty} e^{i\xi.X(t)}dt \overset{F}{\ge} E\int_{\tau}^{\infty} e^{i\xi.X(t)}\,dt.$$

or

$$E\int_{\tau}^{\infty} e^{i\xi.X(t)}\,dt = E\left(e^{i\xi.X(\tau)}\, E_\tau \int_{\tau}^{\infty} e^{i\xi.(X(t)-X(\tau))}dt\right)$$

où E_τ désigne l'espérance quand τ est fixé (c'est-à-dire l'espérance conditionnelle par rapport à la tribu engendrée par τ). Comme

$$E_\tau \int_{\tau}^{\infty} e^{i\xi.(X(t)-X(\tau))}\,dt = E\int_{0}^{\infty} e^{i\xi.X(s)}ds$$

puisque le processus est à accroissements indépendants et stationnaires, on obtient

$$E\int_\tau^\infty e^{i\xi.X(t)}dt = E(e^{i\xi.X(\tau)}\ \hat{n}(\xi)) = \hat{\mu}(\xi)\ \hat{n}(\xi).$$

L'inégalité

$$\hat{\mu}_0(\xi)\ \hat{n}(\xi) \overset{F}{\geq} \hat{\mu}(\xi)\ \hat{n}(\xi)$$

équivaut a l'énoncé du théorème.

Remarquons que l'inégalité du théoreme vaut a fortiori si μ est l'image par $X(\tau)$ de la probabilité P restreinte à une partie de l'espace de probabilité Ω.

On verra une application dans la prochaine leçon.

TROISIEME LEÇON.- Théorie de Kakutani. Rencontres et points multiples

Capacité et polarité

En 1944, ayant en vue d'étudier l'existence de points doubles pour le mouvement brownien à valeurs dans $\mathbb{R}^d$, Kakutani montrait qu'un ensemble compact F dans $\mathbb{R}^d \setminus \{0\}$ est de capacité newtonienne nulle si et seulement si, presque sûrement, la fonction de Wiener $X(t)$ à valeurs dans $\mathbb{R}^d$ ne prend aucune valeur dans F (c'est-à-dire si F est "polaire" pour le mouvement brownien):

$$P(X(\mathbb{R}^+) \cap F = \emptyset) = 1.$$

Entre 1944 et 1958, en collaboration avec Dvoretzky, Erdös, S.J. Taylor, il établissait les résultats suivants:

- pour $d \geq 4$, $X(t)$ n'a pas de point double p. s.:

$$P(\exists\, t_1 < t_2 \ ;\ X(t_1) = X(t_2)) = 0$$

- pour $d = 3$, $X(t)$ a p. s. des points doubles, et pas de point triple:

$$P(\exists\, t_1 < t_2 \ ;\ X(t_1) = X(t_2)) = 1$$

$$P(\exists\, t_1 < t_2 < t_3 \ ;\ X(t_1) = X(t_2) = X(t_3)) = 0$$

- pour $d = 2$, $X(t)$ a p. s. des points multiples de multiplicité non dénombrable:

$$P(\exists\, E \ ,\ \text{Card}\, E > \text{aleph}_0 \ ;\ \text{Card}\, X(E) = 1) = 1.$$

L'extension de ces résultats à d'autres processus à accroissements indépendants a été faite par S. J. Taylor, Blumenthal et Getoor, Orey, Hawkes. Nous allons établir une partie de ces résultats. Dans la suite, sauf exception, $X(t)$ sera un processus à accroissement indépendant du type (d,γ), c'est-à-dire à valeurs dans $\mathbb{R}^d$ $(d \geq 3)$, et vérifiant

$$E(e^{i\xi.(X(t)-X(t'))}) = e^{-|t-t'|\psi(\xi)}$$

$$\psi(\xi) = |\xi|^\gamma \psi(\xi') \ ,\quad \text{Inf Re}\ \psi(\xi) > 0 \qquad (0 < \gamma \leq 2).$$

Pour $0 < \gamma < 2$, c'est un processus de Lévy stable d'exposant γ, et on désigne par Γ le support de la distribution de $X(t)$ $(t > 0)$; c'est un cône convexe de sommet 0 et d'intérieur non vide, ou c'est R^d.

En général, E désignera une partie compacte de $R^+ \setminus \{0\}$, F une partie compacte de $\overset{\circ}{\Gamma} \setminus \{0\}$. On utilisera le lemme simple suivant.

Lemme. *Pour avoir* $P(X(E) \cap F \neq \emptyset) > 0$, *il faut et suffit que* $\text{mes}_d(X(E) - F) > 0$ p. s.

Preuve. Posons $X(t) = X(a) + X_a(t-a)$, $a < \inf E$. Alors

$$P(X(E) \cap F \neq \emptyset) = P(X(a) \in F - X_a(E - a))$$

et comme la distribution de $X(a)$ est équivalente à la mesure de Lebesgue dans Γ (à vérifier), et que la v. a. $X(a)$ et l'ensemble aléatoire $X_a(E - a)$ sont indépendants, cette probabilité est positive si et seulement si

$$P(\text{mes}_d(F - X_a(E-a))) > 0$$

ce qu'on peut écrire, avec $a < b < \inf E$,

$$P(\text{mes}_d(F - X_a(b) - X_b(E-b))) > 0$$

soit - les distributions de $X_a(b)$ et $X(b)$ étant équivalentes -

$$P(\text{mes}_d(F - X(b) - X_b(E-b))) > 0$$

soit

$$P(\text{mes}_d(F - X(E)) > 0$$

Désignons par I l'intervalle $[a, a+1]$ $(a > 0$ donné).

THEOREME 1. *Soit* $X(t)$ *un processus du type* (d,γ), *admettant* Γ *comme cône des valeurs, et* F *une partie compacte de* $\overset{\circ}{\Gamma}$ *(ou* $R^d \setminus \{0\}$ *si* $\Gamma = R^d$*). Alors*

$$\text{Cap}_{d-\gamma} F > 0 \iff P(X(I) \cap F \neq \emptyset) > 0.$$

Preuve. Nous allons donner une preuve circulaire qui établit immédiatement, dans ce cas particulier, l'implication

$$\text{Cap}_\alpha^{(2)} F > 0 \implies \text{Cap}_\alpha^{(0)} F > 0$$

dans les notations de la première leçon, quand $d-2 \leq \alpha < d$

$(\gamma = d-\alpha)$.

Supposons $\mathrm{Cap}_\alpha^{(2)} F > 0$. Soit $T \in \mathcal{D}'(F)$, $T \neq 0$, une distribution d'énergie finie:

$$\int |\hat{T}(-\xi)|^2 |\xi|^{\alpha-d} d\xi < \infty.$$

Soit $\mu = \mu_\omega$ l'image de la mesure de Lebesgue dt sur I par l'application $t \to X(t)$; ainsi

$$\hat{\mu}(\xi) = \int_I e^{i\xi.X(t)} dt$$

$$|\hat{\mu}(\xi)|^2 = \iint_{I^2} e^{i\xi.(X(t)-X(t'))} dt\, dt'$$

$$E(|\hat{\mu}(\xi)|^2) = \iint_{I^2} e^{-|t-t'|\psi(\xi)} dt\, dt'$$

$$\leq \frac{2}{\mathrm{Re}\ \psi(\xi)} \leq \frac{C}{|\xi|^\gamma}.$$

Comme $\alpha+\gamma = d$, on a

$$E \int |\hat{\mu}(\xi)\ \hat{T}(-\xi)|^2 d\xi < \infty$$

donc p. s. la convolution $\mu * \tilde{T}$ a une densité dans L^2, donc p. s. son support $X(I) - F$ est de mesure de Lebesgue positive. D'après le lemme, on a donc

$$\mathrm{Cap}_\alpha^{(2)} F > 0 \Longrightarrow P(X(I) \cap F \neq \emptyset) > 0.$$

Supposons $P(X(I) \cap F \neq \emptyset) > 0$. Soit τ le temps d'arrêt défini par

$$\tau = \inf\{t | X(t) \in F\}.$$

Choisissons $\varepsilon > 0$ tel que $P(\varepsilon < \tau < \infty) > 0$ (en fait, tout $\varepsilon > 0$ convient), et soit μ l'image de la probabilité P, restreinte a l'évènement $(\varepsilon < \tau < \infty)$. En vertu du théorème qui termine la leçon 2 (avec $\tau_o = \varepsilon$ et en utilisant la remarque), le potentiel d'ordre $\alpha = d - \gamma$ de μ et borné. Donc

$$P(X(I) \cap F \neq \emptyset) > 0 \Longrightarrow \mathrm{Cap}_\alpha^{(0)} F > 0.$$

Comme $\mathrm{Cap}_\alpha^{(2)} F \geq \mathrm{Cap}_\alpha^{(0)} F$, la preuve est terminée.

C'est le théorème de Kakutani pour $\gamma = 2$.

Points doubles et points triple du mouvement brownien

Restreignons nous pour un moment au mouvement brownien à valeurs dans $\mathbb{R}^d$. Considérons trois intervalles disjoints sur R^+, soit I, J, K. Le problème des points doubles (resp. triples) se ramène à décider si $P(X(I) \cap X(J) = \emptyset)$ (resp. $P(X(I) \cap X(J) \cap \cap X(K) = \emptyset)$) est nul ou non. Comme $X(I)$, $X(J)$, $X(K)$ sont, à des translations près, des ensembles indépendants, il s'agit de décider si $X(I)$ (resp. $X(I) \cap X(J)$) est p. s. polaire ou non.

THEOREME 2. *Si* $d \geq 4$, $X(I)$ *est p. s. polaire. Si* $d = 3$, $X(I)$ *est p. s. non-polaire et* $X(I) \cap X(J)$ *est p. s. polaire*.

Preuve. Supposons $d = 4$. Il suffit (voir 1ère leçon) de montrer que p. s. $\text{mes}_2 X(I) < \infty$ (cela entraîne $\text{Cap}_2\, X(I) = 0$). Montrons que

$$(*) \qquad E(\text{mes}_4(X(I))_\varepsilon) = O(\varepsilon^2) \qquad (\varepsilon \to 0).$$

Pour cela, on découpe I en $O(\frac{1}{\varepsilon^2})$ intervalles I_j de longueur $|I_j| = \varepsilon^2$, et on observe que

$$E(\text{mes}_4(X(I_j))_\varepsilon) = E(\text{mes}_4(X([0,\varepsilon^2]))_\varepsilon = c\ \varepsilon^4$$

parce qu'une dilatation temporelle de rapport ε^2 équavaut (en distribution) a une dilatation spatiale de rapport ε. D'où (*), qui entraîne (Fatou) que p. s.

$$\underline{\lim}_{\varepsilon \to 0}\ \varepsilon^{-2}\ \text{mes}_4(X(I))_\varepsilon < \infty,$$

d'où $\text{mes}_2\ X(I) < \infty$ p. s.

Supposons maintenant $d = 3$. On obtient alors

$$E(\text{mes}_3(X(I))_\varepsilon + \text{mes}_3(X(J))_\varepsilon) = O(\varepsilon)$$

donc p. s.

$$\underline{\lim}_{\varepsilon \to 0}\ \varepsilon^{-1}(\text{mes}_3(X(I))_\varepsilon + \text{mes}_3(X(J))_\varepsilon) < \infty$$

donc (voir leçon 1 : $E = X(I)$, $F = X(J)$, $d = 3$, $\alpha = \beta = 2$)

$$\underline{\lim}_{\varepsilon \to 0}\ \varepsilon^{-2}\ \text{mes}_3(X(I) \cap (X(J) + x)) < \infty$$

pour presque tout $x \in \mathbb{R}^3$. Comme les distributions de $X(J)$ et $X(J) + x$ sont équivalentes, on a p. s.

$$\overline{\lim_{\varepsilon\to 0}}\ \varepsilon^{-2}\ \mathrm{mes}_3(X(I)\cap X(J)) < \infty,$$

d'où $\mathrm{mes}_1(X(I)\cap X(J)) < \infty$, d'où $\mathrm{Cap}_1(X(I)\cap X(J)) = 0$.

Reste à montrer que, en supposant toujours $d = 3$, $X(I)$ est p. s. non polaire. L'image par X de la mesure de Lebesgue dt sur I est la mesure aléatoire μ dont la transformée de Fourier est

$$\hat{\mu}(\xi) = \int_I e^{i\xi . X(t)}\, dt.$$

On a (calcul déjà fait)

$$E(|\hat{\mu}(\xi)|^2) = O(|\xi|^{-2})$$

donc p. s.

$$\int_{R^3} |\hat{\mu}(\xi)|^2\ |\xi|^{-2}\ d\xi < \infty$$

donc p. s. $\mathrm{Cap}_1\ X(I) > 0$, ce qui termine la preuve.

Pour le cas $d = 2$, il est facile de voir par récurrence que, quels que soient les intervalles disjoints $I_1, I_2, \ldots, I_n, \ldots$ l'ensemble

$$X(I_1)\cap X(I_2)\ \ldots \cap X(I_n)$$

a p. s. la dimension 2. Comme corollaire, p. s.

$$\bigcap_{n=1}^{\infty} X(I_n) \neq \emptyset.$$

Condition suffisante pour que $X(E)$ recontre F avec probabilité positive

On peut chercher à étendre le théorème 1 dans deux directions:

1. en remplaçant l'intervalle I par un compact $E \subset \mathbb{R}^+ \setminus \{0\}$

2. en considérant d'autres processus que les processus (d,γ).

On va se borner a la première direction.

Cherchons donc, d'abord, une condition suffisante pour

$$P(X(E)\cap F \neq \emptyset) > 0,$$

$X(t)$ et F satisfaisant aux hypotheses du théorème 1. En reprenant la première partie de la preuve, on voit qu'il suffit que F porte une distribution $T \neq 0$, et E une mesure $d\sigma \neq 0$, telles que

$$(*) \qquad \iiint |\hat{T}(-\xi)|^2 \, e^{-|t-t'|\psi(\xi)} \, d\sigma(t)\, d\sigma(t')\, d\xi < \infty.$$

THEOREME 3. *Soit* X(t) *un processus* (d,γ), E compact $\subset R^+ \setminus \{0\}$, F *compact intérieur au cône des valeurs de* X(t). *Si, pour un* $\alpha > 0$, *on a l'une des deux hypothèses*

$$(H_1) \qquad \begin{cases} \text{mes}_\alpha \; E > 0 \\ \text{Cap}_{d-\gamma\alpha} \; F > 0, \end{cases}$$

$$(H_2) \qquad \begin{cases} \text{Cap}_\alpha \; E > 0 \\ \text{mes}_{d-\gamma a} \; F > 0 \end{cases}$$

on a

$$P(X(E) \cap F \neq \emptyset) > 0.$$

Preuve. Supposons (H_1). D'après le lemme de Frostman, on peut choisir la mesure positive $d\sigma$ de façon que

$$\int_{2^{-j} \leq |t-t'| \leq 2^{-j+1}} d\sigma(t)\, d\sigma(t') \leq C\, 2^{-j\alpha}$$

Il en résulte

$$\iint_{E^2} e^{-|t-t'|\psi(\xi)} \, d\sigma(t)\, d\sigma(t') \leq c' \, |\xi|^{-\gamma\alpha}$$

En choisissant pour T une distribution d'énergie finie par rapport au noyau $|x|^{-d+\gamma\alpha}$, on a bien (*) d'où la conclusion.

Supposons maintenant (H_2). D'après le lemme de Frostman, on peut choisir pour T une mesure positive $d\nu$ de façon que

$$\nu(B) \leq (\text{diam } B)^{d-\gamma\alpha}.$$

Il en résulte

$$\int |\hat{\nu}(-\xi)|^2 \, e^{-|t-t'|\psi(\xi)} d\xi \leq \int |\hat{\nu}(-\xi)|^2 \, e^{-d|t-t'||\xi|^\gamma} \, d\xi$$
$$\leq C\, |t-t'|^{-\alpha},$$

En choisissant pour $d\sigma$ une mesure d'énergie finie par rapport au noyau $|t|^{-\alpha}$, on a encore (*), d'où la conclusion.

Enonçons une autre conséquence simple de la formule (*).

THEOREME 4. *Si* E *est un fermé non dénombrable, il existe un fermé* F *de mesure de Lebesgue nulle tel que*

$$P(X(E) \cap F \neq \emptyset) > 0.$$

Preuve. On choisit pour $d\sigma$ une mesure continue. Alors le facteur de $|\hat{T}(-\xi)|^2 \, d\xi$ dans (*) est une fonction de ξ qui tend vers 0 à l'infini.

Remarquons que si E est au contraire un fermé dénombrable, on a, pour tout fermé F de mesure de Lebesgue nulle,

$$P(X(E) \cap F = \emptyset) = 1.$$

Cella traduit simplement le fait que la distribution de $X(t)$ est absolument continue.

Condition nécessaire pour que X(E) recontre F avec probabilité positive

Cherchons maintenant une condition nécessaire pour

$$P(X(E) \cap F \neq \emptyset) > 0,$$

en utilisant le théorème de la leçon 2.

Supposons d'abord que la mesure de Lebesgue de E est positive: mes E > 0. Comme corollaire des théorèmes 1 et 2, on a tout de suite:

THEOREME 5. *Si* mes E > 0, *on peut remplacer dans l'énoncé du théorème 1* X(I) *par* X(E).

Supposons maintenant que la mesure de Lebesgue de E est nulle: mes E = 0. Soit $[a,b]$ le plus petit segment contenant E, et $I_1, I_2, \ldots, I_n, \ldots$ les intervalles ouverts contenus dans $[a,b]$ et contigüs à E. Soit $h(t)$ $(t \geq 0)$ une fonction positive concave, telle que $h(0) = 0$. Nous faisons l'hypothèse:

$$BT(h) : \Sigma\, h(|I_n|) < \infty$$

(ce genre de conditions a été utilisé par Besicovitch et Taylor, en vue d'obtenir des majorations de dimensions). Posons enfin

$$n = \int_0^\infty \mu_t \, dh(t)$$

où μ_t est la distribution de $X(t)$, c'est-à-dire

$$\hat{n}(\xi) = \int_0^\infty e^{-t\psi(\xi)} \, dh(t).$$

THEOREME 6. *Supposons* BT(h) *et*

$$P(X(E) \cap F \neq \emptyset) > 0.$$

Alors F *porte une mesure positive* μ, *non nulle, telle que*

$$\sup_x \mu * n(x) < \infty$$

n *étant le noyau défini ci-dessus.*

Preuve. Désignons par ρ_0 la mesure, portée par $[b,\infty[$, définie par

$$d\rho_0(t) = dh(t - b) \qquad (t \geq b)$$

et par ρ_n la mesure, portée par $I_n = [a_n, b_n]$, définie par

$$d\rho_n(t) = 1_{[a_n,b_n]}(t)dh(t-a_n).$$

D'après l'hypothèse BT(h), la somme des masses totales des ρ_n $(n \geq 1)$ est finie. Soit $\rho = \rho_0 + \rho_1 + \rho_2 + \dots \rho_n + \dots$ Si $t \in E$ et $\theta > 0$, la ρ-mesure de l'intervalle $[t, t+\theta]$ est la somme des ρ-mesures des intervalles I_n inclus dans $[t, t+\theta]$, plus éventuellement la ρ-mesure de $I_m \cap [t, t+\theta]$, si I_m est l'intervalle (s'il existe) qui contient $t+\theta$ (I_m existe $\Longleftrightarrow$ $t+\theta \notin E$). Ainsi (E étant de mesure nulle!),

$$\rho([t, t+\theta]) = \Sigma \, h(\ell_n)$$

avec $\Sigma \, \ell_n = \theta$. D'après l'hypothèse de concavité,

$$\rho([t, t+\theta]) \geq h(\theta).$$

Considérons le temps d'arrêt

$$\tau = \inf\{t \in E; X(t) \in F\}$$

et $\varepsilon > 0$ tel que $P(\varepsilon < \tau < \infty) > 0$. Avec les notations de la leçon 2, on a

$$E\int_\varepsilon^\infty e^{i\xi.X(t)}d\rho(t) \overset{F}{\geq} E\int_\tau^\infty e^{i\xi.X(t)}d\rho(t)$$
$$= E(e^{i\xi.X(\tau)}E_\tau\int_\tau^\infty e^{i\xi.(X(t)-X(\tau))}\,d\rho(t))$$
$$= E(e^{i\xi.X(\tau)}\int_\tau^\infty e^{-(t-\tau)\psi(\xi)}\,d\rho(t)).$$

Pour la démonstration du théorème, on peut remplacer $X(t)$ par le processus symétrisé,c'est-à-dire remplacer $\psi(\xi)$ par $\mathrm{Re}\,\psi(\xi)$ (l'hypothèse pour le processus donné entraîne l'hypothèse pour le symétrisé, et la conclusion pour le symétrysé entraîne la conclusion pour le processus donné). Supposons donc $\psi(\xi) > 0$. Alors, au moyen de deux intégrations par parties et de la majoration (*), on a

$$\int_\tau^\infty e^{-(t-\tau)\psi(\xi)}\,d\rho(t) \overset{F}{\geq} \int_0^\infty e^{-t\psi(\xi)}\,dh(t) = \hat{n}(\xi).$$

Désignons par μ la distribution de $X(\tau)$ sur l'événement $(\varepsilon < \tau < \infty)$. On obtient

$$\hat{\mu}(\xi)\,\hat{n}(\xi) \overset{F}{\leq} \int_\varepsilon^\infty e^{-t\psi(\xi)}\,d\rho(t)$$

c'est-à-dire

$$\mu * n \leq \int_\varepsilon^\infty \mu_t\,d\rho(t).$$

Comme $\hat{\mu}_t = e^{-t\psi} \in L^1(R^n)$, la mesure μ_t a une densité bornée. Par abus de notation, désignons par $\mu_t(x)$ la valeur de cette densité au point x. Comme

$$\hat{\mu}_t(\xi) = \hat{\mu}_1(t^{1/\gamma}\,|\xi|),$$

on a

$$\mu_t(x) = t^{-d/\gamma}\,\mu_1(t^{-1/\gamma}\,x)$$

donc

$$\sup_x \int_\varepsilon^\infty \mu_t(x)\,d\rho(t) < \infty.$$

Donc

$$\sup_x \mu * n(x) < \infty$$

et $\mu \in M^+(F)$, ce qui démontre le théorème.

COROLLAIRES

1. Si E est de mesure de Lebesgue nulle, il existe $F \subset \mathbb{R}^d$, $\text{Cap}_{d-\gamma} F > 0$, tel que

$$P(X(E) \cap F = \emptyset) = 1.$$

2. Si on choisit $E = E(\xi_1, \xi_2, \ldots)$ (défini dans la 1ère leçon), avec $(\xi_1 \xi_2 \ldots \xi_n)^\alpha = n^{-2}\, 2^{-n}$, et si $\text{Cap}_{d-\gamma\alpha} F = 0$, on a

$$P(X(E) \cap F = \emptyset) = 1.$$

Question: a-t-on la même conclusion si on suppose

$$(\xi_1 \xi_2 \ldots \xi_n)^\alpha = 2^{-n},$$

c'est-à-dire si $E = E(\xi)$ avec $\xi^\alpha = \frac{1}{2}$? Ce serait alors une bonne réciproque au théorème 2 (hypothèse (H_1), cas $0 < \alpha < 1$). Le corollaire 1 est une bonne réciproque pour le cas $\alpha = 1$.

Conditions pour que $X(E)$ recontre $X(F)$

Revenons maintenant sur la question des points doubles, traitée par le théorème 2. Au lieu de deux intervalles I et J disjoints, considérons maintenant deux compacts E et F, portés par des intervalles disjoints de $\mathbb{R}^+ \setminus \{0\}$, et un processus du type (d,γ). Nous allons établir ceci.

THEOREME 7. *Si* $X(t)$ *est un processus* (d,γ), E *et* F *compacts portés par des intervalles disjoints de* $\mathbb{R}^+ \setminus \{0\}$, *on a les implications*

$$\text{Cap}_{\frac{d}{\gamma}}(E \times F) > 0 \Longrightarrow P(X(E) \cap X(F) \neq \emptyset) > 0 \Longrightarrow \text{mes}_{\frac{d}{\gamma}}(E \times F) > 0$$

avec pour corollaire

$$\dim E \times F > \frac{d}{\gamma} \Longrightarrow P(X(E) \cap X(F) \neq \emptyset) > 0 \Longrightarrow \dim E \times F \geq \frac{d}{\gamma}.$$

Preuve de la première implication. Supposons $\text{Cap}_{\frac{d}{\gamma}}(E \times F) > 0$. Il existe une mesure $\sigma \in M_1^+(E \times F)$ telle que

$$\iiiint \frac{d\sigma(s,t)\, d\sigma(s',t')}{(|s-s'| + |t-t'|)^{d/\gamma}} < \infty.$$

Soit μ l'image de σ par l'application

$$(s,t) \to X(s) - X(t).$$

Sa transformée de Fourier est

$$\hat{\mu}(\xi) = \iint_{E\times F} \exp(i\xi.(X(s)-X(t)))d\sigma(s,t)$$

donc

$$|\hat{\mu}(\xi)|^2 = \iiiint \exp(i\xi.(X(s)-X(s')-(X(t)-(X(t'))d\sigma(s,t)d\sigma(s't')$$

et les accroissements $X(s) - X(s')$ et $X(t) - X(t')$ sont indépendants puisque E et F sont portés par des intervalles disjoints. Donc

$$E(|\hat{\mu}(\xi)|^2) = \exp(-|s-s'| + |t-t'|))\psi(\xi)$$

et, compte tenu de l'hypothèse sur σ, on obtient

$$E \int_{R^d} |\hat{\mu}(\xi)|^2 \, d\xi < \infty$$

ce qui entraîne p. s. le support de μ, $X(E) - X(F)$, a sa mesure de Lebesgue positive, d'où (voir lemme au début de la leçon),

$$P(X(E) \cap X(F) \neq \emptyset) > 0.$$

Preuve de la seconde implication. Si I et J sont deux intervalles disjoints de longueur commune ℓ, on pose

$$\phi(I,J) = E(\mathrm{mes}_d(X(I)-X(J))).$$

On vérifie que $\phi(I,J) < \infty$; c'est évident dans le cas $\gamma = 2$, et çà nécessite un retour à la définition des processus de Lévy pour $0 < \gamma < 2$, et la décomposition de $X(t)$ en $X_1(t) + X_2(t)$ (somme de processus indépendants correspondant à la décomposition de la mesure de Lévy $d\nu$ en $d\nu_1 + d\nu_2$, $d\nu_1$ étant portée par une boule de centre 0, et $d\nu_2$ par le complémentaire); le détail est dans le séminaire d'Orsay 1983, p. 88. Cela étant, il est clair que $\phi(I,J)$ ne dépend que de ℓ. Comme une dilatation des temps dans le rapport λ équivaut (en distribution) à une dilatation de l'espace dans le rapport $\lambda^{1/\gamma}$, qui multiplie la mesure par $\lambda^{d/\gamma}$, on a

$$\phi(I,J) = C\,\ell^{d/\gamma}.$$

Supposons $\mathrm{mes}_{d/\gamma}(E \times F) = 0$. On peut donc recouvrir $E \times F$ une infinité de fois par des pavés $I_m \times J_m$ ($|I_m| = |J_m| = \ell_m$) de façon que $\sum_1^\infty \ell_m^{d/\gamma} < \infty$. Donc

$$E(\sum_1^\infty \mathrm{mes}_d(X(I_m) - X(J_m))) < \infty,$$

donc p. s. la série converge, et, comme n'importe quel reste de cette série majore $mes_d(X(E) - X(F))$, on a p. s.

$$mes_d(X(E) - X(F)) = 0$$

ce qui équivaut à

$$P(X(E) \cap X(F) \neq \emptyset) = 0.$$

Cela termine la preuve du théorème 7.

Dimension de $X(E)$

En vue de la prochaine leçon, voici un résultat simple sur la dimension de $X(E)$, pour les deux types de processus introduits dans la deuxième leçon.

THEOREME 8. 1) *Soit* $X(t)$ *un processus* (d,γ) *(Wiener pour* $\gamma = 2$, *Lévy pour* $0 < \gamma < 2$*),* E *un compact de* $\mathbb{R}^+$*. Alors*

$$mes_\alpha E = 0 \Longrightarrow mes_{\gamma\alpha} X(E) = 0 \quad \text{p. s.}$$

$$\left.\begin{array}{l} Cap_\alpha E > 0 \\ \gamma\alpha < d \end{array}\right\} \Longrightarrow Cap_{\gamma\alpha} X(E) > 0 \quad \text{p. s.}$$

En conséquence

$$\dim X(E) = \inf(d,\gamma\alpha) \quad \text{p. s.}$$

2) *Soit maintenant* $X(t)$ *un processus gaussien* (n,d,γ) *(mouvement brownien à* n *paramètres pour* $\gamma = 1$, *mouvement brownien fractionnaire à* n *paramètres pour* $0 < \gamma < 2$, $\gamma \neq 1$*), et soit* E *un compact de* $\mathbb{R}^n$*. Alors*

$$mes_\alpha E = 0 \Longrightarrow mes_{2\alpha/\gamma} X(E) = 0 \quad \text{p. s.}$$

$$\left.\begin{array}{l} Cap_\alpha E > 0 \\ 2\alpha/\gamma < d \end{array}\right\} \Longrightarrow Cap_{2\alpha/\gamma} X(E) = 0 \quad \text{p. s.}$$

En conséquence

$$\dim X(E) = \inf(d, 2\alpha/\gamma) \quad \text{p. s.}$$

Preuve. Donnons la d'abord pour le mouvement brownien ordinaire à valeurs dans $\mathbb{R}^d$. Si $mes_\alpha E = 0$, on a $E \subset \overline{\lim\limits_{n\to\infty}}\, I_n$ avec $\sum_1^\infty |I_n|^\alpha < \infty$. Par homogénéité, on a pour tout $\beta > 0$

$$(*) \qquad E((\operatorname{diam} X(I_n))^\beta) = c_\beta\, |I_n|^{\beta/2}$$

et on vérifie aisément que $c_\beta < \infty$. En choisissant $\beta = 2\alpha$, on a

$$E\left(\sum_1^\infty (\operatorname{diam} X(I_n))^\beta\right) < \infty$$

d'où

$$\sum_1^\infty (\operatorname{diam} X(I_n))^\beta < \infty \quad \text{p. s.}$$

d'où, puisque $X(E)$ est contenu dans $\bigcup_N^\infty X(I_n)$ quel que soit N,

$$\operatorname{mes}_\beta X(E) = 0.$$

Si d'autre part $\operatorname{Cap}_\alpha E > 0$ et $2\alpha > d$, on vérifie que l'image par $X(.)$ d'une mesure $\alpha \in M_1^+(E)$ d'énergie finie par rapport a $|t|^{-\alpha}$ est p. s. d'énergie finie par rapport à $|x|^{-2\alpha}$, d'où $\operatorname{Cap}_{2\alpha} X(E) > 0$ p. s..

La même preuve convient au cas gaussien (n,d,γ), mutatis mutandis, et la seconde partie de la preuve convient également au cas d'un processus de Lévy (d,γ) $(0 < \gamma < 2)$. La première partie de la preuve s'applique encore dans un processus de Lévy (d,γ) quand $\beta = \alpha\gamma < \gamma$ (c'est-à-dire $\alpha < 1$), mais elle ne convient pas au cas $\beta = \gamma$ (c'est-à-dire $\alpha = 1$), parce qu'alors, pour tout intervalle I, on a

$$E((\operatorname{diam} X(I))^\beta) = \infty.$$

Dans le cas $\alpha = 1$, il convient d'utiliser au lieu de (*) la formule

$$E(\operatorname{mes}_\gamma X(I_n)) = c|I_n|$$

qu'on obtient sans difficulté, en écrivant comme dans la section précédente

$$X(t) = X_1(t) + X_2(t)$$

où $X_1(t)$ et $X_2(t)$ sont des processus de Lévy indépendants, $X_1(t)$ n'ayant que des sauts inférieurs à 1, et $X_2(t)$ n'ayant que des sauts supérieurs ou égaux à 1.

QUATRIEME LEÇON.- Ensembles de Salem. Propriétés de Fourier des mesures images. Densité d'occupation

Ensembles U et M. Deux théorèmes de Salem

Soit F un compact dans $\mathbb{R}^d$. Dans la théorie classique des séries trigonométriques (cas $d = 1$), on dit que F est un ensemble de type U (U pour "unicité") si F ne porte aucune distribution $\neq 0$ dont la transformée de Fourier tende vers 0 à l'infini:

$$\left.\begin{array}{l} T \in \mathcal{D}'(F) \\ \lim\limits_{|\xi|\to\infty} \hat{T}(\xi) = 0 \end{array}\right\} \Longrightarrow T = 0.$$

On dit que F est un ensemble de type M (M pour multiplicité) dans le cas contraire, et un ensemble de type M_o si F porte une mesure positive $\neq 0$ dont la transformée de Fourier tend vers 0 à l'infini:

$$\mu \in M^+(F), \quad \mu \neq 0, \quad \lim_{|\xi|\to\infty} \hat{\mu}(\xi) = 0.$$

L'origine de la terminologie remonte à Cantor; les ensembles de type U (cas $d = 1$) peuvent être définis par la propriété suivante: si une série trigonométrique converge vers 0 en dehors de l'ensemble, elle est identiquement nulle.

A priori, on pourrait penser que, si un ensemble est "mince" au sens de la mesure et de la dimension de Hausdorff (par exemple $\dim F = 0$), il est de type U, et que, s'il est "gros" (par example $\dim F = d$), il est de type M. Il n'en est rien.

Aucune condition sur la mesure de Hausdorff n'entraîne que F soit de type U. En effet, étant donné une fonction $h(t)$ concave, tendant vers 0 quand $t \to 0$ aussi lentement qu'on veut, on peut trouver $F \subset \mathbb{R}$ de type M_o et tel que $\text{mes}_h F = 0$ (Ivašev-Musatov).

Aucune condition sur la mesure de Hausdorff n'entraîne que F soit de type M. En effet, étant donné une fonction $h(t)$ quelconque telle que $t^d = 0(h(t))$ $(t \to 0)$, on peut trouver $F \subset \mathbb{R}^d$ de type U tel que $\text{mes}_h F = \infty$ (Wik, Kaufman). Il s'agit même d'un fait générique, dans le sens suivant. Définissons la distance de deux compacts F et F' dans $\mathbb{R}^d$ comme

$$d(F,F') = \inf\{\sup_{x\in\mathbb{R}^d} (|f(x)-x| + |\log Jf(x)|);\ f \in \text{Diff}\ \mathbb{R}^d;\ f(F) = F'\}.$$

$\text{Diff}\,\mathbb{R}^d$ étant d'ensemble des difféomorphismes de classe C^1 de $\mathbb{R}^d$, et Jf le jacobien de f. Disons que F et F' appartiennent à la même classe si $d(F,F') < \infty$. Chaque classe est un espace métrique complet. Partons d'un ensemble F_0 totalement discontinu avec la propriété suivante: il existe $\varepsilon_j \to 0$ et $\mu_j \to \infty$ tels que, pour chaque j, F_0 est contenu dans une réunion de boules de diamètres ε_j dont les distances mutuelles dépassent $\mu_j\varepsilon_j$ (on vérifie facilement que cela est compatible avec $\text{mes}_h F_0 = \infty$, quelle que soit la fonction $h(t)$ donnée). Alors, au sens de Baire, quasi tout F dans la classe de F_0 est un ensemble de type U (c'est-à-dire que les F de type M constituent un ensemble maigre), et de plus $\text{mes}_h F = \infty$.

L'ensemble triadique de Cantor ($E(\xi)$ avec $\xi = \frac{1}{3}$) est de type U (Rajchman); les ensembles $E(\xi)$ avec ξ irrationnel sont de type M (Nina Bari). La classification des $E(\xi)$ selon le type U ou M est l'objet d'un théorème fameux de Salem et Zygmund: $E(\xi)$ est de type U si et seulement si ξ^{-1} est un nombre de Pisot, c'est-à-dire un entier algébrique dont tous les conjugués (sauf lui-même) ont leur module strictement inférieur à 1. Cela concerne le cas $d = 1$.

Si $d > 1$, tout ensemble F contenu dans une réunion finie d'hyperplans est de type U; par example, la frontière d'un cube est de type U. Au contraire, une sphère est de type M, puisque la transformée de Fourier de la mesure de masse 1 équirépartie sur S^{d-1} satisfait

$$\hat{\mu}(\xi) = O(|\xi|^{-\frac{d-1}{2}}) \qquad (\xi \to \infty)$$

Après ces préliminaires, venons-en aux deux théorèmes de Salem que cette leçon va illustrer.

S1. Si $\underline{\lim}_{\varepsilon\to 0} (N_\varepsilon(F)/\log \frac{1}{\varepsilon}) = 0$, F est de type U, au sens fort que voici: pour chaque $T \in \mathcal{D}'(F)$,

$$\overline{\lim}_{|\xi|\to\infty} |\hat{T}(\xi)| = \sup_{\xi} |\hat{T}(\xi)|.$$

Si $\underline{\lim}_{\varepsilon\to 0} (N_\varepsilon(F)/\log \frac{1}{\varepsilon}) < \infty$, F est de type U, et il existe un $c > 0$ tel que, pour chaque $T \in \mathcal{D}'(F)$,

$$\overline{\lim}_{|\xi|\to\infty} |\hat{T}(\xi)| \geq c \sup_{\xi} |\hat{T}(\xi)|.$$

(le second énoncé est de Kahane et Katznelson, et se trouve dans le

Journal d'Analyse Mathématique de Jérusalem 23 (1970), 185-197, avec les références).

S2. Pour tout $\beta \in]0,1[$, il existe un ensemble $F \subset \mathbb{R}$, de dimension β, et portant une mesure positive $\mu \neq 0$ telle que

$$\hat{\mu}(\xi) = O(|\xi|^{-\beta/2+\varepsilon}) \qquad (\xi \to \infty)$$

pour $\varepsilon > 0$.

Nous appellerons ensembles de Salem tous les ensembles $F \subset \mathbb{R}^d$ ayant cette propriété: il existe $\mu \in M_1^+(F)$ telle que

$$\hat{\mu}(\xi) = O(|\xi|^{-\frac{1}{2}\dim F+\varepsilon}) \qquad (\xi \to \infty)$$

pour tout $\varepsilon > 0$. Remarquons que pour aucun $\varepsilon > 0$ il n'existe de distribution $T \in \mathcal{D}'(F)$ non nulle et telle que

$$\hat{T}(\xi) = O(|\xi|^{-\frac{1}{2}\dim F-\varepsilon}) \qquad (\xi \to \infty);$$

en effet, cela entrainerait

$$\int |\hat{T}(\xi)|^2 |\xi|^{\beta-d}\, d\xi < \infty$$

avec $\beta > \dim F$, contrairement a la définition de la dimension comme dimension capacitaire.

Nous venons de voir que les spheres S^{d-1} sont des ensembles de Salem. On peut le vérifier aussi pour des frontières de convexes assez ronds. Mais nous avons vu aussi que les frontières de cubes sont des ensembles de type U, donc à l'opposé des ensembles de Salem.

La construction de Salem est probabiliste, et utilise un peu de théorie des nombres. En fait, en dehors de l'exemple mentionné (frontière de convexe rond), et des cas $d = 0$ et $d = \dim F$, des constructions explicites seraient très laborieuses. Au contraire, comme nous allons le voir, tous les ensembles $X(E)$, où E est un compact donné, et X un processus de Lévy (d,γ), ou le processus du mouvement brownien, ou d'un mouvement brownien fractionnaire, sont des ensembles de Salem. Ils nous fourniront d'ailleurs, dans le cas de la dimension 0, une réciproque du théorème 1, montrant le rôle critique de la fonction log.

Dans le cas où $X(E)$ a pour dimension d, on peut étudier la manière dont $X(E)$ occupe l'espace, et plus précisément la "densité d'occupation" (sur cette notion et ses applications, voir la mise au point fait par Geman et Horowitz en 1980 dans Annals of Probability). Par exemple, si $\dim E > \frac{1}{2}$ et X est le mouvement brownien linéaire $(d = 1)$, p. s. $X(E)$ a des points intérieurs (R. Kaufman 1975). Nous allons montrer que la méthode de Fourier, inspirée de Salem, donne rapidement de tels résultats.

Ainsi les méthodes probabilistes et la méthode de Baire donnent des résultats en sens opposé, ce qu'on peut interpréter ainsi: la méthode de Baire introduit des résonnances, donc créé des bosses dans les spectres, tandis qu'au contraire, les méthodes probabilistes suppriment les résonnaces et lissent les spectres.

Proprietés de Fourier des ensembles $X(E)$. Cas du processus (d,γ) (Wiener ou Lévy)

Dans toute la suite E est un compact porté par $\mathbb{R}^+\setminus\{0\}$ (ou quelquefois $\mathbb{R}\setminus\{0\}$), et $X(t)$ est un processus de Lévy du type (d,γ) ou un mouvement brownien d-dimensionnel (type $(d,2)$), ou un mouvement brownien fractionnaire à valeurs dans $\mathbb{R}^d$

$$E(\|X(t) - X(t')\|^2_{\mathbb{R}^d}) = |t-t'|^\gamma$$

defini sur $\mathbb{R}$ ou quelquefois sur $\mathbb{R}^n$ (types $(1,d,\gamma)$ ou (n,d,γ)).

On suppose la fonction $h(t)$ concave, ou bien convexe avec $h(2t) = O(h(t))$, et $\text{mes}_h E > 0$. Soit $\sigma \in M^+(E)$ une mesure, non nulle, telle que $\sigma(I) \leq h(|I|)$ pour tout intervalle I (ou boule I dans le cas de $\mathbb{R}^n$). On désigne par μ (mesure aléatoire) l'image de σ par $X(t)$, de sorte que

$$\hat{\mu}(\xi) = \int e^{i\xi.X(t)}\, d\sigma(t).$$

L'étude consiste à évaluer soigneusement, pour p entier ≥ 1,

$$E(|\hat{\mu}(\xi)|^{2p}) = \underbrace{\int\cdots\int}_{2p} e^{i\xi.(X(t_1)+\ldots+X(t_p)-X(t'_1)\ldots-X(t'_p))}\, d\sigma(t_1)\ldots d\sigma(t'_p)$$

On obtiendra une majoration de la forme

$$E(|\hat{\mu}(\xi)|^{2p}) \leq (\varepsilon(\xi,p))^{2p}.$$

En choisissant $\xi = n$ et $p = p_n$ et en ajoutant, on obtient

$$\sum_{n\in Z^d\setminus\{0\}} E(|\hat{\mu}(n)/\varepsilon(n,p_n)|^{2p_n}) \frac{1}{|n|^{d+1}} < \infty$$

d'où p.s.

$$|\hat{\mu}(n)| \leq \varepsilon(n,p_n)\ |n|^{\frac{d+1}{2p_n}}$$

pour $|n|$ assez grand. A partir de là, et d'estimations analogues pour $\hat{\mu}(\alpha n)$ avec $\alpha > 0$ donné, on obtient une majoration presque sûre de $|\hat{\mu}(\xi)|$ lorsque $|\xi|$ est assez grand.

Le cas du mouvement brownien (type $(d,2)$) avec $h(.)$ concave se trouve traité en détail dans SRSF (dernier chapitre). Le cas des processus (d,γ) se traite de la même façon, et on obtient que p. s.

$$E(|\hat{\mu}(\xi)|^{2p}) \leq (Cp\sigma(E)\ h(|\xi|^{-\gamma}))^p$$

d'où p. s. (avec une autre constante C)

$$(*) \qquad |\hat{\mu}(\xi)| \leq C(\alpha(E)\ h(|\xi|^{-\gamma})\log|\xi|)^{1/2}$$

pour $|\xi|$ assez grand, les constantes ne dépendant que du processus, c'est-à-dire de d et de la fonction $\psi(\xi)$. Voici deux applications.

THEOREME 1. *Si* E *est de dimension* α, *et* $\alpha\gamma \leq d$, *son image* $X(E)$ *par un processus* (d,γ) *(Lévy ou Wiener a valeurs dans* $\mathbb{R}^d$*) est p. s. un ensemble de Salem de dimension* $\alpha\gamma$.

Preuve. On sait déjà (troisième leçon, théorème 8) que p. s.

$$\dim X(E) \leq \alpha\gamma.$$

Comme on a $\text{mes}_h\ E > 0$ si $h(t) = t^{\alpha-\varepsilon}$, il existe p. s. une mesure $\mu \in M^+(X(E))$ telle que

$$\hat{\mu}(\xi) = O(|\xi|^{-\frac{1}{2}\gamma\ (\alpha-2\varepsilon)})$$

cela quel que soit $\varepsilon > 0$, donc $X(E)$ est p. s. un ensemble de Salem de dimension $\alpha\gamma$.

Remarquons que si E est de dimension α et vérifie $\text{mes}_h\ E > 0$ avec $h(t) = t^\alpha/\log\frac{1}{t}$ $(t < \frac{1}{2})$, on a p. s.

$$\hat{\mu}(\xi) = O(|\xi|^{-\frac{1}{2}\alpha\gamma}) \qquad (|\xi| \to \infty).$$

THEOREME 2. *Si* $\text{mes}_h E > 0$ *avec* $h(t) = O((\log\frac{1}{t})^{-1})$ $(t \to 0)$, *p. s. son image* $X(E)$ *par un processus* (d,γ) *est un ensemble de type* M_0. *Si* $\text{mes}_h E$ *est assez grand, avec* $h(t) = (\log\frac{1}{t})^{-1}$ $(t \to 0)$, *p. s.* $X(E)$ *n'est pas un ensemble de type* U *au sens fort.*

Preuve. La première partie est immédiate. Pour la seconde, il suffit d'avoir $\sigma(E) > C^{-1}$ dans la formule (*) pour conclure

$$\overline{\lim_{|\xi|\to\infty}} \; |\hat{\mu}(\xi)| < \hat{\mu}(0) = \sigma(E).$$

En application du théorème 2, choisissons $E = E(\xi_1,\xi_2,\ldots,\xi_n,\ldots)$ avec $\xi_n = \rho_n^{2^n}$ $(\rho_n < 1)$. Si $\lim_{n\to\infty} \rho_n = 1$, on est dans le premier cas; si $\rho_n = \rho < 1$, ρ étant assez voisin de 1, on est dans le second. Etant donné une fonction $A(\varepsilon) \to \infty$ $(\varepsilon \to 0)$ on peut choisir $\rho_n \to 1$ de telle sorte que

$$N_\varepsilon(E) = O(A(\varepsilon^{1/3}) \log \frac{1}{\varepsilon}) \qquad (\varepsilon \to 0).$$

Si $X(t)$ est la fonction de Wiener, on a, compte-tenu du module de continuité,

$$N_\varepsilon((X(E))) = O(N_{\varepsilon^3}(E)) = O(A(\varepsilon) \log \frac{1}{\varepsilon}) \qquad (\varepsilon \to 0).$$

Donc il est faux que

$$\underline{\lim}_{\varepsilon\to 0} \; (N_\varepsilon(F)/A(\varepsilon) \log \frac{1}{\varepsilon}) < \infty$$

entraîne que F est de type U. De même, en choisissant $\rho_n = \rho$ assez proche de 1, on a

$$N_\varepsilon(E) = O(\log \frac{1}{\varepsilon})$$

$$N_\varepsilon(X(E)) = O(\log \frac{1}{\varepsilon})$$

donc il est faux que

$$\underline{\lim}_{\varepsilon\to 0} \; (N_\varepsilon(F)/\log \frac{1}{\varepsilon}) < \infty$$

entraîne que F soit un ensemble de type U au sens fort. Donc l'énoncé S1 est un énoncé précis, et l'estimation (*) est également une estimation précise.

En application de la première partie du théorème 2, si l'on choisit $E = E(\xi_1,\ldots \xi_n,\ldots)$ avec $\xi_n = \rho_n^{2^n}$, $\lim \rho_n = 1$ et $\lim \xi_n = 0$ (conditions évidement compatibles), $X(E)$ est p. s. un ensemble de type M_o. Il a de plus (à cause de la condition $\lim \xi_n = 0$) une propriété arithmétique intéressante: il est indépendant sur les rationnels. Il est facile de vérifier qu'un ensemble de Salem de dimension positive n'est pas indépendant sur les rationnels. Pour les détails là-dessus, voir SRSF.

Propriétés de Fourier des ensembles $X(E)$. Cas des processus gaussiens (n,d,γ)

Considérons maintenant le cas où $X(t)$ est un processus de mouvement brownien fractionnaire défini sur $\mathbb{R}^n$ $(t \in \mathbb{R}^n)$ a valeurs dans $\mathbb{R}^d$, donc

$$E(|X(t') - X(t)|^2) = d|t'-t|^\gamma \qquad (t,t' \in \mathbb{R}^n)$$
$$E(e^{i\xi.(X(t')-X(t))}) = e^{-\frac{1}{2}|\xi|^2|t'-t|^\gamma} \qquad (\xi \in \mathbb{R}^d)$$

avec $0 < \gamma < 2$. Le mouvement brownien correspond à $\gamma = 1$. La fonction déterminante de Hausdorff $h(t)$ est supposée soit concave, soit convexe avec $h(2t) = O(h(t))$ $(t \to 0)$, et E est un compact de $\mathbb{R}^n$ tel que $\text{mes}_h\, E > 0$. On désigne toujours par σ une mesure positive $\neq 0$, portée par E, telle que $\sigma(I) \leq h(\text{diam}\, I)$ pour toute boule I de $\mathbb{R}^n$. L'image de σ par $X(t)$ est la mesure aléatoire μ. On va de nouveau estimer $E(|\hat{\mu}(\xi)|^{2p})$ (p entier ≥ 1, $\xi \in \mathbb{R}^d$).

Posons

$$\psi(\underline{t},\underline{t}') = d^{-1}E(|X(t_1) +\ldots+ X(t_p) - X(t_1') -\ldots- X(t_p')|^2)$$

ou $\underline{t} = (t_1,\ldots,t_p)$ et $\underline{t}' = (t_1',\ldots,t_p')$. Ainsi

$$(*) \quad E(|\hat{\mu}(\xi)|^{2p}) = \underbrace{\int\cdots\int}_{2p} e^{-\frac{1}{2}|\xi|^2\psi(\underline{t},\underline{t}')} d\sigma(t_1)\ldots d\sigma(t_p)d\sigma(t_1')\ldots d\sigma(t_p').$$

L'estimation consistera d'abord à minorer $\psi(\underline{t},\underline{t}')$. Pour cela, écrivons

$$\psi(\underline{t},\underline{t}') = c\int_{\mathbb{R}^n} |e^{iu.t_1}+\ldots+e^{iu.t_p}-e^{iu.t_1'}-\ldots-e^{iu.t_p'}|^2 \frac{du}{|u|^{n+\gamma}}$$

c étant défini par

$$c\int_{\mathbb{R}^n} |e^{iu.t} - 1|^2 \frac{du}{|u|^{n+\gamma}} = |t|^\gamma.$$

Etant donné $\underline{t}' = (t_1',\ldots,t_p') \in (\mathbb{R}^n)^p$ et $\varepsilon > 0$, désignons par $F(\underline{t}',\varepsilon)$ l'ensemble des $t \in \mathbb{R}^n$ tels que

$$\inf_j |t - t_j'| \leq \varepsilon$$

et par $G(\underline{t}',\varepsilon)$ l'ensemble des $\underline{t} = (t_1,t_2,\ldots,t_p) \in (\mathbb{R}^n)^p$ tels que, pour tout k, t_k appartienne a $F(\underline{t}',\varepsilon)$. Il est immédiat que

$$\sigma(F(\underline{t}',\varepsilon)) \leq p\, h(2\varepsilon),$$

d'où

$$(\ddagger)\qquad \underbrace{\int\cdots\int}_{p}{}_{G(t',\varepsilon)} d\sigma(t_1)\ldots d\sigma(t_p) \le (ph(2\varepsilon))^p .$$

Supposons $\underline{t} \notin G(t',\varepsilon)$, donc, pour un k convenable,

$$\inf_j |t_k - t_j'| \ge \varepsilon .$$

Désignons par $\delta(.)$ une fonction de classe C^∞, à support dans la boule $B(0,1)$ de $\mathbb{R}^n$, partout ≥ 0, et égale à 1 en 0. Soit $\gamma(.)$ sa cotransformée de Fourier, et soit $\delta_\varepsilon(t) = \varepsilon^{-n}\delta(t/\varepsilon)$. Ainsi

$$\delta(t) = \int_{\mathbb{R}^n} e^{iu.t}\,\gamma(u)du,$$

$$\delta_\varepsilon(t) = \int_{\mathbb{R}^n} e^{iu.t}\,\gamma(\varepsilon u)du,$$

$$\int_{\mathbb{R}^n} (e^{iu.t_1}+\ldots+e^{iu.t_p}-e^{iu.t_1'}-\ldots-e^{iu.t_p'})e^{-iu.t_k}\,\gamma(\varepsilon u)du =$$

$$\delta_\varepsilon(t_1-t_k) +\ldots+ \delta_\varepsilon(t_p-t_k)-\delta_\varepsilon(t_1'-t_k) -\ldots- \delta_\varepsilon(t_p'-t_k) \ge \varepsilon^{-n}$$

donc, en appliquant l'inégalité de Schwarz,

$$\varepsilon^{-2n} \le c^{-1}\,\psi(\underline{t},\underline{t}') \int_{\mathbb{R}^n} |u|^{n+\gamma}\,|\gamma^2(\varepsilon u)|du = c_1^{-1}\varepsilon^{-2n-\gamma}\,\psi(\underline{t},\underline{t}')$$

où c_1 ne dépend que de n et de γ (par l'intermédiaire de la fonction $\delta(.)$ choisie).

On a donc l'implication

$$(\ddagger\ddagger)\qquad \underline{t} \notin G(\underline{t}',\varepsilon) \Longrightarrow \psi(\underline{t},\underline{t}') \ge c_1\varepsilon^\gamma .$$

Etant donné $\xi \in \mathbb{R}^d$, choisissons $\varepsilon > 0$ de sorte que $\varepsilon^\gamma\,|\xi|^2 = 1$, et majorons l'intégrale de (*) en intégrant d'abord par rapport à $d\sigma(t_1)\ \ldots\ d\sigma(t_p)$. On décompose l'intégrale ainsi:

$$\underbrace{\int\cdots\int}_{p} e^{-\frac{1}{2}|\xi|^2\psi(\underline{t},\underline{t}')}\, d\sigma(t_1)\ldots d\sigma(t_p) = \int_{G(\underline{t}',\varepsilon)} +$$

$$+ \sum_{\nu=1}^{\infty} \int_{G(\underline{t}',\varepsilon 2^\nu)\setminus G(\underline{t}',\varepsilon 2^{\nu-1})} .$$

D'après ($\ddagger$) et ($\ddagger\ddagger$) le second membre est majoré par

$$(p\ h(2\varepsilon))^p + \sum_{\nu=1}^{\infty} \exp(-\tfrac{1}{2}\,c_1\ e^{(\nu-1)\gamma})(ph(2^{\nu+1}\ \varepsilon))^p$$

ce qui, grâce a l'hypothèse $h(2\varepsilon) = o(h(\varepsilon))$ $(\varepsilon \to 0)$, es majoré par

$$(C\ p\ h(|\xi|^{-2/\gamma}))^p .$$

Ensuite, l'intégration par rapport à $d\sigma(t'_1) \ldots d\sigma(t'_p)$ multiplie par $(\sigma(E))^p$, et on obtient finalement

$$E(|\hat{\mu}(\xi)|^{2p}) \leq (C\,p\,\sigma(E)\,h(|\xi|^{-2/\gamma}))^p$$

ou $C = C(n,\gamma,h(.))$.

Dès lors on procede par la même méthode que dans la section précédente, et on obtient:

THEOREME 3. *Si* $X(t)$ *est un processus gaussien* (n,d,γ), *et si* E *est un compact de* $\mathbb{R}^n$ *tel que* $\text{mes}_h E > 0$, $h(.)$ *ayant les propriétés de la leçon 1, p. s.* $X(E)$ *porte une mesure positive* $\mu \neq 0$ *telle que*

$$|\hat{\mu}(\xi)| \leq C(h(|\xi|^{-2/\gamma}) \log |\xi|)^{1/2}$$

quand ξ *est assez grand,* C *ne dépendant que de* n, d, γ *et* $h(.)$.

Corollaire (cf. théorème 8, leçon 3). $X(E)$ est p. s. un ensemble de Salem de dimension $\frac{2}{\gamma}$ dim E, lorsque $\frac{2}{\gamma}$ dim E $\leq$ d. C'est un ensemble de mesure de Lebesgue positive si $\frac{2}{\gamma}$ dim E > d.

Les processus (n,d,γ) permettent d'obtenir des ensembles de Salem de dimension ≤ 2.

Dans le cas $\gamma d < 2$ dim E, nous allons voir qu'on a un résultat beaucoup plus fort que le corollaire, à savoir que $X(E)$ a un intérieur non vide.

Densité d'occupation. Cas des processus (d,γ)

On examine pour terminer le cas où dim $X(E) = d$. On va voir qu'en général le mesure μ, image de σ par $X(.)$, a p. s. une densité continue. Il en résultera que p. s. $X(E)$ a des points intérieurs. De façon précise, cela aura lieu 1°) pour les processus (d,γ), quand γ dim E > d (ce qui impose $d = 1$, donc on se bornera à ce cas) 2°) pour les processus (n,d,γ) quand 2 dim E > γd.

Formellement, la densité de μ, que nous écrivons $\mu(.)$ par abus de notation, s'obtient par la formule

$$\mu(x) = (2\pi)^{-d} \int e^{-i\xi . x}\, \hat{\mu}(\xi) d\xi.$$

donc

$$\mu(x) - \mu(y) = (2\pi)^{-d} \int (e^{-i\xi.x} - e^{-i\xi.y})\hat{\mu}(\xi)d\xi$$

et

$$E(|\mu(x)-\mu(y)|^{2p}) = (2\pi)^{-2dp} E \underbrace{\int\cdots\int}_{2p} \prod_{j=1}^{2p} (e^{-i\xi_j.x} - e^{-i\xi_j.y}) \prod_{j=1}^{2p} \hat{\mu}(\xi_j)d\xi_1\ldots d\xi_{2p}$$

avec

$$E(\prod_{j=1}^{2p} \hat{\mu}(\xi_j)) = E \underbrace{\int\cdots\int}_{2p} e^{i(\xi_1.X(t_1)+\ldots+\xi_{2p}.X(t_{2p}))} d\sigma(t_1)\ \ldots\ d\sigma(t_{2p}).$$

Considérons d'abord le cas d'un processus $(1,\gamma)$, et d'une fonction $h(.)$ concave. Ainsi $X(.)$ est défini sur $\mathbb{R}^+$, à valeurs dans $\mathbb{R}$, E est un compact de $\mathbb{R}^+$, $\sigma \in M^+(E)$ et $\sigma(I) \leq h(|I|)$ pour tout intervalle I. Ecrivons

$$\xi_1 X(t_1) + \ldots + \xi_{2p}X(t_{2p}) = \zeta_1 X(s_1) + \ldots + \zeta_{2p}X(s_{2p})$$

où $s_1 \leq s_2 \leq \ldots \leq s_{2p}$ est la suite positive $t_1, t_2, \ldots, t_{2p}$ réarrangée dans l'ordre croissant, et $\zeta_j = \xi_k$ si $s_j = t_k$. Posons $s_o = 0$, donc $X(s_o) = 0$. Utilisant la transformation d'Abel, on a

$$E(e^{i(\xi_1 X(t_1)+\ldots+\xi_{2p}X(t_{2p}))}) = E \exp i \sum_{j=1}^{2p} (\zeta_j + \ldots + \zeta_{2p})(X(s_j)-X(s_{j-1}))$$

$$= \prod_{j=1}^{2p} e^{-(s_j - s_{j-1})\psi(\zeta_j+\ldots+\zeta_{2p})}$$

dont le module est

$$\exp(-\sum_{j=1}^{2p} (s_j - s_{j-1})|\zeta_j + \ldots + \zeta_{2p}|^{\gamma})$$

si on suppose (comme on le peut, quitte à modifier un facteur) $\mathrm{Re}\ \psi(\xi) = |\xi|^{\gamma}$. Donc

$$|E(\prod_1^{2p} \hat{\mu}(\xi_j))| \leq \sum_{(P)} \int\cdots\int_{0 \leq s_1 \leq \ldots \leq s_{2p}} \exp(-\sum_{j=1}^{2p} (s_j - s_{j-1})|\zeta_j+\ldots+\zeta_{2p}|^{\gamma})d\sigma(s_1)\ldots d\sigma(s_{2p})$$

la somme $\sum_{(P)}$ étant prise pour toutes les permutations P de $(\xi_1, \ldots, \xi_{2p})$ sous la forme $(\zeta_1, \ldots, \zeta_{2p})$.

Quels que soient $s_j \geq 0$ et $A > 0$, on a

$$\int_0^{\infty} e^{-sA}\, d\sigma(s-s_j) \leq 2h(A^{-1}).$$

En effet, en intégrant par parties, l'intégrale de droite est majorée par

$$\int_0^{\infty} Ae^{-sA}\, h(s)ds$$

qu'on majore en décomposant sous la forme $\int_0^{A^{-1}} + \int_{A^{-1}}^{\infty}$, et en utilisant les inégalités $h(s) \leq h(A^{-1})$ pour $s \leq A^{-1}$ et $h(s) \leq \leq A\, s\, h(A^{-1})$ pour $s \geq A^{-1}$. Utilisant cette inégalité, on obtient

$$|E(\prod_1^{2p} \hat{\mu}(\xi_j))| \leq 2^{2p} \sum_{(P)} h(|\zeta_j + \ldots + \zeta_{2p}|^{-\gamma})$$

d'où

$$E(|\mu(x)-\mu(y)|^{2p}) \leq 2^{2p}(2p)! \underbrace{\int\cdots\int}_{2p} \prod_{j=1}^{2p} (e^{-i\zeta_j x} - e^{-i\zeta_j y}) \prod_{j=1}^{2p} h(|\zeta_j + \ldots + \zeta_{2p}|^{-\gamma}) d\zeta_1 \ldots d\zeta_{2p}.$$

Bornons-nous, pour simplifier, au cas

$$h(t) = 1 \wedge t^{\alpha} \qquad (0 < \alpha < 1, \quad \gamma\alpha > 1)$$

et choisissons $0 < \gamma < 1-\gamma\alpha$. Utilisant les majorations

$$\left|e^{-i\zeta_j x} - e^{-i\zeta_j y}\right| \leq C\, |\zeta_j|^{\delta}\, |x-y|^{\delta} \qquad (C = C(\delta))$$

pour $j=1,\ldots,2p$, on obtient

$$E(|\mu(x)-\mu(y)|^{2p}) \leq (Cp|x-y|^{\delta})^{2p} \underbrace{\int\cdots\int}_{2p} \prod_{j=1}^{2p} |\zeta_j|^{\delta}(1 \wedge |\zeta_j + \ldots + \zeta_{2p}|^{-\gamma\alpha}) d\zeta_1 \ldots d\zeta_{2p}$$

et, compte tenu de la condition sur δ, l'intégrale est finie. Quitte à changer $C = C(\delta)$, on obtient

$$E(|\mu(x) - \mu(y)|^{2p}) \leq (Cp\, |x-y|^{\delta})^{2p}.$$

Fixons un intervalle I réel de longueur 1. Pour chaque entier positif p, considérons le partage de I en 2^p intervalles égaux, et désignons pas I_p l'ensemble de leurs extrémités. Quitte à changer encore C, on a

$$E \sum_{(x,y)\in I_p \times I_p} |\mu(x)-\mu(y)|^{2p} \leq 2^{-p}(Cp\, |x-y|^{\delta})^{2p}$$

donc

$$E \sum_{p=1}^{\infty} \sum_{(x,y)\in I_p \times I_p} \left(\frac{|\mu(x)-\mu(y)|}{Cp\, |x-y|^{\delta}}\right)^{2p} < \infty.$$

Donc il existe p. s. $p_0 = p_0(\omega)$ tel que pour tout $p \geq p_0$ on ait

$$(*) \qquad |\mu(x) - \mu(y)| \leq Cp\, |x-y|^{\delta} \quad \text{quand} \quad (x,y) \in I_p \times I_p.$$

Si $(x,y) \in I \times I$ et $|x-y| \sim 2^{-\mu}$ ($\sim$ signifiant que le rapport des deux membres est compris entre $\frac{1}{2}$ et 2), et si $\mu > p_0$, on applique (*) sous la forme

$$|\mu(x_\mu) - \mu(y_\mu)| \leq 2C\mu 2^{-\mu\delta}$$

$$\left.\begin{array}{l}|\mu(x_{n+1}) - \mu(x_n)| \leq Cn2^{-n\delta} \\ |\mu(y_{n+1}) - \mu(y_n)| \leq Cn2^{-n\delta}\end{array}\right\} \quad (n = \mu, \ \mu+1, \ldots)$$

où x_n (resp. y_n) est le point de I_n le plus proche de x (resp. y), d'où

$$(\ddagger) \qquad |\mu(x) - \mu(y)| \leq C_\delta \ |x-y|^\delta \log \frac{1}{|x-y|}$$

quand $|x-y|$ est assez petit ($|x-y| \leq 2^{-p_0(\omega)}$).

Les calculs qui précèdent sont formels, puisqu'on sait seulement que p. s. $\hat{\mu} \in L^2$ dans le cas $\gamma\alpha > 1$. Si cependant on pose

$$\mu_\varepsilon(x) = (2\pi)^{-1} \int e^{-i\xi x} e^{-\varepsilon^2\xi^2} \hat{\mu}(\xi) d\xi$$

on peut faire tous les calculs en remplaçant μ par μ_ε, et on obtient l'existence presque sûre de $p_0 = p_0(\omega)$ tel que pour tout $p \geq p_0$ on ait

$$\lim_{\varepsilon \to 0} \sum_{N=0}^{\infty} \sum_{p=1}^{\infty} \sum_{(x,y)\in I_p \times I_p} \left(\frac{|\mu_\varepsilon(x) - \mu_\varepsilon(y)|}{Cp \ |x-y|^\delta}\right)^{2p} < \infty$$

et il en résulte que p. s. une suite convenable de $\mu_\varepsilon(x)$ ($\varepsilon = \varepsilon_j \to 0$) converge uniformément vers une fonction $\mu(.)$ qui vérifie l'inégalité ($\ddagger$) quand $|x-y|$ est assez petit. Enonçons le résultat.

<u>THEOREME 4</u>. <u>*Si*</u> $X(t)$ <u>*est un processus*</u> $(1,\gamma)$ *(Lévy ou Wiener, $1 < \gamma \leq 2$)* <u>*et si*</u> E <u>*est un compact de*</u> $\mathbf{R}^+$ <u>*tel que*</u>

$$\gamma \dim E > 1,$$

$X(E)$ <u>*a p. s. des points intérieurs. De plus, la mesure*</u> μ <u>*image de*</u> σ <u>*par*</u> $X(.)$ <u>*est absolument continue et sa densité appartient aux classes*</u> Lip δ <u>*pour tout*</u> $\delta < \gamma \dim E-1$ ($\sigma \in M^+(E)$, $\sigma(I) \leq C_\alpha |I|^\alpha$ <u>*pour toute boule*</u> I <u>*et tout*</u> $\alpha < \dim E$).

Pour $\gamma = 2$, c'est le résultat de R. Kaufman annoncé dans l'introduction de cette leçon. Selon Geman et Horowicz, $\mu(.)$ est la "densité d'occupation".

Il est à peu près clair, en reprenant la démonstration, que l'hypothèse

$$\int_1^\infty \log |\xi| h(|\xi|^{-\gamma}) d\xi < \infty$$

suffit à assurer que p. s. la densité $\mu(.)$ soit continue. Enonçons le résultat.

THEOREME 5. *Si* $X(t)$ *est un processus* $(1,\gamma)$ $(1 < \gamma \leq 2)$ *et si* E *est un compact de* $\mathbb{R}^+$ *tel que* $\mathrm{mes}_h E > 0$ *avec*

$$\int_0^1 t^{-1/\gamma}\, h(t) \log \frac{1}{t} \frac{dt}{t} < \infty,$$

$X(E)$ *a p. s. des points intérieures.*

Passons maintenant au cas des processus gaussiens (n,d,γ).

Densité d'occupation. Cas des processus gaussiens (n,d,γ)

De nouveau le point crucial est l'estimation de

$$e^{-\psi(\underline{t},\underline{\xi})} = E(e^{i(\xi_1 . X(t_1)+\ldots+\xi_{2p}.X(t_{2p}))}).$$

On pose ici $\underline{t} = (t_1,\ldots,t_{2p}) \in (\mathbb{R}^n)^{2p}$, $\underline{\xi} = (\xi_1,\ldots,\xi_{2p}) \in (\mathbb{R}^d)^{2p}$. On suppose $\sigma \in M^+(E)$, E compact $\subset \mathbb{R}^n$, et $\sigma(I) \leq h(\mathrm{diam}\, I)$ pour tout borélien $I \subset \mathbb{R}^n$. On peut écrire

$$\begin{aligned}\psi(\underline{t},\underline{\xi}) &= d^{-1}\, E(|\xi_1 . X(t_1) +\ldots+ \xi_{2p}.X(t_{2p})|^2) \\ &= c \int_{\mathbb{R}^n} |\xi_1 e^{iu.t_1} +\ldots+ \xi_{2p} e^{iu.t_{2p}} - (\xi_1+\ldots+\xi_{2p})|^2 \frac{du}{|u|^{n+\gamma}}\end{aligned}$$

le constante c étant définie par

$$c \int_{\mathbb{R}^n} |e^{iu.t} - 1|^2 \frac{du}{|u|^{n+\gamma}} = |t|^\gamma .$$

La clé sera la minoration de $\psi(\underline{t},\underline{\xi})$. Posons $t_o = 0$, $\xi_o = -(\xi_1 +\ldots+ \xi_{2p})$,

$$\psi(\underline{t},\underline{\xi}) = c \int_{\mathbb{R}^n} \Big|\sum_{j=0}^{2p} \xi_j e^{iu.t_j}\Big|^2 \frac{du}{|u|^{n+\gamma}} .$$

Lemme. *Il existe une constante* $c(n,\gamma)$ *telle que, pour tout* $j \in \{0,1,\ldots 2p\}$;

$$\psi(\underline{t},\underline{\xi}) \geq c(n,\gamma)\, |\xi_j|^2 \Big(\inf_{\substack{k\neq j \\ k\in\{0,1,\ldots 2p\}}} |t_k - t_j|\Big)^\gamma .$$

Preuve. On se ramène tout de suite au cas $j = 0$. Supposons donc $|t_k| \geq \varepsilon$ $(k = 1,2,\ldots 2p)$; il s'agit de montrer que

$$(*) \qquad \psi(\underline{t},\underline{\xi}) \geq c(n,\gamma)\, |\xi_0|^2\, \varepsilon^\gamma .$$

Soit, comme plus haut, $\delta(.)$ une fonction de classe C^∞ à support dans la boule unité de $\mathbb{R}^n$, égale à 1 en 0, soit $\gamma(.)$ sa co-transformée de Fourier, et $\delta_\varepsilon(t) = \varepsilon^{-n}\,\delta(t/\varepsilon)$. Ainsi

$$\delta_\varepsilon(t) = \int_{\mathbb{R}^n} e^{iu.t}\,\gamma(\varepsilon u)du$$

$$\xi_0 = \int_{\mathbb{R}^n} \sum_{j=0}^{2p} \xi_j\, e^{iu.t_j}\,\varepsilon^{-n}\,\gamma(\varepsilon u)du$$

et l'inégalité de Schwarz donne

$$|\xi_0|^2 \leq \int_{\mathbb{R}^n} \Big|\sum_{j=0}^{2p} \xi_j\, e^{iu.t_j}\Big|^2 \frac{du}{|u|^{n+\gamma}} \int_{\mathbb{R}^n} |u|^{n+\gamma}\varepsilon^{-2n}|\gamma^2(\varepsilon u)|du$$

d'où l'inégalité (*), avec un $c(n,\gamma)$ convenable.

VARIANTE. On a

$$\psi(\underline{t},\underline{\xi}) \geq c(n,\gamma)\,|\xi_j^*|^2\,\varepsilon_j^\gamma$$

en posant

$$\xi_j^* = \sum_k \delta\Big(\frac{\xi_k - \xi_j}{\varepsilon_j}\Big)\,\xi_k .$$

<u>Corollaire</u>

Quel que soit $k = 0,1,\ldots,$

$$\psi(\underline{t},\underline{\xi}) \geq \frac{1}{2p} \sum_{\substack{j=0\\ j\neq k}}^{2p} |\xi_j^*|^2\,\varepsilon_j^\gamma .$$

Nous allons choisir les ε_j en fonction de $\underline{t}$, de façon à majorer de façon convenable

$$E(|\mu(x)-\mu(y)|^{2p}) = (2\pi)^{-2dp} \underbrace{\int\cdots\int}_{2p} \underbrace{\int\cdots\int}_{2p} \prod_{j=1}^{2p} (e^{-i\xi_j.x} - e^{-i\xi_j.y})$$

$$e^{-\psi(\underline{t},\underline{\xi})}\, d\xi_1\ldots d\xi_{2p}\, d\sigma(t_1)\ldots d\sigma(t_{2p}).$$

Comme σ est une mesure continue, on peut se restreindre aux $\underline{t}$ tels que les $|t_j - t_k|$ soient tous distincts. Posons

$$\rho(t_j) = \inf_{\substack{k\neq j\\ k\in\{0,1,\ldots 2p\}}} |t_k - t_j| .$$

Quitte à réordonner les t_j, on peut supposer (condition $(^*_{**})$)

$$\rho(t_1) = \sup_{k\geq 1} \rho(t_k) = |t_1| \quad \text{ou} \quad |t_2 - t_1|$$

(si $\rho(t_1) = |t_1|$, on choisit t_2 de façon arbitraire)

$$\rho(t_3) = \sup_{k\geq 3} \rho(t_k) = |t_3| \quad \text{ou} \quad |t_2-t_3| \quad \text{ou} \quad |t_4-t_3|$$

(si $\rho(t_3) = |t_3|$ ou $|t_3-t_2|$, on choisit t_4 de façon arbitraire),...

$$\rho(t_{2p-1}) = \sup_{k\geq 2p-1} \rho(t_k) = |t_{2p-1}| \quad \text{ou} \quad |t_2-t_{2p-1}| \quad \text{ou...ou} \quad |t_{2p}-t_{2p-1}|.$$

Permutons maintenant la suite $t_0,t_2,t_4,\dots,t_{2p}$, sous la forme $t_{P0}, t_{P2}, t_{P4},\dots,t_{P(2p)}$, de la manière que voici: $t_{P2} = t_2$, t_{P4} est aussi proche que possible de t_{P2}, t_{P6} aussi proche possible de t_{P4}, et ainsi de suite. On choisit alors

$$\varepsilon_1 = \rho(t_1), \quad \varepsilon_3 = \rho(t_3), \dots, \varepsilon_{2p-1} = \rho(t_{2p-1}),$$

$$\varepsilon_{P2} = |t_{P2}-t_{P4}|, \quad \varepsilon_{P4} = |t_{P4}-t_{P6}|, \dots, \varepsilon_{2p} = |t_{P(2p)}-t_{P0}|$$

et on pose en tous cas $\varepsilon_j = |\theta_j - t_j|$,

$$\theta_1 = 0 \text{ ou } t_2, \quad \theta_2 = 0 \text{ ou } t_2 \text{ ou } t_4, \text{ etc...}$$

$$\theta_{P2} = t_{P4}, \quad \theta_{P4} = t_{P6},\dots,\theta_{P(2p)} = t_{P0}.$$

Puis on utilise la majoration

$$(\phi) \qquad |e^{-i\xi_j.x} - e^{-i\xi_j.y}| \leq C_\delta \, |\xi_j|^\delta \, |x-y|^\delta$$

quand j est impair, et on majore par 2 quand j es pair. Observons que $\xi_j^* = \xi_j$ quand j est impair, parce que la boule $B(t_1,\rho(t_1))$ ne contient que t_1 parmi les points $t_1,t_2,\dots,t_{2p}$, la boule $B(t_3,\rho(t_3))$ ne contient que t_3, etc... De plus

$$\xi^*_{P2} - \xi_{P2} = \text{combinaison linéaire de } \xi_1,\xi_3,\dots,\xi_{2p-1}$$

$$\xi^*_{P4} - \xi_{P4} = \text{combinaison linéaire de } \xi_1,\xi_3,\dots,\xi_{2p-1},\xi_{P2},\dots$$

$$\xi^*_{P(2p)} - \xi_{P(2p)} = \text{combinaison linéaire de } \xi_1,\xi_3,\dots,\xi_{2p-1},\xi_{P2},\dots,\xi_{P(2p-2)}.$$

Nous avons

$$E(|\mu(x)-\mu(y)|^{2p}) \leq (2\pi)^{-2pd}(2p)! \underbrace{\int\dots\int}_{2p} \underbrace{\int\dots\int}_{2p} (2C_\delta|x-y|^\delta)^p \prod_{j=1}^{p} |\xi_{2j-1}|^\delta \, e^{-\psi(\underline{t},\underline{\xi})} \, d\xi_1\dots d\xi_{2p} d\sigma(t_1)\dots d\sigma(t_{2p})$$

l'intégrale étant restreinte à l'ensemble des $\underline{t}$ vérifiant la condition $(^{*}_{**})$. Au lieu d'intégrer par rapport à $d\xi_1 \dots d\xi_{2p}$, on peut intégrer par rapport à $d\xi_1 \dots d\xi_{2p-1}\, d\xi_{P2} \dots d\xi_{P(2p)}$, ou mieux par rapport a $d\xi_1 \dots d\xi_{2p-1}\, d\xi^*_{P2} \dots d\xi^*_{P(2p)}$ en utilisant la minoration (voir corollaire)

$$\psi(\underline{t},\underline{\xi}) \geq \frac{1}{2p}\Big(\sum_{j=1}^{p} |\xi_{2j-1}|^2 |\theta_{2j-1} - t_{2j-1}|^\gamma + \sum_{j=1}^{p} |\xi^*_{P(2j)}|^2\ |\theta_{P(2j)} - t_{P(2j)}|^\gamma\Big).$$

On obtient

$$E(|\mu(x)-\mu(y)|^{2p}) \leq (C\, p^{2+d+\delta} |x-y|^\delta\, JJ^*)^p \qquad (C = C(\delta))$$

où

$$J = \sup_{\theta\in\mathbb{R}^n} \int_E \frac{d\sigma(t)}{|t-\theta|^{-\frac{1}{2}\gamma(d+\delta)}}$$

$$J^* = \sup_{\theta\in\mathbb{R}^n} \int_E \frac{d\sigma(t)}{|t-\theta|^{-\frac{1}{2}\gamma d}}$$

et J et J^* sont finies si $\frac{1}{2}\gamma(d+\delta) < \dim E$ et si la mesure σ est bien choisie. En procédant à partir de là comme pour le théorème 4, on obtient ceci.

<u>THEOREME 6</u>. <u>*Si*</u> $X(t)$ <u>*est un processus gaussien*</u> (n,d,γ), <u>*et si est un compact de*</u> $\mathbb{R}^n$ <u>*tel que*</u>

$$\dim E > \frac{1}{2}\gamma\, d,$$

$X(E)$ <u>*a p. s. des points intérieurs. De plus, si*</u> μ <u>*est l'image d'une mesure*</u> σ <u>*comme dans le théorème 4,*</u> μ <u>*est absolument continue p. s. et sa densité est de classe*</u> $C^{\delta/2}$ <u>*pour tout*</u> $\delta < \frac{2}{\gamma}\dim E - d$.

(Dans le cas du brownien, le théorème 4 est donc meilleur que le théorème 6).

Au lieu de la majoration $(^{\phi}_{\phi})$, on peut utiliser, pour j impair,

$$\int |e^{-i\xi_j \cdot x} - e^{-i\xi_j \cdot y}|\ e^{-|\xi_j|^2 |t_j - \theta_j|^\gamma}\, d\xi_j\, d\sigma(t_j) = o(\log^q \frac{1}{|x-y|})$$

sous la condition

$$h(t) = o(t^{\frac{1}{2}\gamma d} (\log \frac{1}{t})^{-q-1}).$$

L'analogue du théorème 5 qu'on obtient ainsi est le suivant.

THEOREME 7. *Si* $\mathrm{mes}_h\, E = \infty$, *avec* $h(t) = t^{\frac{1}{2}\gamma d}(\log \frac{1}{t})^{-3-d}$, X(E) *a p. s. des points intérieurs:*

Recent Progress in Fourier Analysis
I. Peral and J.-L. Rubio de Francia (Editors)

KATO'S SQUARE ROOTS OF ACCRETIVE OPERATORS AND CAUCHY KERNELS ON LIPSCHITZ CURVES ARE THE SAME

C. Kenig and Y. Meyer
University of Minnesota and École Polytechnique

1. Introduction

Kato's definition of the square root of an m-accretive operator can be extended to a slightly more general situation. Let H be a Banach space, $\mathcal{B}(H,H)$ be the algebra of bounded linear operators on H and T be a densely defined closed operator such that a constant C would exist with the following property:

(1) for all $\varepsilon \geq 0$, $(1+\varepsilon T)^{-1} \in \mathcal{B}(H,H)$

and

(2) $\|(1+\varepsilon T)^{-1}\| \leq C$.

Then the domain of T is $V = (1+T)^{-1}H$ and we can define

(3) $(T+1)^{-1/2} = \frac{1}{\pi}\int_0^\infty (T+1+\lambda)^{-1}\lambda^{-1/2}\,d\lambda$.

We call W the range of $(T+1)^{-1/2}$. Obviously $W \supset V$ and

(4) $$\sqrt{T} = \frac{2}{\pi} T \int_0^\infty (1+t^2T)^{-1}\,dt$$

is a closed operator whose domain is W.

Moreover

$$(\sqrt{T})^2 = T.$$

The proofs of these remarks follow closely Kato's approach and will be sketched later on for the reader's convenience.

We now return to a concrete situation. Let $H^1(\mathbb{R}) \subset L^2(\mathbb{R})$ be the usual Sobolev space and $D = -i\frac{d}{dx} : H^1(\mathbb{R}) \to L^2(\mathbb{R})$. In what follows H will be $L^2(\mathbb{R};dx)$ and A and B will be the pointwise multiplication operators by $a \in L^\infty(\mathbb{R})$ and $b \in L^\infty(\mathbb{R})$.

It will be furthermore assumed that for some $\delta > 0$

(5) $\operatorname{Re} a(x) \geq \delta$ and $\operatorname{Re} b(x) \geq \delta$ almost everywhere.

With these notations we have

THEOREM 1. *The operator* $T = BDAD$ *satisfies* (1) *and* (2) *and the domain of* $\sqrt{BDAD}$ *is* $H^1(\mathbb{R})$.

Moreover

(6) $$\sqrt{BDAD} = J(A,B)D$$

where $J(A,B) : L^2(\mathbb{R}) \to L^2(\mathbb{R})$ *is an isomorphism.*

THEOREM 2. *Let* $\phi : \mathbb{R} \to \mathbb{R}$ *be a Lipschitz function*, $z(x) = x+i\phi(x)$ *and* C_ϕ *be the Cauchy operator whose kernel is*

$$\frac{1}{\pi i}\ \text{p.v.}\ \frac{1}{z(y) - z(x)} .$$

Then with Theorem 1's notations,

(7) $$J(A,A) = C_\phi \quad \textit{when} \quad a(x) = \frac{1}{1 + i\phi'(x)} .$$

This means that the boundedness of the Cauchy kernel on Lipschitz curves is a special case of a more general theorem concerning square roots of second order differential operators.

This paper will be organized as follows. The second section will contain general remarks on square roots of unbounded operators. The 3rd section will give the computation of the resolvent in the abstract situation. In Section 4, a fundamental estimate is proved for a truncated version of $J(A,B)$. In Section 6 Theorem 1 will be completely proved and the next section will be devoted to Theorem 2's proof. In the last section a more general symbolic calculus on unbounded operators will be discussed.

2. Square roots of unbounded operator (the abstract approach)

Let H be a Banach space, $T \in C(H,H)$ be a densely defined closed operator and $G \subset \mathbb{C}$ be the resolvent set of T.

We assume that G contains $]-\infty,0[$. In other words for $t \in \mathbb{R}$ $(1 + t^2T)^{-1} \in \mathcal{B}(H,H)$, the algebra of bounded operators on H. Moreover it is assumed that a constant C exists such that

(8) $$|(1 + t^2T)^{-1}| \leq C \quad \text{for all} \quad t \in \mathbb{R}.$$

Here and below, if $L \in \mathcal{B}(H,H)$, $|L|$ will always denote the operator norm of L. If $x \in H$, $|x|$ will denote the corresponding norm of x. Other norms will be specified by indices.

Let $T_\varepsilon = T + \varepsilon$, $\varepsilon > 0$, and

(9) $$T_\varepsilon^{-1/2} = \frac{1}{\pi}\int_0^\infty (T_\varepsilon + \lambda)^{-1}\,\lambda^{-1/2}\,d\lambda.$$

Then following Kato ([4], ch. V, §3, sect. 11) we have $T_\varepsilon^{-1/2} \in \mathcal{B}(H,H)$, $|T_\varepsilon^{-1/2}| \le \frac{C}{\sqrt{\varepsilon}}$ and $T_\varepsilon^{-1/2}T_\varepsilon^{-1/2} = T_\varepsilon^{-1}$. Moreover Range$(T_\varepsilon^{-1/2} - T_\eta^{-1/2}) \subset$ Domain (T) for all $\varepsilon > 0$ and $\eta > 0$.

Then $T_\varepsilon^{1/2}$ is next defined by $T_\varepsilon^{1/2} = T_\varepsilon T_\varepsilon^{-1/2}$. The domain of $T_\varepsilon^{1/2}$ is the range of $T_\varepsilon^{-1/2}$ and does not depend on $\varepsilon > 0$. The domain of $T_\varepsilon^{1/2}$ contains the domain of T and we have $T_\varepsilon = T_\varepsilon^{1/2}T_\varepsilon^{1/2}$.

Lemma 1. *We define*

$$T\int_0^\infty (T + \lambda)^{-1}\,\lambda^{-1/2}\,d\lambda$$
$$= \int_0^1 T(T + \lambda)^{-1}\,\lambda^{-1/2}\,d\lambda + T\int_1^\infty (T + \lambda)^{-1}\,\lambda^{-1/2}\,d\lambda =$$

(10) $$2 - \int_0^1 (T + \lambda)^{-1}\,\lambda^{1/2}\,d\lambda + T\int_1^\infty (T + \lambda)^{-1}\,\lambda^{-1/2}\,d\lambda.$$

Then the domain of this operator $T\int_0^\infty (T + \lambda)^{-1}\,\lambda^{-1/2}\,d\lambda$ *coincides with the domain of* $\sqrt{T_\varepsilon}$ *and we have*

$$T_\varepsilon^{1/2} = \frac{1}{\pi}T\int_0^\infty (T + \lambda)^{-1}\,\lambda^{-1/2}\,d\lambda + R_\varepsilon$$

where $|R_\varepsilon| \le C\sqrt{\varepsilon}$.

Before proving the lemma, observe that $\int_1^\infty (T + \lambda)^{-1}\,\lambda^{-1/2}\,d\lambda$ is bounded on H while $\int_0^1 (T + \lambda)^{-1}\,\lambda^{-1/2}\,d\lambda$ does not make sense. Nevertheless $T\int_0^1 (T + \lambda)^{-1}\,\lambda^{-1/2}\,d\lambda$ makes sense if it is defined as $\int_0^1 T(T + \lambda)^{-1}\lambda^{-1/2}\,d\lambda$. The operator $T(T + \lambda)^{-1}$ is indeed uniformly bounded on H for $0 < \lambda \le 1$.

The proof of Lemma 1 is straightforward.

We have $T_\varepsilon^{1/2} = \frac{1}{\pi}(T + \varepsilon)\int_0^\infty (T + \varepsilon + \lambda)^{-1}\lambda^{-1/2}\,d\lambda$ $= \frac{1}{\pi}T\int_0^\infty (T + \varepsilon + \lambda)^{-1}\lambda^{-1/2}\,d\lambda + R_\varepsilon^{(1)}$. We have $R_\varepsilon^{(1)} = \varepsilon\,T_\varepsilon^{-1/2}$ and therefore $|R_\varepsilon^{(1)}| \le C\sqrt{\varepsilon}$. Then the other error terms will have the general form $T\int_0^\infty (T + \lambda)^{-1}\omega(\lambda)d\lambda$ where $\int_0^\infty |\omega(\lambda)|d\lambda \le C\sqrt{\varepsilon}$ and where $T\int_0^\infty (T + \lambda)^{-1}\omega(\lambda)d\lambda$ is by definition

$$\int_0^1 T(T + \lambda)^{-1}\omega(\lambda)d\lambda + T\int_1^\infty (T + \lambda)^{-1}\omega(\lambda)d\lambda = \int_0^\infty T(T + \lambda)^{-1}\omega(\lambda)d\lambda.$$

Since $\|T(T+\lambda)^{-1}\| \le C$, the operator norm of $T\int_0^\infty (T+\lambda)^{-1}\omega(\lambda)d\lambda$ is bounded by $C\|\omega\|_1 \le C'\sqrt{\varepsilon}$.

Returning to Lemma 1, $T\int_0^\infty (T+\varepsilon+\lambda)^{-1}\lambda^{-1/2}d\lambda =$ $T\int_0^\infty (T+\varepsilon+\lambda)^{-1}(\varepsilon+\lambda)^{-1/2}d\lambda + R_\varepsilon^{(2)}$ and the estimate $|R_\varepsilon^{(2)}| \le C\sqrt{\varepsilon}$ follows from the preceding observations. Finally $T\int_\varepsilon^\infty (T+\lambda)^{-1}\lambda^{-1/2}d\lambda = T\int_0^\infty (T+\lambda)^{-1}\lambda^{-1/2}d\lambda + R_\varepsilon^{(3)}$ and $|R_\varepsilon^{(3)}| \le C\sqrt{\varepsilon}$.

Lemma 1 shows that $T_\varepsilon^{1/2}$ converges in norm to $[\frac{1}{\pi}T\int_0^\infty (T+\lambda)^{-1}\lambda^{-1/2}d\lambda]$ as ε tends to 0. The latter operator will be defined as $T^{1/2}$. It is closed and $T^{1/2}T^{1/2} = T$ as one easily checks. Finally the domain of $T^{1/2}$ is $(T+1)^{-1/2}H$.

3. Resolvent identities for DAD and BDAD

Let H be a Hilbert space on the field of complex numbers and $D : H^1 \to H$ be a selfadjoint operator with domain H^1.

The operators A and B will be assumed to be bounded on H, with $\|A\| \le M$, $\|B\| \le M$ and to satisfy $\mathcal{R}e \langle Ax,x\rangle \geqq \|x\|^2$, $\mathcal{R}e \langle Bx,x\rangle \geqq \|x\|^2$ for all $x \in H$.

We then would like to write the resolvents of DAD and BDAD as Taylor series expansions.

For doing it we first observe that $A^{-1} = 1-\alpha$, $B^{-1} = 1-\beta$ where $\alpha \in \mathcal{B}(H,H)$, $\beta \in \mathcal{B}(H,H)$ and $\|\alpha\| \le \sqrt{1-\frac{1}{M^2}}$, $\|\beta\| \le \sqrt{1-\frac{1}{M^2}}$. In order to check that observation, we write $\|A(x)-x\|^2 = \|A(x)\|^2 - 2\,\mathcal{R}e\langle A(x),x\rangle + \|x\|^2 \le \|A(x)\|^2 - \|x\|^2 \le \|A(x)\|^2(1-\frac{1}{M^2})$.

If $A(x) = y$, then $\alpha(y) = A(x) - x$ and $\|\alpha(y)\| \le \|y\|\sqrt{1-\frac{1}{M^2}}$.

Finaly it is well known that a bounded accretive operator on a Hilbert space H is an isomorphism from H to itself. This implies that y is any element of H.

Our first task will be to compute the domain of the accretive operator DAD.

We have $1 + DAD = (1+D^2)^{1/2}S(1+D^2)^{1/2}$ where $S = (1+D^2)^{-1} + RAR$ and $R = D(1+D^2)^{-1/2}$. Then $S \in \mathcal{B}(H,H)$ and $\mathcal{R}e\langle Sx,x\rangle$

$\geq |(1 + D^2)^{-1/2}x|^2 + |D(1 + D^2)^{-1/2}x|^2 = |x|^2$. Therefore $S : H \to H$ is an isomorphism and the domain of DAD is $V = (1 + D^2)^{-1/2}S^{-1}(1 + D^2)^{-1/2}H$. It coincides with the linear subspace $V \subset H$ of all $x \in H^1$ (H^1 is defined as the domain of D) such that $Ax \in H^1$.

We now define for $t \in \mathbf{R}$, $S_t = (1 + t^2D^2)^{-1} + R_tAR_t$ where $R_t = \dfrac{tD}{\sqrt{1 + t^2D^2}}$ and we then have the following identity.

Lemma 2. *If* $P_t = (1 + t^2D^2)^{-1}$, *then*

$$(11) \qquad S_t^{-1} = 1 - R_t(1 - \alpha P_t)^{-1}\alpha R_t .$$

The proof of this identity is left to the reader.

We now want to study the operator $1 + t^2DAD : V \to H$. The special case $t = 0$ being obvious, we shall assume that $t \neq 0$. Then $V = (1 + t^2DAD)^{-1}H$ which means that DAD is m-accreative (in Kato's terminology).

Lemma 3. *Define* $\pi_t = (1 + t^2DAD)^{-1} : H \to H$. *Then* $|\pi_t| \leq 1$.

We have indeed for $x \in V = \text{Range}\ \pi_t$, $\mathcal{R}e \langle x, (1 + t^2DAD)x\rangle \geq |x|^2 + t^2|Dx|^2 \geq |x|^2$. But $x = \pi_t(y)$ and $y = (1 + t^2DAD)x$. Therefore $|x|^2 \leq \mathcal{R}e \langle x,y\rangle = |\mathcal{R}e \langle x,y\rangle| \leq |x||y|$ which yields $|\pi_t| \leq 1$.

Lemma 4. *Define* $P_t = (1 + t^2D^2)^{-1}$ *and* $Q_t = tDP_t = tD(1+t^2D^2)^{-1}$. *Then*

$$(12) \quad \pi_t = (1 + t^2DAD)^{-1} = P_t - Q_t(1 - \alpha P_t)^{-1}\alpha Q_t .$$

$$(13) \quad \pi_t tDA = Q_t(1 - \alpha P_t)^{-1} .$$

$$(14) \quad tAD\pi_t = (1 - P_t\alpha)^{-1}Q_t .$$

The first identity comes from (11). In (13) the domain of the left hand side is a priori the dense subspace $A^{-1}(H^1) \subset H$. By an abuse of notation (13) means that the left hand side of (13) is indeed continuous and extends to the operator given by the right hand side.

There is no difficulty in defining $tAD\pi_t$ since the domain of DAD is indeed $\pi_t(H)$.

The two identities (13) and (14) follow easily from (12).

Lemma 5. *With the preceding notations,* $1 + t^2 BDAD : V \to H$ *is an (algebraic) isomorphism and* $|(1 + t^2 BDAD)^{-1}| \leq 2M^2$.

In fact $1 + t^2 BDAD = B\{1 + t^2 DAD - \beta\}$ where $|\beta| < 1$. Using the definition of π_t we have $1 + t^2 BDAD = B(1 - \beta\pi_t)(1 + t^2 DAD)$. Since $|\pi_t| \leq 1$, $1 - \beta\pi_t : H \to H$ is an isomorphism and so is $1 + t^2 BDAD : V \to H$. We have

$$(15) \qquad (1 + t^2 BDAD)^{-1} = \sum_{k \geq 0} \pi_t (\beta\pi_t)^k B^{-1}$$

and the second information in Lemma 5 follows from (15).

Let again V be the domain of DAD. The subspace $DV \subset H$ will be called V_0 and will be dense in H in the concrete case discussed in §4.

In other words $f \in V_0$ means that $Af = g \in H^1$ (H^1 is the domain of D) and $f = Du$ where $u \in H^1$.

The following lemma provides the Taylor expansion of $J(A,B)$ on V_0.

Lemma 6. *Let* u *belong to* $V = \text{Domain (DAD)}$ *and* $f = Du$. *Then*

$$\sqrt{BDAD}\, u = J(A,B)f = \frac{2}{\pi}\int_0^\infty (1 + t^2 BDAD)^{-1} BDAf\, dt,$$

$$J(A,B)f = \sum_{p=0}^{\infty} J_p(A,B)f$$

and

$$J_p(A,B)f = \lim_{\varepsilon\downarrow 0, R\uparrow +\infty} \frac{2}{\pi}\int_\varepsilon^R (\pi_t\beta)^p Q_t (1 - \alpha P_t)^{-1} f \frac{dt}{t}.$$

Limits and integrals exist in the norm of H *when* $f \in V_0 = D(V)$.

The operators $J_p(A,B) : V_0 \to H$ will be studied first. We assumed $f = Du$, $u \in V$ and we observe that the operators $(1 - \alpha P_t)^{-1} tD : H \to H$ are uniformly bounded when $\|\alpha\| < 1$. Therefore $|Q_t(1 - \alpha P_t)^{-1} f| \leq C(\alpha) \frac{\|u\|}{t}$ which gives the convergence when $R \to +\infty$.

For proving the convergence when $\varepsilon \to 0$, we use the second information that $Af = g \in H^1$ and (13) gives $Q_t(1 - \alpha P_t)^{-1} f = \pi_t tD(Af) = \pi_t(tDg)$ whose L^2 norm is $0(t)$.

Finally for $f \in V_0$, $|(\pi_t\beta)^p Q_t (1 - \alpha P_t)^{-1} f| \leq C(\alpha)\|\beta\|^p \frac{t}{1 + t^2}$ $(|u| + |Dg|)$ which gives all required convergences in Lemma 6. Then

$$\sum_{p=o}^{\infty} J_p(A,B)f = \frac{2}{\pi}\int_0^{\infty} (1 - \pi_t\beta)^{-1}Q_t(1 - \alpha P_t)^{-1}f \frac{dt}{t}$$

$$= \frac{2}{\pi}\int_0^{\infty} (1 + t^2 \text{ BDAD})^{-1}\text{BDAD } u \, dt = \sqrt{\text{BDAD}}\, u.$$

Indeed the domain of $J(A,B)$ contains V_0 but cannot be proved to be H by the abstract approach.

Indeed even if $B = 1$ a counter example will be given where A satisfies all the preceding properties and where J is not continuous on H.

This means that the abstract approach has to be completed by real variable estimates in which the specific meaning of A, B and D will be crucial.

4. Real variable estimates

There is indeed a "square function" which yields the deep real variable information.

Lemma 7. *If* $D = -i\frac{d}{dx} : H^1 \to L^2(\mathbb{R};dx)$ *and if* $\alpha : L^2 \to L^2$ *is the operator of pointwise multiplication by a function in* $L^{\infty}(\mathbb{R};dx)$, *then for all* $f \in L^2(\mathbb{R})$,

$$(17) \qquad \|(\int_0^{\infty} |Q_t(\alpha P_t)^q f|^2 \frac{dt}{t})^{1/2}\| \leq C(1 + q)\|\alpha\|^q\|f\|$$

where C *is a numerical constant and* $q \in \mathbb{N}$ *is an arbitrary integer*.

The proof of lemma 7 is given in [2] or in [3].

Using (17) and a simple minded Cauchy-Schwarz inequality we obtain

Lemma 8. *If the operators* $L_t \in \mathcal{B}(H,H)$, $t \geq 0$, *depend continuously on* t *and satisfy* $\|L_t\| \leq C_0$, *then there exists a constant* C_1 *such that for every* $R > 0$, *every* $\varepsilon > 0$, $p \in \mathbb{N}$, $q \in \mathbb{N}$, *and every pair* α, β *of operators of pointwise multiplication by functions in* $L^{\infty}(\mathbb{R})$, *we have*

$$(18) \qquad \|\int_{\varepsilon}^{R} (P_t\beta)^p Q_t L_t Q_t(\alpha P_t)^q \frac{dt}{t}\| \leq C_1(1 + p)(1 + q)\|\alpha\|^q\|\beta\|^p.$$

It is indeed surprising that $Q_t L_t Q_t$ cannot be replaced by $L_t Q_t$ (or by $Q_t L_t$) as counter-examples show (section 8). The two operators Q_t are necessary for cancelling L_t or which we have no information. But if $L_t = 1$, one Q_t is enough.

Lemma 9. *With the preceding notations, we have*

$$(19)\qquad \left|\int_\varepsilon^R (P_t\beta)^p Q_t(\alpha P_t)^q \frac{dt}{t}\right| \le C(1+p^2)(1+q^2)|\alpha|^q|\beta|^p.$$

For proving (19), the identity

$$(20)\qquad t\frac{\partial}{\partial t}(-Q_t + 2P_tQ_t) = Q_t - 8Q_t^3$$

is used together with integration by parts. Since $t\frac{\partial}{\partial t}P_t = 2Q_t^2$, all the $p+q$ terms which come out contain at least two operators Q_t. This means that Lemma 8 and Lemma 9 follow from the square function estimate by formal operator theory.

We are now interested in studying the operator

$$L_{p,q}^{(\varepsilon,R)}(\alpha,\beta) = \int_\varepsilon^R (\pi_t\beta)^p Q_t(\alpha P_t)^q \frac{dt}{t}$$

in which $\alpha : L^2(\mathbb{R}) \to L^2(\mathbb{R})$ and $\beta : L^2(\mathbb{R}) \to L^2(\mathbb{R})$ are pointwise multiplications by L^∞ functions.

THEOREM 3. *There exists a numerical constant* C *such that*

$$|L_{p,q}^{(\varepsilon,R)}(\alpha,\beta)| \le C(1+p^2)(1+q^2)|\alpha|^q|\beta|^p(1-|\alpha|)^{-1}$$

for all $R > \varepsilon > 0, p,q \in \mathbb{N}$, β *and* α *such that* $\|\alpha\| < 1$.

For convenience we shall drop out all indices in $L_{p,q}^{\varepsilon,R}(\alpha,\beta)$ and call this operator L_p. Then L_p will be compared to

$$T_{j,p} = \int_\varepsilon^R (P_t\beta)^{p-j}(\pi_t\beta)^j Q_t(\alpha P_t)^q \frac{dt}{t}, \qquad 0 \le j \le p.$$

Indeed $T_{p,q} = L_p$ while $T_{o,p}$ is given by Lemma 9. We would like to prove that $T_{j+1,p} = T_{j,p} + R_{j,p}$ where $R_{j,p}$ can be studied directly by Lemma 8. This is quite easy since (12) yields $\pi_t = P_t + Q_tL_t$ where $|L_t| \le \frac{1}{2}|\alpha|(1-\|\alpha|)^{-1}$. We therefore define

$$R_{j,p} = \int_\varepsilon^R (P_t\beta)^{p-j-1}Q_tL_t\beta(\pi_t\beta)^j Q_t(\alpha P_t)^q \frac{dt}{t}$$

and observe that

$$|L_t\beta(\pi_t\beta)^j| \le \frac{1}{2}|\alpha|(1-\|\alpha|)^{-1}|\beta|^{j+1}.$$

Lemma 8 can be applied to $R_{j,p}$ and yields

$$|R_{j,p}| \le C(1+p)(1+q)\|\beta\|^p\|\alpha\|^q(1-|\alpha|)^{-1}.$$

This ends the proof of Theorem 3.

Lemma 10. *There is a numerical constant* C *such that (keeping Lemma 6's notations) for every* $R > \varepsilon > 0$,

$$(21) \qquad \left\| \int_\varepsilon^R (BDAD + \lambda)^{-1}\, BDA\ \lambda^{-1/2}\, d\lambda \right\| \le C\|A\|^6\|B\|^6$$

and

$$(22) \qquad \left\| \int_\varepsilon^R AD(BDAD + \lambda)^{-1}\lambda^{-1/2}\, d\lambda \right\| \le C\|A\|^6\|B\|^6.$$

These estimates follow directly from Theorem 3 and Lemma 6.

5. The domain of $\sqrt{BDAD}$ is H^1

Lemma 11. *Keeping the preceding notations, there is a constant* C *such that for all* $\varepsilon > 0$

$$(23) \qquad \|(BDAD + \varepsilon^2)^{-1/2}BDA\| \le C\|A\|^6\|B\|^6.$$

$$(24) \qquad \|D(BDAD + \varepsilon^2)^{-1/2}\| \le C\|A\|^6\|B\|^6.$$

$$(25) \qquad \|(BDAD + \varepsilon^2)^{1/2}(D + i\varepsilon)^{-1}\| \le C\|A\|^6\|B\|^6.$$

$$(26) \qquad \|(D + i\varepsilon)(BDAD + \varepsilon^2)^{-1/2}\| \le C\|A\|^6\|B\|^6.$$

Before proving these estimates, some remarks should be made on the domains of these operators.

When S and T are unbounded operators, the domain of $S \circ T$ is, by definition, the subspace of the domain of T consisting of those x for which $T(x)$ belongs to the domain of S. If T and S are closed operators, most of the time $S \circ T$ is not closed and can be therefore extended to a larger domain. For example, the domain of $(BDAD + \varepsilon^2)^{-1/2}BDA$ is, by definition, $A^{-1}(H^1)$. Since by (23) this operator is bounded, it can be extended to the whole of $L^2(\mathbb{R})$.

The domain of $D(BDAD + \varepsilon^2)^{-1/2}$ certainly contains the domain of $(BDAD)^{1/2}$ and is therefore dense in $L^2(\mathbb{R})$.

The following observation is due to Kato [4].

Lemma 12. *Let* H *be a Hilbert space,* $\Delta \subset H$ *be a dense subspace,* $L \in \mathcal{B}(H,H)$ *be a continuous operator and* $M \in C(H,H)$ *be a closed operator. Assume that* ML *is defined on* Δ *and satisfies*

$|ML(x)| \leq C|x|$ *for some constant* C *and every* $x \in \Delta$. *Then the domain of* ML *is* H.

The proof is immediate. Let $\xi \in H$, $x_k \in \Delta$ be such that $x_k \to \xi$ $(k \to +\infty)$. Then $z_k = (ML)(x_k)$ is a Cauchy sequence is H whose limit will be called ζ. On the other end $y_k = L(x_k) \longrightarrow \to L(\xi) = \eta$.

Therefore $y_k \to \eta$ and $M(y_k) \to \zeta$. Since M is closed, this implies that $\eta \in$ Domain (M) and $M(\eta) = \zeta$. Then Lemma 11 implies the following

Lemma 13. *The domain of* $D(BDAD + \varepsilon^2)^{-1/2}$ *is* $L^2(\mathbb{R})$. *In other words the range of* $(BDAD + \varepsilon^2)^{-1/2}$ *which is the domain of* $\sqrt{BDAD}$ *is contained in* H^1.

Similarly $(BDAD + \varepsilon^2)^{1/2}(D + i\varepsilon)^{-1}f$ makes sense if $(D + i\varepsilon)^{-1}f \in V =$ Domain (DAD). This condition on f can be rewritten $f \in U_\varepsilon S_\varepsilon^{-1}(H^1)$ where U_ε is the unitary operator $(D + i\varepsilon)(D^2 + \varepsilon^2)^{-1/2}$. Therefore $(BDAD + \varepsilon^2)^{1/2}(D + i\varepsilon)^{-1}$ is defined on a dense subspace of L^2. Since $(BDAD + \varepsilon^2)^{1/2}$ is closed, Lemma 12 applies and gives.

Lemma 14. *The domain of* $(BDAD + \varepsilon^2)^{1/2}(D + i\varepsilon)^{-1}$ *is* $L^2(\mathbb{R})$. *In other words* $(BDAD + \varepsilon^2)^{1/2}$ *is defined on* H^1 *and the domain of* $\sqrt{BDAD}$ *is therefore* H^1.

It is now time to return to the proof of Lemma 11.

We define $\omega_\varepsilon(\lambda) = \lambda^{-1/2} - (\lambda + \varepsilon^2)^{-1/2} \in L^1\,[0,+\infty[$ and we have

$$(BDAD + \varepsilon^2)^{-1/2} = \frac{1}{\pi}\int_0^\infty (BDAD + \varepsilon^2 + \lambda)^{-1}\lambda^{-1/2}\,d\lambda$$

$$= \frac{1}{\pi}\int_0^\infty (BDAD + \varepsilon^2 + \lambda)^{-1}(\varepsilon^2 + \lambda)^{-1/2}\,d\lambda$$

$$+ \frac{1}{\pi}\int_0^\infty (BDAD + \varepsilon^2 + \lambda)^{-1}\omega_\varepsilon(\lambda)\,d\lambda.$$

In other words

$$(27) \qquad (BDAD + \varepsilon^2)^{-1/2} = \frac{1}{\pi}\int_{\varepsilon^2}^\infty (BDAD + \lambda)^{-1}\lambda^{-1/2}\,d\lambda + R_\varepsilon$$

and the two terms in the right hand side of (27) are operator valued Bochner integrals.

Let us compute $(BDAD + \varepsilon^2)^{-1/2}BDA\,f$ when $f \in L^2$ and

$Af \in H^1$. Then

$$g_\varepsilon = \int_{\varepsilon^2}^{+\infty} (BDAD + \lambda)^{-1} BDAf \, \lambda^{-1/2} \, d\lambda$$

is an L^2-valued Bochner integral and (21) implies $|g_\varepsilon| \leq C(A,B)|f|$.

We then have to take care of the error term $R_\varepsilon(BDAf)$. The following observation will be used.

Lemma 15. *There exists* $C = C(A,B)$ *such that, for every* $\lambda > 0$

(28) $$|(BDAD + \lambda)^{-1} BDA| \leq \frac{C}{\sqrt{\lambda}}$$

and

(29) $$\|AD(BDAD + \lambda)^{-1}| \leq \frac{C}{\sqrt{\lambda}}$$

This follows immediately from (13), (14) and (15).

We now return to the error term $R_\varepsilon BDA$. It will be treated as an operator valued Bochner integral. Using (28) and the definition of R_ε we are led to computing

$$\int_0^\infty \frac{1}{\sqrt{\lambda + \varepsilon^2}} \left(\frac{1}{\sqrt{\lambda}} - \frac{1}{\sqrt{\lambda + \varepsilon^2}}\right) d\lambda$$

which is finite and does not depend on $\varepsilon > 0$.

The proof of (23) is now complete.

The proof of (24) is almost identical and left to the reader.

We now would like to prove (25).

This time the testing function f should belong to $U_\varepsilon S_\varepsilon^{-1}(H^1)$ where $U_\varepsilon = (D + i\varepsilon)(D^2 + \varepsilon^2)^{-1}$. In that case $(D + i\varepsilon)^{-1} f \in V =$ $=$ Domain (DAD) $=$ Domain $(BDAD) \subset$ Domain $(BDAD + \varepsilon^2)^{1/2}$.

Then $AD(D + i\varepsilon)^{-1} f \in H^1$ and by (23)

(30) $$|(BDAD + \varepsilon^2)^{-1/2} BDAD(D + i\varepsilon)^{-1} f| \leq C(A,B)|D(D + i\varepsilon)^{-1} f| \leq C(A,B)|f|.$$

On the other hand, the easy estimates $|(BDAD + \varepsilon^2)^{-1/2}| \leq \frac{C}{\varepsilon}$ and $|(D + i\varepsilon)^{-1}| \leq \frac{1}{\varepsilon}$ give

(31) $$|(BDAD + \varepsilon^2)^{-1/2} \varepsilon^2 (D + i\varepsilon)^{-1}\| \leq C.$$

Then (30) and (31) together give

$$\|(BDAD + \varepsilon^2)^{-1/2}(BDAD + \varepsilon^2)(D + i\varepsilon)^{-1} f\| \leq C(A,B)|f|$$

which is precisely the required inequality (25). The proof of (26) is even simpler since

$$\varepsilon\|(BDAD + \varepsilon^2)^{-1/2}| \leq C.$$

6. Definition and invertibility of J(A,B)

We start with the operator valued truncated singular integrals

$$J_{\varepsilon,R} = \frac{1}{\pi}\int_{\varepsilon}^{R} (BDAD + \lambda)^{-1} BDA\, \lambda^{-1/2}\, d\lambda$$

and observe by Lemma 10 that $\|J_{\varepsilon,R}\| \leq C$.

Moreover if $f \in V_0 = DV$, $J_{\varepsilon,R}(f)$ converges in L^2 to

$$\frac{1}{\pi}\int_{0}^{\infty} (BDAD + \lambda)^{-1} BDAf\, \lambda^{-1/2}\, d\lambda.$$

Therefore the same holds for every f in L^2 and we express this fact by writing $J_{\varepsilon,R} \underset{s}{\to} J(A,B)$ or more formally

$$J(A,B) = \frac{1}{\pi}\int_{0}^{\infty} (BDAD + \lambda)^{-1}\, BDA\, \lambda^{-1/2}\, d\lambda.$$

Then if $f \in V = \text{Domain }(BDAD)$, we have

$$J(A,B)Df = \lim_{\varepsilon\downarrow o, R\uparrow+\infty} J_{\varepsilon,R}(Df) = \sqrt{BDAD}\, f.$$

Since $J(A,B)$ is continuous on L^2 and $\sqrt{BDAD}$ is continuous on H^1, this equality holds for all $f \in H^1$.

For proving the invertibility of $J(A,B)$ we shall exhibit a family M of invertible operators such that $M_\varepsilon \underset{s}{\to} J(A,B)$ and $M_\varepsilon^{-1} \underset{s}{\to} J_1(A,B)$ as $\varepsilon \to 0$.

More precisely

Lemma 16. *With the preceding notations we have*

$$(32) \qquad (BDAD + \varepsilon^2)^{1/2}(D + i\varepsilon)^{-1} \underset{s}{\to} J(A,B)$$

and

$$(33) \qquad (D + i\varepsilon)(BDAD + \varepsilon^2)^{-1/2} \underset{s}{\to} J_1(A,B).$$

We recall that $T_\varepsilon \underset{s}{\to} T$ means that for every $f \in L^2$, $|T_\varepsilon(f) - T(f)| \to 0$ as $\varepsilon \to 0$. By Banach-Steinhaus theorem it suffices to prove $|T_\varepsilon| \leq C$ and the required convergence when f belongs to some dense subset of L^2.

In the case of (32) the dense subset will consist of those $f \in S(\mathbb{R})$ such that $\hat{f} = 0$ in some neighborhood of 0. Then $f = Dg$ where g has the same properties and therefore $(D + i\varepsilon)^{-1}f$

$= g + \varepsilon r_\varepsilon$ where the H^1-norm of the function r_ε is uniformly bounded. On the other hand we know that $\sqrt{BDAD + \varepsilon^2} = \sqrt{BDAD} + R_\varepsilon$ where $|R_\varepsilon| \le C_\varepsilon$. Altogether this gives

$$\sqrt{BDAD + \varepsilon^2}\,(D + i\varepsilon)^{-1} f \to \sqrt{BDAD}\, g = J(A,B)Dg.$$

The proof of (33) will now be given.

As in the proof of Lemma 14, we have

$$(BDAD + \varepsilon^2)^{-1/2} = \frac{1}{\pi}\int_{\varepsilon^2}^{\infty} (BDAD + \lambda)^{-1}\, \lambda^{-1/2}\, d\lambda + R_\varepsilon$$

where

$$R_\varepsilon = \frac{1}{\pi}\int_0^{\infty} (BDAD + \varepsilon^2 + \lambda)^{-1} \omega_\varepsilon(\lambda)\, d\lambda.$$

We now observe that

$$(35) \qquad |D(BDAD + \lambda)^{-1}BD| \le C(A,B)$$

and

$$(36) \qquad |(BDAD + \lambda)^{-1}BD| \le \frac{C(A,B)}{\sqrt{\lambda}}\,.$$

The second observation has already been used (Lemma 15) and the first one is also a consequence of (13), (14) and (15).

Let us first prove that $(D + i\varepsilon)R_\varepsilon \underset{s}{\to} 0$. In fact Lemma 15 and straightforward estimates give $|(D + i\varepsilon)R_\varepsilon| \le C$. If $(D + i\varepsilon)R_\varepsilon$ is next tested on $f = BDg$, $g \in H^1$, inequalities (35) and (36) can be used and give

$$|(D + i\varepsilon)R_\varepsilon\, BDg| \le C\int_0^{\infty} \omega_\varepsilon(\lambda)\, d\lambda = C\varepsilon.$$

For comparing the main term

$$(D + i\varepsilon)\int_{\varepsilon^2}^{\infty} (BDAD + \lambda)^{-1}\, \lambda^{-1/2}\, d\lambda$$

to

$$D\int_0^{\infty} (BDAD + \lambda)^{-1}\, \lambda^{-1/2}\, d\lambda$$

we have to study two error terms $R_\varepsilon^{(1)}$ and $R_\varepsilon^{(2)}$ where

$$R_\varepsilon^{(1)} = i\varepsilon\int_{\varepsilon^2}^{\infty} (BDAD + \lambda)^{-1}\, \lambda^{-1/2}\, d\lambda$$

and

$$R_\varepsilon^{(2)} = D\int_0^{\varepsilon^2} (BDAD + \lambda)^{-1}\, \lambda^{-1/2}\, d\lambda.$$

The operator norm of $R_\varepsilon^{(1)}$ is uniformly bounded since $|(BDAD + \lambda)^{-1}| \le C\lambda^{-1}$ and its action on $f = BDg$, $g \in H^1$, can be estimated using (21). We have indeed $|R_\varepsilon^{(1)}(BDg)| \le C\varepsilon|g|$ for $g \in H^1$ and therefore $R_\varepsilon^{(1)} \underset{s}{\to} 0$.

The second error term is uniformly bounded in norm by (22) and its action on BDg can be estimated using (35). We then have $R_\varepsilon^{(2)} \underset{s}{\to} 0$.

Once that (32) and (33) are proved, (34) is obvious.

7. The Cauchy kernel on Lipschitz curves

Let $\phi : \mathbf{R} \to \mathbf{R}$ be a Lipschitz function: $\|\phi'\|_\infty \leq M < +\infty$ and let $\Gamma = \{x + i\phi(x);\ \ -\infty < x < +\infty\}$ be the graph of ϕ.

Our first task will be to define a space $V(\Gamma) \subset L^2(\Gamma)$ of testing functions $f : \Gamma \to \mathbf{C}$ with the following properties:

(37) $V(\Gamma)$ is dense in $L^2(\Gamma) = L^2(\Gamma; ds)$ where s is the arc-length on Γ;

(38) each function $f \in V(\Gamma)$ is holomorphic in some neighborhood of Γ.

Then the operator $\partial = -i \frac{d}{dz}$ will be defined on $V(\Gamma)$ using the analytic continuation of $f \in V(\Gamma)$ to a neighborhood of Γ.

Together with the space $V(\Gamma)$ of testing functions, an operator algebra $A(\Gamma)$ will be constructed. Elements of $A(\Gamma)$ will act as operators from $V(\Gamma)$ to itself. This algebra will be commutative and will contain both the operator ∂ and the Cauchy operator C_Γ.

Let $M' > M$ and G be the union of the two open cones G_+ and G_- in the complex plane respectively defined by $\zeta = \xi + i\eta$, $|\eta| < M'\xi$ or $|\eta| < -M'\xi$.

The algebra $H(G)$ will consist of all holomorphic functions $\phi : G \to \mathbf{C}$ such that

$$(39) \qquad |\phi(z)| \leq C(1 + |z|)^m$$

for some constants $C \geq 0$ and $m \geq 0$.

Since G has two components, $H(G) = H(G_+) + H(G_-)$ where $\phi \in H(G_+)$ vanishes by definition on G_- and viceversa.

For each $\phi \in H(G)$, two Fourier-Laplace transforms are defined by

$$(40) \qquad \Phi^+(z) = \frac{1}{2\pi} \int_0^{+\infty} e^{iz\xi} \phi(\xi)\, d\xi \quad \text{which makes sense if } \operatorname{Im} z > 0$$

and

$$(41) \qquad \Phi^-(z) = \frac{1}{2\pi} \int_{-\infty}^{0} e^{iz\xi} \phi(\xi)\, d\xi \quad \text{which makes sense if } \operatorname{Im} z < 0.$$

Since $\phi \in H(G)$, the path of integration $[0,+\infty[$ can be moved to any half line $z = te^{i\theta}$, $t \geq 0$ and $-\theta_o < \theta < \theta_o < \frac{\pi}{2}$ (where $\tan \theta_o = M'$). Therefore $\Phi^+(z)$ has an analytic continuation to the open sector $u > -M'|v|$ of the complex plane and similarly $\Phi^-(z)$ can be analytically continued to $u < M'|v|$.

Lemma 17. *Let* ϕ_1 *and* ϕ_2 *be two analytic functions in* $H(G_+)$ *and let* $\phi_3 = \phi_1\phi_2$. *Denote by* Φ_1^+, Φ_2^+ *and* Φ_3^+ *the corresponding Fourier-Laplace transforms defined by (40). If* $z_1 = x_1 + iy_1$ *is below* Γ $(y_1 < \phi(x_1))$ *and if* $z_2 = x_2 + iy_2$ *is above* Γ $(y_2 > \phi(x_2))$ *we then have the Plancherel identity*

$$\int_\Gamma \Phi_2^+(z_2 - z)\Phi_1^+(z - z_1)\, dz = \Phi_3^+(z_2 - z_1). \tag{42}$$

For a Lipschitz curve Γ, the open set above Γ will be denoted Ω^+ and the open set below Γ will be Ω^-. For $z_2 \in \Omega^+$, the sector $S^+(z_2)$ is defined by $v - y_2 > M'|u - x_2|$ if $z = u + iv$, $z_2 = x_2 + iy_2$. For $z_1 \in \Omega^-$, the sector $S^-(z_1)$ is similarly defined by $v - y_1 < -M'|u - x_1|$. Then $S^+(z_2) \subset \Omega^+$ when $z_2 \in \Omega^+$ and similarly $S^-(z_1) \subset \Omega^-$ when $z_1 \in \Omega^-$. Moreover $\Phi_2^+(z_2 - z)$ is holomorphic away from $S^+(z_2)$ and similarly $\Phi_1^+(z - z_1)$ is holomorphic away from $S^-(z_1)$. These two functions are holomorphic in an open set containing Γ.

Moreover estimates like $|\Phi^+(z)| \leq C_\varepsilon(|z|^{-1} + |z|^{-m-1})$ when $z = u + iv$, $v \geq -(M' - \varepsilon)|u|$ are easily proved. Therefore the integral (42) in absolutely convergent at ∞.

For z_1 kept fixed in Ω^-, both sides of (42) are holomorphic functions of $z_2 \in \Omega^+$. The identity (42) for general z_2 will be obtained by analytic continuation from the case where $z_2 = z_1 + iT$ and $T \geq T_o$ is large enough. Then z_1 and z_2 will be kept fixed while Γ will be deformed to a horizontal line without meeting $S^+(z_2)$ or $S^-(z_1)$. During this change of contour, both sides of (42) are constant by Cauchy's theorem. In the case when Γ is a horizontal line (42) is the ordinary Plancherel's formula.

The space $V(\Gamma) \subset L^2(\Gamma;ds)$ of testing functions is defined as the linear span of all $\Phi^+(z - z_o)$ associated to $\phi \in H(G^+)$ and $z_o \in \Omega^-$ and of all $\Phi^-(z - z_o)$ associated to $\phi \in H(G^-)$ and $z_o \in \Omega^+$. Observe that $\Phi^+(z - z_o)$ is defined and holomorphic away from $S^-(z_o)$ while $\Phi^-(z - z_o)$ will be holomorphic away from

$S^+(z_o)$. Both functions are holomorphic on some open set containing Γ.

For example the rational functions $\frac{P(z)}{Q(z)}$ vanishing at ∞ and continuous on Γ belong to $V(\Gamma)$. This proves that $V(\Gamma)$ is dense in $L^2(\Gamma) = L^2(\Gamma;ds)$.

Let us introduce the Hardy spaces $H^2(\Omega^+)$ and $H^2(\Omega^-)$.

A function $F \in H^2(\Omega^+)$ if F is holomorphic on Ω^+ and satisfies

$$\sup_{\varepsilon>0} \int_\Gamma |F(z + i\varepsilon)|^2\, dx < +\infty.$$

The generalized Hardy space $H^2(\Omega^-)$ is defined similarly. If $F \in H^2(\Omega^+)$, then $\lim_{\varepsilon \downarrow 0} F(z + i\varepsilon)$ exists in $L^2(\Gamma;ds)$ and almost everywhere in such a way that $H^2(\Omega^+)$ can be realized as a closed subspace of $L^2(\Gamma;ds)$. This closed subspace is characterized by the property that

$$\int_\Gamma \frac{F(z)}{z - z_o}\, dz = 0$$

for all $z_o \in \Omega^-$. Similar results apply to $H^2(\Omega^-)$. Then $\Phi^+(z - z_o) \in H^2(\Omega^+)$ when $z_o \in \Omega^-$ and similarly $\Phi^-(z - z_o)$ $\in H^2(\Omega^-)$ when $z_o \in \Omega^+$. This means that $V(\Gamma)$ admits a decomposition $V(\Gamma) = V^+(\Gamma) + V^-(\Gamma)$ where $V^+(\Gamma) = V(\Gamma) \cap H^2(\Omega^+)$ and $V^-(\Gamma) = V(\Gamma) \cap H^2(\Omega^-)$. The Cauchy operator C_Γ is then defined by $C_\Gamma(f) = f^+ - f^-$ when $f \in V(\Gamma)$ and $f = f^+ + f^-$, $f^+ \in V^+(\Gamma)$ and $f^- \in V^-(\Gamma)$.

More generally an operator algebra $A(\Gamma)$ will now be introduced. Operators $T \in A(\Gamma)$ will be defined by symbols $\tau \in H(G)$ and will act on $V(\Gamma)$. This action will first be described when the symbol τ belongs to $H(G^+)$. Then the kernel K_τ^+ is defined by $K_\tau^+(z) = \frac{1}{2\pi} \int_0^\infty \tau(\xi) e^{i\xi z}\, d\xi$ and the operator T_τ is given by

$$T_\tau f(z) = \lim_{\varepsilon \downarrow 0} \int_\Gamma K_\tau^+(z + i\varepsilon - \omega) f(\omega)\, d\omega \tag{43}$$

for $f \in V(\Gamma)$.

It is easily checked that $T_\tau f = 0$ if $f \in V^-(\Gamma)$. While if $f = \Phi^+(z - z_o)$ and $z_o \in \Omega^-$, Lemma 17 gives $T_\tau f = \Phi_1^+(z - z_o)$ where $\Phi_1^+(z)$ is the Fourier Laplace transform of $\phi_1 = \tau\phi$ ($\Phi^+(z)$ being the Fourier-Laplace transform of ϕ). Therefore the linear operator T_τ, $\tau \in H(G^+)$ can be described by the following recipe:

(44) $T_\tau(V^+(\Gamma)) \subset V^+(\Gamma)$ and $T_\tau(V^-(\Gamma)) \subset V^-(\Gamma)$

(45) if $\tau \in H(G^+)$, $T_\tau = 0$ on $V^-(\Gamma)$ and vice-versa

(46) if Φ^+ is the Fourier-Laplace transform of $\phi \in H(G^+)$, then $T_\tau(\Phi^+(z - z_0)) = \Phi_1^+(z - z_0)$ for all $z_0 \in \Omega^-$, Φ_1^+ being the Fourier-Laplace transform of $\phi_1 = \tau\phi \in H(G^+)$ and similarly for $H(G^-)$.

This recipe gives the crucial information that

$$T_{\tau_2} \circ T_{\tau_1} = T_{\tau_2\tau_1}.$$

The Cauchy operator C_Γ belongs to $A(\Gamma)$ and the corresponding symbol is $\tau = 1$ on G^+ and $\tau = -1$ on G^-. The operator $\partial = -i \frac{d}{dz}$ also belongs to $A(\Gamma)$ and the corresponding symbol is $\tau(\zeta) = \zeta$. Finally if $T_k \in A(\Gamma)$ and $T \in A(\Gamma)$ we say that T_k converges to T $(k \to +\infty)$ if the symbols τ_k of T_k satisfy uniform estimates

$$|\tau_k(\zeta)| \leq C(1 + |\xi|)^m$$

and if $\tau_k(\zeta) \to \tau(\zeta)$ uniformly on compact subsets of G.

It is quite easy to check that if T_k converges to T in this sense, then $T_k(f) \to T(f)$ in $L^2(\Gamma;ds)$ for all fixed f in $V(\Gamma)$.

Consider for example the symbols $\tau_k(\zeta) = \frac{2}{\pi} \zeta \int_{1/k}^{k} (1+t^2\zeta^2)^{-1} dt \in H(G)$.

They satisfy $|\tau_k(\zeta)| \leq C$ on G and converge uniformly on compact subsets of G to the symbol of C_Γ. Therefore on the operator level we have

(47) $$C_\Gamma = \lim_{k\to+\infty} \frac{2}{\pi} \partial \int_{1/k}^{k} (1 + t^2\partial^2)^{-1} dt.$$

Let us now relate ∂ to $AD : H^1(\mathbb{R}) \to L^2(\mathbb{R})$ and the Cauchy operator C_Γ to $J(A,A)$ defined by $\sqrt{ADAD} = J(A,A)D$.

For doing it, the canonical mapping $X : L^2(\Gamma;ds) \to L^2(\mathbb{R};dx)$ will be used. This mapping X is defined on $f \in V_\Gamma$ by $(Xf)(x) = f(x + i\phi(x))$.

If A is the operator of pointwise multiplication by $\frac{1}{1 + i\phi'(x)}$ we have, for every $f \in V_\Gamma$

$$ADXf = \frac{-i}{1 + i\phi'(x)} \frac{d}{dx} f(x + i\phi(x)) = -if'(x + i\phi(x)) = X\partial f.$$

Therefore $\partial = X^{-1}ADX$ and

$$\frac{2}{\pi} \int_{1/k}^{k} (1 + t^2\partial^2)^{-1} \partial \, dt = X^{-1}L_k X$$

where

$$L_k = \frac{2}{\pi} \int_{1/k}^{k} (1 + t^2 ADAD)^{-1} AD \, dt \underset{s}{\rightarrow} J(A,A)A^{-1}.$$

We then obtain

$$C_\Gamma = X^{-1}J(A,A)A^{-1}X$$

which relates the Cauchy operator C_Γ acting on functions defined on the curve to the operator $J(A,A)$ acting on functions defined on the real axis. This change of parametrization is responsible for the operators X and A^{-1}. Theorem 2 is now completely proved.

8. Pseudo-differential operators

One of the main goals of A.P. Calderón was to construct new algebras of pseudo-differential operators in which multiplications by non smooth functions is allowed.

An example of such an algebra is given by the following theorem.

THEOREM 4. *Let* $M \geq \delta > 0$ *and* $\varepsilon > 0$ *be positive numbers. Then a constant* $C(M,\delta,\varepsilon)$ *exists with the following properties. If* $a \in L^\infty(\mathbb{R})$ *satisfies* $\mathrm{Re}\, a(x) \geq \delta$ *almost everywhere and* $\|a\|_\infty \leq M$, *if* A *denotes the operator of pointwise multiplication by* $a(x)$, *if* $D = -i\frac{d}{dx}$ *and if* F *is holomorphic and bounded in the open sectors* $G_\varepsilon = |\eta| < (\frac{1}{\delta}\sqrt{M^2 - \delta^2} + \varepsilon)|\xi|$, $\zeta = \xi + i\eta \in \mathbb{C}$, *then* $F(AD)$ *is bounded on* $L^2(\mathbb{R})$ *with a norm not exceeding* $C(M,\delta,\varepsilon) \sup_{G_\varepsilon} |F|$.

This symbolic calculus needs some explanations. It means the existence of a mapping $\mathcal{T} : H^\infty(G_\varepsilon) \rightarrow \mathcal{L}(L^2(\mathbb{R}), L^2(\mathbb{R}))$ with the following properties

(48) $\mathcal{T}$ is linear and continuous

(49) for $F_1 \in H^\infty(G_\varepsilon)$ and $F_2 \in H^\infty(G_\varepsilon)$,

$$\mathcal{T}(F_1 F_2) = \mathcal{T}(F_1) \circ \mathcal{T}(F_2)$$

(50) for $\zeta_0 \notin \overline{G}_\varepsilon$, $\mathcal{T}(\frac{1}{\zeta - \zeta_0}) = (AD - \zeta_0)^{-1}$

(51) if $F_k \in H^\infty(G_\varepsilon)$, $\|F_k\|_\infty \leq C$ and $F_k \rightarrow F$ uniformly on compact subsets of G_ε, then $\mathcal{T}(F_k) \underset{s}{\rightarrow} \mathcal{T}(F)$.

When a is real valued, Theorem 4 is easily reduced to the self adjoint case. Indeed a bi-lipschitzian increasing mapping $h = \mathbb{R} \to \mathbb{R}$ exists with the following property:

(52) $$AD = V_h D V_h^{-1} \quad \text{when} \quad V_h(f) = f \circ h.$$

Therefore $F(AD) = V_h F(D) V_h^{-1}$ and the functional calculus is reduced to the self-adjoint case. When a is complex valued, this cheap trick does not work but can be used to transform AD into A_1D where $A_1D = V_h(AD)V_h^{-1}$ and A_1 is the pointwise multiplication operator by $a_1(x) = \frac{a(h(x))}{h'(x)}$.

The choosing correctly the bi-lipschitzian mapping h, the problem will be reduced to the case when $a_1(x) = \frac{1}{1 + i\phi'(x)}$ where $\phi : \mathbb{R} \to \mathbb{R}$ is Lipschitz.

Then the required operator theory will be constructed on $L^2(\Gamma;ds)$ where Γ is the graph of the function ϕ. It suffices to prove that $F(\partial) : L^2(\Gamma;ds) \to L^2(\Gamma;ds)$ is bounded when F is holomorphic and bounded in the open sectors G_ε defined by $|\eta| < (\|\phi'\|_\infty + \varepsilon)|\xi|$, $\varepsilon > 0$.

Then $F(\partial) \in A(\Gamma)$ and the corresponding symbol is precisely $F(\zeta) \in H^\infty(G_\varepsilon)$.

It suffices now to consider the two cases when either $F = 0$ on $\xi < 0$ or $F = 0$ on $\xi > 0$. The general case will obviously follow from the preceding ones.

Define $\sigma = \|\phi'\|_\infty + \frac{\varepsilon}{2}$ and observe, that if $F = 0$ on $\xi > 0$, the corresponding kernel $K(z)$ is holomorphic on $y > -\sigma|x|$ and satisfies $|K(z)| \leq \frac{C}{|z|}$.

Then it is proved in [1] that the "convolution operator" $\int_\Gamma K(z - w)f(w)dw$, $z \in \Omega^+$, $f \in H^2(\Omega^+)$ maps $H^2(\Omega^+)$ into itself.

Since our symbol $F(\zeta)$ vanishes on $\xi < 0$, the corresponding operator $F(\partial)$ annihilates $H^2(\Omega^-)$. Finally we have $L^2(\Gamma) = H^2(\Omega^+) + H^2(\Omega^-)$ by the L^2-boundedness of the Cauchy operator C_Γ. Therefore $F(\partial)$ is continuous on $L^2(\Gamma)$.

The case when $F = 0$ on $\xi > 0$ is similar and Theorem 4 is completely proved.

There are some similar results of a more superficial nature. For example, with Theorem 4's notations, if θ is holomorphic and bounded on $\mathrm{Re}\, z > 0$, then $\theta(DAD)$ is bounded on L^2.

This remark is an abstract property shared by all m-accretive operators T as von Neumann's theorem shows. Indeed if T is m-accretive, $\frac{T-1}{T+1}$ is a contraction and if $g \in H^\infty(\Delta)$ where is the unit disc, then $g(\frac{T-1}{T+1})$ is bounded on L^2.

The fact that $T = DAD$ would be m-accretive is due to the following abstract properties of A:

(53) $\qquad \mathrm{Re}\, \langle Ax,x\rangle \geq |x|^2$ and $A \in \mathcal{B}(H,H)$.

Nevertheless Theorem 4 is not an abstract theorem. For proving this remark, for every $\varepsilon > 0$, an operator $A : L^2(\mathbb{R}) \to L^2(\mathbb{R})$ satisfying $\|A - 1\| \leq \varepsilon$ will be constructed such that the conclusion of Theorem 4 would fail.

Indeed it suffices to show that this conclusion fails for $F(\zeta) = \mathrm{sign}\,(\mathrm{Re}\,\zeta)$ when $A_\lambda^{-1} = 1 + \lambda\alpha$ where $\alpha : L^2 \to L^2$ is fixed (with norm 1) and $|\lambda| = \varepsilon$.

Then a simple calculation gives as in [2]

$$L = \int_{|\lambda|=\varepsilon} F(A_\lambda D)\, \frac{d\lambda}{\lambda^2} = \mathrm{p.v.}\ \frac{1}{\pi i} \int_{-\infty}^{+\infty} (1 + itD)^{-1}\alpha(1 + itD)^{-1}\, \frac{dt}{t}.$$

We want to produce α such that L is unbounded. We first conjugate by the Fourier transformation $\mathcal{F}$. Then if $\tilde{\alpha} = \mathcal{F}\alpha\mathcal{F}^{-1}$ and $\tilde{L} = \mathcal{F}L\mathcal{F}^{-1}$, we have

$$\tilde{L} = \mathrm{p.v.}\ \frac{1}{\pi i} \int_{-\infty}^{+\infty} M_t \tilde{\alpha} M_t\, \frac{dt}{t}$$

where $M_t : L^2 \to L^2$ is the pointwise multiplication operator defined by $M_t f(\xi) = \frac{1}{1 + it\xi} f(\xi)$.

Let $K(\xi,\eta) \in \mathcal{S}'(\mathbb{R}^2)$ be the distribution kernel of $\tilde{\alpha}$. Then the distribution kernel of $\tilde{L}$ is

$$K(\xi,\eta)\, \frac{|\xi| - |\eta|}{\xi - \eta} = K_1(\xi,\eta).$$

A kernel $K(\xi,\eta)$ will now be constructed such that the corresponding operator would be bounded on L^2 but such that this property would fail with K_1.

Our kernel $K(\xi,\eta)$ will be supported by $\xi < 0$, $\eta > 0$ and defined by

$$K(\xi,\eta) = (\log |\tfrac{\xi}{\eta}|)^{-1} |\xi\eta|^{-1/2}.$$

The change of variables $-\xi = e^u$, $\eta = e^v$ reduces K to $\frac{1}{u - v}$.

The same change of variables reduces K_1 to $-\frac{1}{u-v}\tanh(\frac{u-v}{2})$ which is not bounded on L^2. We have here used the property that the two kernels $K(x,y)$ and $K(h(x),h(y))h'(x)^{1/2}h'(y)^{1/2}$ define two operators with the same norm when h is an increasing absolutely continuous homeomorphism of the line.

The counterexample can also be applied to proving that $\sqrt{DAD}$ is not bounded on H^1 when A is a general operator satisgying $\|A - I\| < \varepsilon$. The real variable estimates actually play a rôle in Theorems 1 and 4.

References

[1] R.R. Coifman, Y. Meyer. Fourier analysis of multilinear convolutions. Euclidean Harmonic Analysis. Proceedings 1979. Edited by J.J. Benedetto. Lecture notes no. 779 (Springer).

[2] R.R. Coifman, A. McIntosh, Y. Meyer. L'intégrale de Cauchy définit un opérateur bourné sur L^2 pour les courbes Lipschitziennes. Annals of Mathematics Vol. 116, no 2, (1982) 361-387.

[3] R.R. Coifman, Y. Meyer, E.M. Stein. Un nouvel espace fonctionnel adapté à l'étude des opérateurs définis par des intégrales singulières. Proceedings of the Italian-American conference (Cortona, July 1982).

[4] T. Kato. Perturbation theory for linear operators. Springer Verlag (1966).

Recent Progress in Fourier Analysis
I. Peral and J.-L. Rubio de Francia (Editors)

CONTINUITE SUR LES ESPACES DE HOLDER ET DE SOBOLEV DES OPERATEURS DEFINIS PAR DES INTEGRALES SINGULIERES

Y. Meyer
École Polytechnique, Palaiseau

Le programme de Calderón est la recherche d'hypothèses optimales entraînant la continuité L^2 ou L^p, $1 < p < +\infty$, d'opérateurs intervenant dans des équations aux dérivées partielles à coefficients peu réguliers. L'opérateur qui permet de résoudre cette équation s'écrit $T(a)$ où a désigne l'ensemble des coefficients $a_\alpha(x)$ de notre équation aux dérivées partielles. Lorsque $a_\alpha \in \mathcal{D}(\mathbb{R}^n)$, $T(a)$ est (du moins dans les problèmes décrits dans [1] et [5]) un opérateur pseudo-différentiel classique d'ordre 0. On cherche alors à affaiblir le plus possible les conditions de régularité portant sur les $a_\alpha(x)$ et entraînant la continuité de $T(a)$ sur L^2 ou sur d'autres espaces fonctionnels. L'étude de ces hypothèses optimales a conduit à abandonner la représentation des opérateurs par des symboles et à retourner à leur description par les noyaux-distributions. Oublions provisoirement la dépendance (non-linéaire dans les exemples les plus intéressants) de $T(a)$ en les coefficients a et posons le problème un peu fou de caractériser tous les opérateurs continus sur les espaces classiques à l'aide de criteres simples sur leur noyau-distribution.

Soit T un opérateur linéaire continu, défini sur l'espace $\mathcal{D}(\mathbb{R}^n)$ des fonctions de test, à valeurs dans l'espace $\mathcal{D}'(\mathbb{R}^n)$ des distributions. Un théorème célèbre de L. Schwartz affirme l'existence d'une unique distribution $K(x,y)$ appartenant à $\mathcal{D}'(\mathbb{R}^n \times \mathbb{R}^n)$ et telle que $\langle Tu,v\rangle = \langle K,u\otimes v\rangle$ pour toute $u \in \mathcal{D}(\mathbb{R}^n)$ et toute $v \in \mathcal{D}(\mathbb{R}^n)$; on a noté par $\langle .\ ,\ .\rangle$ la dualité entre $\mathcal{D}$ et $\mathcal{D}'$. Nous nous proposons de donner des conditions suffisantes portant sur $K \in \mathcal{D}'(\mathbb{R}^n \times \mathbb{R}^n)$ et entraînant la continuité de l'opérateur correspondant T sur différents espaces fonctionnels.

Ces conditions se divisent en deux groupes. Nous ferons a priori une hypothèse très faible (et très facile à vérifier dans la pratique) sur la taille et la régularité de $K(x,y)$ lorsque cette distribution est restreinte à l'ouvert Ω de $\mathbb{R}^n \times \mathbb{R}^n$ défini par $x \neq y$ $(x \in \mathbb{R}^n, \; y \in \mathbb{R}^n)$.

Nous pourrons alors donner la condition du second groupe : c'est une condition sur les oscillations de la distribution $K(x,y)$ autour de la diagonale Δ de $\mathbb{R}^n \times \mathbb{R}^n$, nécessaire et suffisante pour obtenir la continuité désirée sur l'espace fonctionnel considéré.

Il est temps de préciser, d'une part, les espaces fonctionnels utilisés, d'autre part, les conditions portant sur le noyau-distribution $K(x,y)$ de $T : \mathcal{D}(\mathbb{R}^n) \to \mathcal{D}'(\mathbb{R}^n)$.

1. Espaces de Hölder homogènes et inhomogènes

Si $0 < s < 1$, nous désignerons par $\Lambda^s(\mathbb{R}^n)$ l'espace des fonctions continues, modulo les fonctions constantes, vérifiant pour une certaine constante $C \geq 0$, tout $x \in \mathbb{R}^n$ et tout $y \in \mathbb{R}^n$, $|f(x) - f(y)| \leq C|x-y|^s$. Si $s = 1$, nous désignerons, dans cet exposé, par $\Lambda^1(\mathbb{R}^n)$ la classe de Zygmund des fonctions continues (modulo les fonctions affines) vérifiant, pour une certaine constante $C \geq 0$, tout $x \in \mathbb{R}^n$ et tout $y \in \mathbb{R}^n$, $|f(x+y) + f(x-y) - 2f(x)| \leq C|y|$.

Enfin si $s > 1$, on pose $s = m+r$ où $0 < r \leq 1$ et $m \in \mathbb{N}$. On écrit $f \in \Lambda^s(\mathbb{R}^n)$ si (et seulement si) $\partial^\alpha f \in \Lambda^r$ pour tous les multi-indices $\alpha \in \mathbb{N}^n$ de longueur $|\alpha| = m$. Appelons δ_t, $t > 0$, les dilatations définies par $\delta_t(x) = tx$, $x \in \mathbb{R}^n$ et posons $\delta_t(f) = f \circ \delta_t^{-1}$. Alors, pour tout $s > 0$ et toute fonction $f \in \Lambda^s$, $\|\delta_t(f)\|_{\Lambda^s} = t^{-s}\|f\|_{\Lambda^s}$.

Si $m < s < m+1$, Λ^s est un espace de fonctions modulo les polynômes de degré $\leq m$ et si $s \geq 1$ est un entier, Λ^s est un espace de fonctions modulo les polynômes de degré $\leq s$.

Ces espaces Λ^s, $s > 0$, seront appelés les espaces de Hölder homogènes. Les espaces de Hölder inhomogènes $C^s(\mathbb{R}^n)$ sont définis comme suit : si $0 < s < 1$, $C^s = \Lambda^s \cap L^\infty$ et donc C^s n'est plus un espace de fonctions continues modulo les constantes. On pose $\|f\|_{C^s} = \|f\|_{\Lambda^s} + \|f\|_\infty$. On procède de même si $s = 1$ et si $s > 1$, on pose $s = m+r$ où $m \in \mathbb{N}$ et $0 < r \leq 1$. Alors $f \in C^s$ signifie que $\partial^\alpha f \in C^r$ pour tous les multi-indices $\alpha \in \mathbb{N}^n$ tels que $|\alpha| \leq m$.

Cette notation contredit les définitions usuelles lorsque s est entier. L'espace C^s usuel $(s \in \mathbb{N})$ est estrictement inclus dans celui que nous venons de définir (par exemple l'espace C^1 usuel est contenu dans la classe de Zygmund...). Les espaces C^s que nous venons de définir sont préservés par les opérateurs pseudo-différentiels classiques d'ordre 0 et sont des algèbres de Banach.

2. Espaces de Beppo Levi et espaces de Sobolev

Désignons par $\mathcal{S}_o(\mathbb{R}^n) \subset \mathcal{S}(\mathbb{R}^n)$ le sous-espace des fonctions f telles que $\hat{f}(\xi)$, transformée de Fourier de f, soit nulle au voisinage de 0. Nous munissons, pour tout $s \in \mathbb{R}$, $\mathcal{S}_o(\mathbb{R}^n)$ de la structure pré-hilbertienne définie par $\langle f,g\rangle_s = \int_{\mathbb{R}^n} \hat{f}(\xi)\overline{\hat{g}}(\xi)|\xi|^{2s}d\xi$ et appelons B^s l'espace de Hilbert associé. Le probleme que nous allons résoudre est de décrire B^s comme un espace fonctionnel. Si $s = 0$, $B^s = L^2(\mathbb{R}^n)$. Il convient ensuite d'étudier le cas $0 < s < \frac{n}{2}$. On note q l'exposant tel que $\frac{1}{2} - \frac{1}{q} = \frac{s}{n}$; alors B^s est canoniquement inclus dans $L^q(\mathbb{R}^n)$ et est donc un espace de fonctions.

Si $-\frac{n}{2} < s < 0$, on définit $p \in]1,2[$ par $\frac{1}{p} - \frac{1}{2} = -\frac{s}{n}$ et on a, canoniquement $L^p(\mathbb{R}^n) \subset B^s(\mathbb{R}^n)$. Si l'on pose alors $t = -s$, les espaces de Hilbert $B^s(\mathbb{R}^n)$ et $B^t(\mathbb{R}^n)$ sont le dual l'un de l'autre pour une dualité qui coincide avec celle reliant $L^p(\mathbb{R}^n)$ et $L^q(\mathbb{R}^n)$ lorqu'on se sert de $L^p \subset B^s$ et de $B^t \subset L^q$.

Il en résulte que B^s est canoniquement un espace de distributions si $-\frac{n}{2} < s < 0$.

Si $s = -\frac{n}{2}$, B^s contient canoniquement l'espace de Hardy $H^1(\mathbb{R}^n)$, généralisé par Stein et Weiss; $H^1(\mathbb{R}^n)$ est dense dans B^s qui est encore un espace de distributions tempérées.

Si $s < -\frac{n}{2}$, B^s est relié a l'espace $H^p(\mathbb{R}^n)$, $0 < p < 1$, de Stein et Weiss (généralisation de l'espace de Hardy) de la façon suivante. On suppose $\frac{s}{n} = \frac{1}{2} - \frac{1}{p}$ et alors $H^p(\mathbb{R}^n)$ est inclus dans $B^s(\mathbb{R}^n)$ qui est un espace de distributions.

En sens contraire, si $s \geq \frac{n}{2}$, B^s n'est plus un espace de distributions. En effet, si $s = \frac{n}{2}$, B^s est canoniquement inclus dans BMO de John et Nirenberg; cet espace n'est pas un espace de distributions mais un espace quotient, modulo les fonctions constantes. Il en est de même pour $B^{n/2}$. Une autre remarque est que BMO est le dual de $H^1(\mathbb{R}^n)$ et que $B^{n/2}$ est (pour la même dualité) le dual de

$H^1(\mathbb{R}^n)$ lorsque cet espace est muni de la norme $B^{-n/2}$. Mais $\mathcal{D}(\mathbb{R}^n)$ n'est pas inclus dans $H^1(\mathbb{R}^n)$ et, pour cette raison, $B^{n/2}$ n'est pas un espace de distributions.

Si $s = \frac{n}{2} + \gamma$, $\gamma > 0$, B^s est canoniquement inclus dans $\Lambda^\gamma(\mathbb{R}^n)$. C'est à dire que, tout comme Λ^γ, B^s est un espace de fonctions modulo les polynômes de degré $\leq\gamma$.

L'espace B^s est homogène au sens que $\|\delta_t f\|_{B^s} = t^{n/2-s}\|f\|_{B^s}$ pour tout $t > 0$, tout $s \in \mathbb{R}$ et $f \in B^s$. L'espace de Sobolev $H^s(\mathbb{R}^n)$ est la version inhomogène de $B^s(\mathbb{R}^n)$. Si $s \geq 0$, $H^s = B^s \cap L^2$ etc...

On peut, si $0 < s < 1$, définir la norme dans B^s sans recourir à la transformation de Fourier. On a, pour une certaine constante $c(s,n) > 0$,

$$c(s,n)\|f\|_{B^s} = \left(\iint |f(x)-f(y)|^2 |x-y|^{-n-2s}\, dx\, dy\right)^{1/2}.$$

Si $s = 1$, $\|f\|_{B^1} = \|\mathrm{Grad}\, f\|_2$ et toutes les autres normes B^s, $s \geq 1$, peuvent se calculer à l'aide de ces deux cas.

3. Les hypothèses a priori sur le noyau-distribution K de T

Notre but est d'étudier la continuité d'un opérateur linéaire continu $T : \mathcal{D}(\mathbb{R}^n) \to \mathcal{D}'(\mathbb{R}^n)$ sur l'espace Λ^s ou l'espace B^s, $s > 0$.

Rappelons que $(\delta_t f)(x) = f(\frac{x}{t})$ et, si $u \in \mathbb{R}^n$, posons $(R_u f)(x) = f(x-u)$. Une observation évidente mais fondamentale est que la norme d'un opérateur linéaire continu $T : \Lambda^s \to \Lambda^s$ (ou de $T : B^s \to B^s$) ne change pas si T est remplacé par $R_u T R_u^{-1}$ ou par $\delta_t\, T\, \delta_t^{-1}$. Le noyau-distribution $K(x,y)$ de T devient alors $K(x+u, y+u)$ ou $\frac{1}{t^n} K(\frac{x}{t}, \frac{y}{t})$.

Il est donc naturel de faire sur le comportement de $K(x,y)$ hors de la diagonale des hypotheses invariantes par ces deux transformations. La plus simple de ces hypothèses est l'existence d'une constante $C \geq 0$ et d'un exposant $\varepsilon \in]0,1]$ tels que la restriction à Ω de $K(x,y)$ soit une fonction localement intégrable vérifiant les deux conditions suivantes

$$(3.1) \qquad |K(x,y)| \leq C|x-y|^{-n} \quad \text{si} \quad x \neq y$$

et

(3.2) $$|K(x',y)-K(x,y)| \leq C|x-x'|^{\varepsilon}\ |x-y|^{-n-\varepsilon}$$

si $|x'-x| \leq \frac{1}{2}|x-y|$, $x \neq y$.

Aucune régularité en y n'est demandée.

Si $\varepsilon > 1$, on écrit $\varepsilon = m + \eta$, $m \in \mathbb{N}$, $0 < \eta \leq 1$, et (3.2) est remplacé par les deux propriétés suivantes

(3.3) $$|\partial_x^{\alpha} K(x,y)| \leq C|x-y|^{-n-|\alpha|}$$

si $0 \leq |\alpha| \leq m$

(3.4) $$|\partial_x^{\alpha} K(x,y) - \partial_x^{\alpha} K(x',y)| \leq C|x-x'|^{\eta}\ |x-y|^{-n-\varepsilon}$$

si $|\alpha| = m$ et $|x'-x| \leq \frac{1}{2}|x-y|$.

Là encore aucune régularité en y n'est demandée.

Ceci nous conduit à la définition suivante.

Définition 1. *Pour tout* $\varepsilon > 0$, *on appelle* L_{ε} *l'espace vectoriel des applications linéaires continues* $T : \mathcal{D}(\mathbb{R}^n) \to \mathcal{D}'(\mathbb{R}^n)$ *dont le noyau-distribution vérifie* (3.1) *et* (3.2) *si* $0 < \varepsilon \leq 1$ *ou* (3.1), (3.3) *et* (3.4) *si* $\varepsilon > 1$.

Nous allons présenter la propriété de "cancellation faible" et d'invariance faible par translation" qui sont nécessaires et suffisantes à la continuité sur $\Lambda^s (0 < s < \varepsilon)$ des opérateurs $T \in L_{\varepsilon}$.

4. La "cancellation faible" et l'"invariance faible par translation"

Soient $K \subset \mathbb{R}^n$ une partie compacte de $\mathbb{R}^n$ et C_{α}, $\alpha \in \mathbb{N}^n$, une suite de constantes positives. L'ensemble des fonctions $\phi \in \mathcal{D}(\mathbb{R}^n)$ dont le support est contenu dans K et qui vérifient $|\partial^{\alpha}\phi| \leq C_{\alpha}$ pour tout $\alpha \in \mathbb{N}^n$ est une partie bornée $\mathcal{B}$ de $\mathcal{D}(\mathbb{R}^n)$ et toute partie bornée $\mathcal{B}_1$ de $\mathcal{D}(\mathbb{R}^n)$ est contenue dans une partie bornée $\mathcal{B}$ décrite par le procédé précédant. Une famille T_i, $i \in I$, d'opérateurs linéaires continus de $\mathcal{D}(\mathbb{R}^n)$ dans $\mathcal{D}'(\mathbb{R}^n)$ est dite bornée si, pour tout couple $\mathcal{B}_1$, $\mathcal{B}_2$ de deux parties bornées de $\mathcal{D}(\mathbb{R}^n)$, il existe une constante $C(\mathcal{B}_1,\mathcal{B}_2)$ telle que $|\langle T_i f,g\rangle| \leq C(\mathcal{B}_1,\mathcal{B}_2)$ pour toute $f \in \mathcal{B}_1$, toute $g \in \mathcal{B}_2$ et tout $i \in I$.

Nous arrivons à la première condition décrivant les oscillations du noyau-distribution $K(x,y)$ de T autour de la diagonale (condition dite de "cancellation").

Définition 2. *Un opérateur linéaire continu* $T : \mathcal{D}(\mathbb{R}^n) \to \mathcal{D}'(\mathbb{R}^n)$ *est d'ordre* 0 *(au sens faible) si l'ensemble des opérateurs* $R_u\ \delta_t T \delta_t^{-1}\ R_u^{-1} : \mathcal{D}(\mathbb{R}^n) \to \mathcal{D}'(\mathbb{R}^n)$ *est borné lorsque* u *décrit* $\mathbb{R}^n$ *et* t *décrit* $]0,+\infty[$.

Cela signifie qu'en posant $\phi^{(u,t)}(x) = \phi(\frac{x-u}{t})$, on doit avoir $|\langle T\ \phi_1^{(u,t)}, \phi_2^{(u,t)}\rangle| \leq C(\phi_1,\phi_2)t^n$ où $C(\phi_1,\phi_2)$ est uniformément bornée lorsque ϕ_1 et ϕ_2 décrivent une partie bornée (arbitraire) de $\mathcal{D}(\mathbb{R}^n)$. Par exemple, un opérateur linéaire continu $T : L^2(\mathbb{R}^n) \to L^2(\mathbb{R}^n)$ est d'ordre 0. La réciproque est évidemment fausse : si $\sigma(x,\xi) \in L^\infty(\mathbb{R}^n \times \mathbb{R}^n)$, cette condition suffit pour que l'opérateur pseudo-différentiel correspondant $\sigma(x,D)$ soit d'ordre 0 mais ne suffit pas pour qu'il soit bornée sur $L^2(\mathbb{R}^n)$. Si $T : \mathcal{D}(\mathbb{R}^n) \to \mathcal{D}'(\mathbb{R}^n)$ est un opérateur linéaire continu, l'adjoint T^* de T est défini par $\langle T^*(u),v\rangle = \langle T(v),u\rangle$ pour toute $u \in \mathcal{D}(\mathbb{R}^n)$ et $v \in \mathcal{D}(\mathbb{R}^n)$. Cette définition est différente de la définition usuelle du cas hilbertien mais ce point n'a aucune importance dans ce qui suit.

Si $m \in \mathbb{N}$, nous désignerons par $\mathcal{D}_m$ le sous-espace de $\mathcal{D}(\mathbb{R}^n)$ défini par $\int x^\alpha\ \phi(x)dx = 0$ pour $|\alpha| \leq m$ (tous les moments d'ordre $\leq m$ sont nuls). On a alors, si $m < \varepsilon \leq m+1$,

Lemme 1. *Si* $T \in L_\varepsilon$ *et si* $\phi \in \mathcal{D}_m$, *la distribution* $T^*(\phi)$ *coincide, en dehors du support de* ϕ, *avec une fonction qui est* $O(|x|^{-n-\varepsilon})$ *a l'infini.*

En effet, en dehors du support de ϕ, on a $T^*\phi(x) = \int K(u,x)\ \phi(u)du$ qu'il suffit d'intégrer par parties.

Désignons par $E_m \subset C^\infty(\mathbb{R}^n)$ le sous-espace des fonctions $f \in C^\infty(\mathbb{R}^n)$ vérifiant $f(x) = O(|x|^m)$ a l'infini. Désignons par $\mathbb{C}_m[X]$ l'espace vectoriel des polynômes de degré $\leq m$ en $x_1,\ldots,x_n$. L'espace dual de $\mathcal{D}_m$ est l'espace quotient $\mathcal{D}'(\mathbb{R}^n) / \mathbb{C}_m[X]$.

On a, avec ces notations, la version duale suivante du lemme 1.

Lemme 2. *Si* $m < \varepsilon \leq m+1$, $f \in E_m$ *et* $g \in \mathcal{D}_m$, *on pose* $\langle Tf\ , g\rangle = \langle T^*g,f\rangle$ *pour tout opérateur* $T \in L_\varepsilon$. *Un tel opérateur* T *définit ainsi une application linéaire de* E_m *dans* $\mathcal{D}'(\mathbb{R}^n)/\mathbb{C}_m[X]$; *cette application linéaire sera encore notée* T, *par abus de langage.*

Le sens de $\langle T^*(g), f\rangle$ est clair lorsque $g \in \mathcal{D}_m$ et $f \in E_m$.

Localement il s'agit de la dualité entre $\mathcal{D}'$ et $\mathcal{D}$ et à l'infini l'intégrale est absolument convergente grâce au lemme 1.

Un exemple très simple permet d'éclairer la définition indirecte donnée par le lemme 2. Si $n = \varepsilon = 1$, nous disposons d'une distribution $K(x,y) \in \mathcal{D}'(\mathbb{R}^2)$ qui a une dérivée partielle $\frac{\partial}{\partial x} K(x,y)$, prise au sens des distributions.

Si $f(x) \in C^\infty(\mathbb{R}) \cap L^\infty(\mathbb{R})$, alors la distribution $S_1(x) = \int \{\frac{\partial}{\partial x} K(x,y)\} f(y)dy$ appartient à $\mathcal{D}'(\mathbb{R})$ et est la dérivée, au sens des distributions, de l'objet $S(x)$ que nous cherchons. De sorte que $S(x)$ n'est pas une distribution mais bien une primitive de $S_1(x)$, modulo des constantes.

Definition 3. *Si* T *appartient à* L_ε, *nous dirons que* T *est faiblement invariante par translation si* $T(x^\alpha) = 0$, *au sens du lemme 2, pour tous les multi-indices* $\alpha \in \mathbb{N}^n$ *tels que* $|\alpha| < \varepsilon$.

Si, par exemple, $T \in L_\varepsilon$ commute avec les translations de $\mathbb{R}^n$, on a obligatoirement $T(x^\alpha) = 0$ pour $|\alpha| < \varepsilon$. Cette remarque justifie notre définition.

Si un symbole $\sigma(x,\xi)$ appartient a la "clase interdite" $S^0_{1,1}$, l'opérateur pseudo-differentiel correspondant $T = \sigma(x,D)$ appartient à L^ε; T est faiblement invariant par translation dès que $\partial^\alpha_\xi \sigma(x,\xi)_{|\xi = 0} = 0$ pour tout $\alpha \in \mathbb{N}^n$.

5. La continuité Λ^s des opérateurs $T \in L_\varepsilon$ lorsque $0 < s < \varepsilon$

Nous devons, puisque $\mathcal{D}(\mathbb{R}^n)$ n'est pas dense dans Λ^s, distinguer le problème de la continuité de $T : \mathcal{D}(\mathbb{R}^n) \to \Lambda^s(\mathbb{R}^n)$ lorsque $\mathcal{D}(\mathbb{R}^n)$ est lui-même muni de la norme Λ^s et le second problème de l'extension de T à Λ^s tout entier. Nous dirons que T est continu pour la norme Λ^s si, pour toute $f \in \mathcal{D}(\mathbb{R}^n)$, $T(f)$ appartient à $\Lambda^s(\mathbb{R}^n)$ et si $\|T(f)\|_{\Lambda^s} \le C\|f\|_{\Lambda^s}$ pour une certaine constante C.

Théorème 1. *Supposons* $0 < s < \varepsilon < 1$ *et* $T \in L_\varepsilon$. *Alors les deux propriétés suivantes sont équivalentes*

(5.1) T *est continu pour la norme* Λ^s

(5.2) T *est d'ordre* 0 *(au sens faible) et* $T(1) = 0$.

Le théorème 1 montre, en particulier, que la continuité Λ^s des opérateurs $T \in L_\varepsilon$ se ramène à deux vérifications algébriques indépendantes de s.

Donnons quelques exemples très superficiels du théorème 1.

Si T est l'opérateur de multiplication ponctuelle par $m(x)$, alors T appartient automatiquement à L_ε pour tout $\varepsilon > 0$ (car la restriction à Ω du noyau-distribution est identiquement nulle). Alors $T(1) = m(x)$ et le théorème 1 nous apprend cette banalité que les seules fonctions avec lesquelles on a le droit de multiplier ponctuellement les fonctions de Λ^s sont les constantes.

Dans le sens opposé, supposons que T commute avec les translations: on a $K(x,y) = S(x-y)$ où $S \in \mathcal{D}'(\mathbb{R}^n)$ et où la restriction de S a $\mathbb{R}^n \setminus \{0\}$ vérifie $|S(u)| \leq C|u|^{-n}$, et $|S(u')-S(u)| \leq$ $\leq C|u'-u|^\varepsilon |u|^{-n-\varepsilon}$ si $|u'-u| \leq \frac{1}{2}|u|$.

On a alors automatiquement $T(1) = 0$ et il est facile de voir que la condition d'être d'ordre 0 équivaut à $\hat{S} \in L^\infty(\mathbb{R}^n)$. En d'autres termes les opérateurs de convolution qui appartiennent à L_ε et sont continus sur Λ^s $(0 < s < \varepsilon)$ sont continus sur L^2 et réciproquement. Nous verrons ensuite qu'il ne s'agit nullement d'un fait général. Par exemple, pour tout symbole $\sigma(x,\xi)$ appartenant à la "classe inderdite" $S^0_{1,1}$, l'opérateur $\sigma(x,D)$ appartient à tous les L_ε. Cet opérateur est borné sur Λ^s $(0 < s < 1)$ si et seulement si $\sigma(x,0) = 0$ mais cette condition n'implique nullement que $\sigma(x,D)$ soit borné sur $L^2(\mathbb{R}^n)$.

Pour mieux comprendre que (5.2) comporte deux conditions, nous allons rappeler qu'il n'y en a plus qu'une lorsque "l'ordre 0" est remplacé par "l'ordre $-\lambda$". Plus précisément, supposons donnée une fonction mesurable $K(x,y) : \mathbb{R}^n \times \mathbb{R}^n \to \mathbb{C}$ et deux exposants λ et ε vérifiant $0 < \lambda < \varepsilon \leq 1$ tels que, pour une certaine constante $C \geq 0$, on ait, pour tout $x \in \mathbb{R}^n$, tout $y \in \mathbb{R}^n$ et tout $x' \in \mathbb{R}^n$ $|K(x,y)| \leq C|x-y|^{-n+\lambda}$ et $|K(x',y)-K(x,y)| \leq C|x-x'|^\varepsilon|x-y|^{-n+\lambda-\varepsilon}$ si $|x'-x| \leq \frac{1}{2}|x-y|$.

Alors on a

<u>Théorème 2</u>. <u>*Sous les hypothèses que nous venons de mentionner, les deux propriétés suivantes sont équivalentes, pour tout exposant*</u> s <u>*tel que*</u> $0 < s < s + \lambda < \varepsilon$:

(5.3) <u>*l'opérateur*</u> T <u>*défini par*</u> $Tf(x) = \int K(x,y)f(y)dy$, $f \in \mathcal{D}(\mathbb{R}^n)$ <u>*est continu de*</u> Λ^s <u>*dans*</u> $\Lambda^{s+\lambda}$

(5.4) $T(1) = 0$ *(modulo les constantes).*

La preuve du théorème 2 est implicite dans la littérature (ou le cas du noyau $|x-y|^{-n+\lambda}$, $0 < \lambda < n$, est traité). De toute façon la démonstration du théorème 1 que nous allons donner conviendra aussi au théorème 2. Nous donnons l'analogue du théorème 1 lorsque $s \geq 1$.

Théorème 3. *Supposons* $0 < s < \varepsilon$ *et désignons par* m_o *la partie entière de* s. *Alors pour tout opérateur* $T \in L_\varepsilon$ *les deux propriétés suivantes sont équivalentes*

(5.5) T *est continu pour la norme* Λ^s

(5.6) T *est d'ordre* 0 *au sens faible et* $T(x^\alpha) = 0$ *pour tous les* $\alpha \in \mathbb{N}^n$ *tels que* $|\alpha| \leq m_o$.

Nous allons maintenant étudier la continuité des opérateurs $T \in L_\varepsilon$ sur les espaces inhomogènes de Hölder.

Rappelons que la transformation de Hilbert définie par $Hf(x) =$ $= \text{v.p.} \frac{1}{\pi} \int_{-\infty}^{+\infty} \frac{f(y)}{x-y}\, dy$ n'est pas continue sur $C^s(\mathbb{R}^n)$; essentiellement parce que H commute avec les dilatations δ_t, $t > 0$, et que H n'est pas continue sur $L^\infty(\mathbb{R})$. Plus généralement si T est un opérateur pseudo-différentiel classique d'ordre 0 dont le symbole est une fonction $\tau(\xi) \in C^\infty(\mathbb{R}^n \setminus \{0\})$, homogène de degré 0, alors T n'est pas borné sur C^s (à moins que T ne soit l'identité).

Nous sommes conduits à la définition suivante.

Définition 4. *Soit* $T : \mathcal{D}(\mathbb{R}^n) \to \mathcal{D}'(\mathbb{R}^n)$ *un opérateur linéaire continu et soit* $\varepsilon > 0$ *un exposant. Nous écrirons* $T \in M_\varepsilon$ *si le noyau-distribution* $K(x,y)$ *de* T *vérifie, dans le complémentaire* $\Omega \subset \mathbb{R}^n \times \mathbb{R}^n$ *de la diagonale*

(5.7) $|K(x,y)| \leq C|x-y|^{-n}$ *et*

$|K(x,y)| \leq C_N|x-y|^{-N}$ *pour tout* $N \geq n$

si $|x-y| \geq 1$

(5.8) $|\partial_x^\alpha K(x,y)| \leq C|x-y|^{-n-|\alpha|}$ *si* $|x-y| \leq 1$ *et*

$|\partial_x^\alpha K(x,y)| \leq C_N|x-y|^{-N}$ *pour tout* $N \geq n + |\alpha|$ *si* $|x-y| \geq 1$

(5.9) *si* $\varepsilon = m + \eta$, $0 < \eta < 1$, *on demande, en outre, que*

$|\partial_x^\alpha K(x,y) - \partial_x^\alpha K(x',y)| \leq C|x-x'|^\eta |x-y|^{-n-\varepsilon}$ *si* $|x-x'| \leq \frac{1}{2}|x-y|$ *et, là encore,* $|x-y|^{-n-\varepsilon}$ *peut être remplacé par* $C_N|x-y|^{-N}$ *pour tout* $N \geq n+\varepsilon$ *dès que* $|x-y| \geq 1$.

Avec ces conditions, nous pouvons énoncer une condition suffisante de continuité des opérateurs $T \in M_\varepsilon$ sur C^s lorsque $0 < s < \varepsilon$.

Théorème 4. *Si* $0 < s < \varepsilon < 1$, *la condition nécessaire et suffisante de continuité sur* C^s *d'un opérateur* $T \in M_\varepsilon$ *est que* T *soit d'ordre* 0 *et que* $T(1)$ *appartienne a* C^s.

Si $1 \leq s \leq \varepsilon$, *une condition suffisante de continuité sur* C^s *d'un opérateur* $T \in M_\varepsilon$ *est que* T *soit d'ordre* 0 *et que, pour toute partie bornée* $\mathcal{B} \subset \mathcal{D}(\mathbb{R}^n)$, *il existe une constante* C *telle que* $\sup_{u\in\mathbb{R}^n} \sup_{f\in\mathcal{B}} \|TR_u f\|_{C^\varepsilon} \leq C$.

6. Enoncés des résultats: continuité sur les espaces de Beppo Levi et de Sobolev

Nous ne disposons pas d'une condition nécessaire et suffisante pour la continuité B^s ou H^s des opérateurs T appartenant à L_ε ou à M_ε.

Voici cependant des conditions suffisantes.

Théorème 5. *Supposons* $0 < s < \varepsilon < 1$ *et* $T \in L_\varepsilon$. *Alors si* T *est d'ordre* 0 *(au sens faible) et si* $T(1) = 0$, T *se prolonge en un opérateur continu sur* B^s.

Nous avons un résultat analogue lorsque $\varepsilon > 1$. On appelle m_0 la partie entière de s et l'on suppose que $0 < s < \varepsilon$, que T appartient à L_ε, que $T(x^\alpha) = 0$ pour $|\alpha| \leq m_0$ et que T est d'ordre 0 (au sens faible). Alors T est continu sur B^s.

Remarquons que ces hypothèses n'impliquent pas la continuité L^2 de T comme nous le verrons sur des contre-exemples.

Théorème 6. *Supposons* $0 < s < \varepsilon < 1$ *et* $T \in M_\varepsilon$. *Si* T *est d'ordre* 0 *(au sens faible) et si* $T(1) = m(x)$ *est un multiplicateur ponctuel de l'espace de Sobolev* H^s, *alors* T *se prolonge en un opérateur continu sur* H^s.

Nous avons un résultat analogue lorsque $\varepsilon > 1$. *On suppose que* T

appartient à M_ε, *que* T *est d'ordre* 0 *au sens faible et que, pour toute partie bornée* $\mathcal{B} \subset \mathcal{D}(\mathbb{R}^n)$, *il existe une constante* C *de sorte que, pour toute translation* R_u *et toute* $f \in \mathcal{B}$, *la norme dans* C^ε *de* $TR_u(f)$ *ne dépasse pas* C.

Alors T *est continu sur* H^s *lorsque* $0 < s < \varepsilon$.

On ne saurait terminer cette présentation générale sans rappeler le célèbre théorème de G. David et J.L. Journé qui (comme nous le verrons) est un corollaire du théorème 5.

Théorème 7. *Soit* $\varepsilon > 0$ *un exposant et* $T : \mathcal{D}(\mathbb{R}^n) \to \mathcal{D}'(\mathbb{R}^n)$ *un opérateur linéaire continu tel que* T *et* T^* *appartiennent à* L_ε. *Alors une condition nécessaire et suffisante pour que* T *soit borné sur* $L^2(\mathbb{R}^n;dx)$ *est que* $T(1) \in BMO$, $T^*(1) \in BMO$ *et que* T *soit d'ordre* 0 *au sens faible*.

7. Exemples et contre-exemples

Le premier exemple a pour rôle d'illustrer le rôle de $T(1)$ dans la continuité C^s.

Soit $a(x) \in C^\varepsilon(\mathbb{R}^n)$, $\varepsilon > 1$, et soit A l'opérateur de multiplication ponctuelle par $a(x)$. Soit T un opérateur pseudo-différentiel classique dont le symbole $\tau(x,\xi)$ appartient à $S^1_{1,0}(\mathbb{R}^n x \mathbb{R}^n)$. On désigne par $\tilde{a}(x)$ la fonction $T(a)$ qui appartient à $C^{\varepsilon-1}(\mathbb{R}^n)$ et l'on appelle $\tilde{A}$ l'opérateur de multiplication ponctuelle par $\tilde{a}(x)$.

Considérons alors les deux opérateurs $L_1 = [T,A]$, commutateur entre T et A et $L_2 = [T,A] + \tilde{A}$. On observe (et ce sera démontré dans un instant) le phénomène curieux suivant. L'opérateur L_1 est continu de $C^s(\mathbb{R}^n)$ dans lui-même pour $0 < s < \varepsilon-1$ mais, si $s > \varepsilon-1$, L_1 est continu de C^s dans $C^{\varepsilon-1}$ et l'on n'a pas mieux en général.

En revanche L_2 est continu de C^s dans C^s lorsque $0 < s < \varepsilon$ et, de nouveau, de C^s dans C^ε si $s > \varepsilon$. C'est-à-dire que si $\varepsilon-1 < s < \varepsilon$, L_2 agit mieux sur C^s que L_1 en un sens que le calcul pseudo-différentiel classique ne peut expliquer. Par exemple, si $a(x) \in C^\infty(\mathbb{R}^n)$ et si toutes les dérivées de $a(x)$ sont bornées, L_2 est exactement d'ordre 0 en général, tout comme L_1.

Désignons par $S(x,y) \in \mathcal{D}'(\mathbb{R}^n \times \mathbb{R}^n)$ le noyau-distribution de

T. Les estimations suivantes sont classiques: si $|x-y| \leq 1$, $|\partial_x^\alpha \partial_y^\beta S(x,y)| \leq C_{\alpha,\beta}|x-y|^{-n-1-|\alpha|-|\beta|}$ et si $|x-y| > 1$, on peut remplacer l'exposant par n'importe quel exposant supérieur.

La restriction à $\Omega = \mathbb{R}^n \times \mathbb{R}^n \setminus \Delta$ du noyau distribution $L_1(x,y)$ de l'opérateur L_1 est $(a(y)-a(x))\, S(x,y)$. Il est alors facile de vérifier que L_1 appartient à la classe M_ε si ε n'est pas entier et, si ε est entier, à toute classe M_γ où $0 < \gamma < \varepsilon$. La même remarque vaut pour L_2 car son noyau-distribution $L_2(x,y)$ coincide avec $L_1(x,y)$ lorsqu'on se restreint à Ω.

Pour calculer $L_2(\phi)$ lorsque $\phi \in \mathcal{D}(\mathbb{R}^n)$, on désigne par Φ l'opérateur de multiplication ponctuelle par ϕ et l'on a $L_2(\phi) = T(a\phi) - aT(\phi) - \phi T(a) = [T,\Phi](a) - aT(\phi)$. Le commutateur $[T,\Phi]$ appartient à une partie bornée de Op $S^0_{1,0}$ lorsque ϕ parcourt l'ensemble $R_u(\mathcal{B})$, $u \in \mathbb{R}^n$, du théorème 4. Or les opérateurs pseudo-différentiels classiques d'ordre 0 opèrent continûment sur les espaces C^s, $s > 0$, tels qu'ils on été définis au paragraphe 1.

Le commutateur L_1 est d'ordre 0 car il est borné sur $L^2(\mathbb{R}^n)$. Ce serait déjà le cas si Grad a appartenait à $L^\infty(\mathbb{R}^n)$ et notre hypothèse est plus précise. De même L_2 est d'ordre 0. Finalement le théorème 4 s'applique à L_2 et fournit la continuité C^s si $0 < s < \varepsilon$. Si $a(x)$ et T sont choisis de sorte que $\tilde{a}(x)$ n'appartienne pas à un meilleur espace que $C^{\varepsilon-1}$, L_1 ne peut être continu sur C^s lorsque $\varepsilon-1 < s < \varepsilon$; plus précisément $L_1(1)$ appartient à $C^{\varepsilon-1}$ seulement.

Le second exemple est l'étude de la continuité H^s des opérateurs L_1 et L_2 et constitue une illustration du théorème 6.

Si $0 < s < \varepsilon-1$, L_1 et L_2 sont tous deux continus sur H^s; si $\varepsilon-1 < s < \varepsilon$, L_1 cesse d'être continu sur H^s et n'est même pas continu de H^s dans $H^{\varepsilon-1}$ tandis que L_2 est encore continu sur H^s. Si enfin $s > \varepsilon$, ni L_1 ni L_2 ne sont continus sur H^s.

Le troisième exemple est dû a A.P. Calderón. On appelle $\phi : \mathbb{R} \to \mathbb{R}$ une fonction lipschitzienne; c'est-à-dire qu'il existe une constante $C \geq 0$ telle que $|\phi(x)-\phi(y)| \leq C|x-y|$ pour tout $x \in \mathbb{R}$ et tout $y \in \mathbb{R}$.

Considérons le noyau-distribution $K(x,y) = \text{v.p.}\ \dfrac{\phi(x)-\phi(y)}{(x-y)^2}$ et l'opérateur $T : \mathcal{D}(\mathbb{R}) \to \mathcal{D}'(\mathbb{R})$ défini par ce noyau-distribution.

Alors T est continu sur $L^2(\mathbb{R})$ (Calderón, 1965), mais, en général, T n'est pas continu sur $H^s(\mathbb{R})$ ou sur $\Lambda^s(\mathbb{R})$ pour $s > 0$.

En revanche le noyau-distribution $K_1(x,y) =$ $= \text{v.p.} \dfrac{\phi(x)-\phi(y)-(x-y)\phi'(y)}{(x-y)^2}$ conduit à un meilleur opérateur T_1 au sens que T_1 est borné sur $H^s(\mathbb{R})$ pour $0 \leq s \leq 1$ et sur $\Lambda^s(\mathbb{R})$ pour $0 < s < 1$.

Là encore, l'explication est que $T(1) = H(\phi') \in BMO$ (et n'appartient pas à un meilleur espace) tandis que $T_1(1) = 0$. La correction faite pour passer de T à T_1 n'a rien changé à la taille ou à la régularité en x du noyau; la régularité en y est complètement détruite mais ce point n'a aucune importance lorsqu'on applique les théorèmes 1 et 5.

Le quatrième exemple concerne la théorie du potentiel dans les domaines lipschitziens. On désigne par $D \subset \mathbb{R}^{n+1}$ un ouvert connexe et borné dont la frontière ∂D est localement le graphe d'une fonction lipschitzienne et l'on appelle $K : L^2(\partial D) \to L^2(\partial D)$ l'opérateur défini par le potentiel de double couche.

En employant des coordonnées locales, ∂D est représenté par un graphe $t = \phi(x)$ où $x \in \mathbb{R}^n$ et où $\phi : \mathbb{R}^n \to \mathbb{R}$ est lipschitzienne. Alors le noyau-distribution $K(x,y)$ de l'opérateur K est donné par

$$K(x,y) = \text{v.p.} \frac{1}{\omega_n} \frac{\phi(x)-\phi(y)-(x-y).\nabla\phi(y)}{\left[|x-y|^2+(\phi(x)-\phi(y))^2\right]^{\frac{n+1}{2}}}$$

où ω_n est la surface de la sphère unité $S^n \subset \mathbb{R}^{n+1}$. Il est bien connu que si ϕ est seulement lipschitzienne, l'opérateur K n'est pas compact. La continuité de K sur $L^2(\mathbb{R}^n;dx)$ a été démontrée par A.P. Calderón en 1977 (Voir aussi Fabes, Jodeit et Rivière) sous l'hypothèse $\|\nabla\phi\|_\infty < \varepsilon$ ($\varepsilon > 0$ étant une mystérieuse constante dont la valeur s'est trouvée devenir $+\infty$ après quelques années).

L'opérateur K est continu sur $H^s(\mathbb{R}^n)$ si $0 \leq s \leq 1$ et sur $\Lambda^s(\mathbb{R}^n)$ si $0 < s < 1$.

En effet $K(1) = 0$ et les théorèmes 1 et 5 s'appliquent. En revanche ces propriétés remarquables de K disparaissent si le numérateur de $K(x,y)$ est remplacé par $\phi(x)-\phi(y)$. Dans le dernier exemple, nous retournons aux opérateurs pseudo-différentiels. Soit $\sigma(x,\xi) \in S^0_{1,1}(\mathbb{R}^n \times \mathbb{R}^n)$ un symbole vérifiant

$$|\partial_\xi^\alpha \, \partial_x^\beta \, \sigma(x,\xi)| \leq C_{\alpha,\beta} \, (1+|\xi|)^{-|\alpha|+|\beta|}.$$

Alors l'opérateur pseudo-différentiel correspondant $\sigma(x,D)$ est borné sur $H^s(\mathbb{R}^n)$ et sur $C^s(\mathbb{R}^n)$ pour tout $s > 0$. Cependant l'opérateur n'est pas continu, en général, sur $L^2(\mathbb{R}^n)$.

Le noyau-distribution $K(x,y)$ de $\sigma(x,D)$ vérifie $|\partial_x^\alpha \, \partial_y^\beta \, K(x,y)| \leq C_{\alpha,\beta}|x-y|^{-n-|\alpha|-|\beta|}$. L'opérateur $\sigma(x,D)$ appartient à M_ε pour tout $\varepsilon > 0$ et les théorèmes 4 et 6 s'appliquent immédiatement.

Voici enfin un contre-exemple dont le rôle est de montrer que dans le théorème de David et Journé, la régularité en y du noyau joue un rôle essentiel. Soit $\psi \in S(\mathbb{R})$ une fonction dont la transformée de Fourier est égale à 1 sur $[4/5, 6/5]$ et à 0 hors de $[3/4, 3/2]$. Formons les noyaus

$$K_N(x,y) = \sum_0^N 2^k \, \psi(2^k(x-y))m_k(y)$$

où $m_k(y) = \exp i(2^k + 2^N)y$.

Désignons par T_N l'opérateur défini par le noyau $K_N(x,y)$. L'adjoint T_N^* est défini par le noyau $L_N(x,y) = = \sum_0^N \bar{m}_k(x)2^k \, \bar{\psi}(2^k(y-x))$. On en déduit immédiatement que le symbole $\tau(x,\xi)$ de T_N^* vérifie $|\tau(x,\xi)| \leq 1$ sur $\mathbb{R}^2$.

L'opérateur T_N^* est (uniformément en N) d'ordre 0 et il en est de même pour T_N.

Par ailleurs, $|\partial_x^j \, K_N(x,y)| \leq C_j|x-y|^{-1-j}$ pour tour entier $j \in \mathbb{N}$. Enfin $T_N(1) = 0$ et $T_N^*(1) = 0$.

Le seule propriété qui fasse défaut est la régularité en y du noyau $K_N(x,y)$. L'opérateur T_N n'est pas (uniformément en N) borné sur $L^2(\mathbb{R})$ car, si $f \in S(\mathbb{R})$ et si $\hat{f}(\xi)$ est portée par $[-\frac{1}{10}, \frac{1}{10}]$, on a (en posant $f_N(y) = \exp(-i2^Ny)f(y)$), $T_N(f_N) = = (\sum_0^N \exp(i2^kx))f(x)$ et $\|T_N(f_N)\|_2 = \sqrt{N+1} \, \|f\|_2 = \sqrt{N+1} \, \|f_N\|_2$.

8. Démonstration de (5.1) $\Rightarrow$ (5.2) dans le théorème 1

Elle repose sur le lemme suivant (dont la démonstration immédiate est laissée au lecteur).

Lemme 3. *Soit* $\varepsilon > 0$ *et* T *un opérateur appartenant à* L_ε. *Alors pour toute fonction* $\phi \in \mathcal{D}(\mathbb{R}^n)$ *et toute fonction* $\psi \in \mathcal{D}_o(\mathbb{R}^n)$, *on a*

$$\lim_{k\to+\infty} \langle T(\delta_k\phi),\psi\rangle = \phi(0) \langle T(1), \psi\rangle.$$

On rappelle que $(\delta_t\phi)(x) = \phi(\frac{x}{t})$.

Supposons donc que $T \in L_\varepsilon$ soit, en fait, continu pour la norme Λ^s $(0 < \varepsilon < 1)$. Soit $\phi \in \mathcal{D}(\mathbb{R}^n)$ telle que $\phi(0) = 1$ et formons $g_k(x) = T(\delta_k\phi)(x)$. Alors $\|g_k\|_{\Lambda^s} \leq C\, k^{-s}$. Il en résulte que si $\psi \in \mathcal{D}_o$, $|\int\psi(x)g_k(x)dx| \leq C(\psi)k^{-s}$ et l'on a donc $T(1) = 0$.

Montrons maintenant que T est d'ordre 0.

Pour alléger un peu l'écriture des démonstrations qui suivent, on désignera systématiquement par Q une boule de centre u arbitraire et de rayon $t > 0$ (arbitraire) de $\mathbb{R}^n$ et par ϕ_Q une famille (indexée par Q) de fonctions de $\mathcal{D}(\mathbb{R}^n)$ vérifiant $|\partial^\alpha\phi_Q| \leq C_\alpha\, t^{-|\alpha|}$. On appellera $|Q|$ le volume de Q si bien que la propriété pour T d'être d'ordre 0 s'écrit plus simplement $|\langle T\, \phi_Q^{(1)}, \phi_Q^{(2)}\rangle| \leq C|Q|$.

On appellera toujours $\tilde{Q}$ la boule double de Q (même centre et rayon double).

Si $x \notin \tilde{Q}$, on a (grâce a la taille du noyau) $|T\phi_Q(x)| \leq C$. Par ailleurs $\|\phi_Q\|_{\Lambda^s} \leq Ct^{-s}$ et la continuité de T sur Λ^s implique $\|T\phi_Q\|_{\Lambda^s} \leq C't^{-s}$. On teste cette norme sur $x_o \in Q$ (Q de centre u) et sur x_1 vérifiant $|x_1-u| = 2t$. On a donc $|T\phi_Q(x_o)-T\phi_Q(x_1)| \leq C't^{-s}|x_1-x_o|^s = C'2^s$. Finalement $|T\phi_Q(x_o)|\leq C''$ et, a fortiori, $|\langle T\phi_Q^{(1)},\phi_Q^{(2)}\rangle| \leq C_{1,2}|Q|$.

Un examen plus attentif de cet argument montre que $\|T(\phi_Q)\|_\infty \leq C$ lorsque $T \in L_\varepsilon$ et lorsque $T : \Lambda^s \to \Lambda^s$ est continu.

Il est donc très naturel que la preuve de la réciproque du théorème 1 passe par la preuve de $\|T(\phi_Q)\|_\infty \leq C$.

Proposition 1. *Soit* $T : \mathcal{D}(\mathbb{R}^n) \to \mathcal{D}'(\mathbb{R}^n)$ *un opérateur linéaire continu appartenant à la classe* L_ε. *Alors si* T *est d'ordre* 0 *et si* $T(1) = 0$, *pour toute famille* ϕ_Q, *il existe une constante* C *telle que* $\|T(\phi_Q)\|_\infty \leq C$.

La preuve de cette proposition 1 est basée sur le lemme suivant (dont la démonstration est laissée au lecteur mais peut également être trouvée dans [6]).

Lemme 4. *Soit* $K \in \mathcal{D}'(\mathbb{R}^n \times \mathbb{R}^n)$ *une distribution dont la restriction à l'ouvert* Ω *vérifie* $|K(x,y)| \leq C|x-y|^{-n}$. *Supposons que l'operateur* $T : \mathcal{D}(\mathbb{R}^n) \to \mathcal{D}'(\mathbb{R}^n)$ *défini par le noyau-distribution* K *soit d'ordre* 0. *Alors, pour toute fonction* $f \in \mathcal{D}(\mathbb{R}^n \times \mathbb{R}^n)$, *nulle sur* $y = x$, *on a*

$$\langle K,f\rangle = \lim_{\varepsilon\downarrow 0} \iint_{|y-x|\geq\varepsilon} K(x,y)f(x,y)dx\, dy.$$

Revenons à la preuve de la proposition 1.

On désigne par u_Q une famille de fonctions de $\mathcal{D}(\mathbb{R}^n)$ telle que $|\partial^\alpha u_Q| \leq C_\alpha\, t^{-|\alpha|}$, que $u_Q(x) = 1$ sur la boule $|x-u| \leq \frac{3}{2}t$ et que u_Q soit nulle hors de $\tilde{Q}$. Soit, par ailleurs, $\theta(x) \in \mathcal{D}(\mathbb{R}^n)$ une fonction arbitraire. On a, dans $\mathcal{D}(\mathbb{R}^n \times \mathbb{R}^n)$,

$$\phi_Q(y)\theta(x) = (\phi_Q(y)-\phi_Q(x))u_Q(y)\theta(x) + \phi_Q(x)u_Q(y)\theta(x).$$

On utilise la dualité entre $\mathcal{D}(\mathbb{R}^n \times \mathbb{R}^n)$ et $\mathcal{D}'(\mathbb{R}^n \times \mathbb{R}^n)$. On obtient $\langle K,\theta(x)\phi_Q(y)\rangle = \langle\theta,T\phi_Q\rangle = I_1 + I_2$ où $I_1 = \langle K,(\phi_Q(y)-\phi_Q(x))u_Q(y)\theta(x)\rangle$ et $I_2 = \langle K,\phi_Q(x)u_Q(y)\theta(x)\rangle$.

Le calcul de I_1 se fait grâce au lemme 4.

Il vient

$$I_1 = \iint K(x,y)(\phi_Q(y)-\phi_Q(x))u_Q(y)\theta(x)dx\, dy$$

et l'intégrale est ici définie comme $\lim_{\varepsilon\downarrow 0} \iint_{|x-y|\geq\varepsilon} \dots\, dx\, dy$. On pose $g(x) = \int K(x,y)(\phi_Q(y)-\phi_Q(x))u_Q(y)dy$. Cette intégrale absolument convergente définit une fonction continue et le théorème de Fubini donne $I_1 = \int g(x)\theta(x)dx$. Si $x \in \tilde{Q}$, $|g(x)| \leq$ $\leq C \int_{|x-y|\leq Ct} |x-y|^{-n} \frac{|x-y|}{t}\, dy \leq C'$.

Pour calculer I_2, on introduit la distribution $T(u_Q)$ et on étudie la restriction de cette distribution à la boule (ouverte Q).

Appelons $\psi \in \mathcal{D}_0$ une fonction portée par Q et posons $v_Q = 1-u_Q$. La condition $T(1) = 0$ s'écrit

$$\langle\psi,T(u_Q)\rangle + \langle\psi,T(v_Q)\rangle = 0. \tag{8.1}$$

La première intégrale vaut, par définition, $\langle K,u_Q(y)\psi(x)\rangle$ et la seconde vaut $\int\{\int\psi(x)K(x,y)dx\}v_Q(y)dy = J(\psi)$. On ne change pas $J(\psi)$ en remplaçant $K(x,y)$ par $K(x,y)-K(u,y)$. On utilise alors $|K(x,y)-K(u,y)| \leq C|u-y|^{-n-\varepsilon}t^{\varepsilon}$ et il vient $|J(\psi)| \leq C_\varepsilon|\psi|_1$.

Tout cela mis ensemble donne $|\langle\psi,T(u_Q)\rangle| \le C_\varepsilon\|\psi\|_1$ chaque fois que ψ est portée par Q et d'intégrale nulle.

La distribution $T(u_Q)$, restreinte à Q, est donc la somme d'une constante λ_Q que l'on ne peut encore majorer et d'une fonction $r_Q \in L^\infty(Q)$ vérifiant $\|r_Q\|_\infty \le C_\varepsilon$.

Nous allons maintenant évaluer λ_Q. Pour cela, on désigne par θ_Q une "bosse ajustée a la boule Q" telle que $\theta_Q \ge 0$, $|\partial^\alpha\theta_Q| \le C_\alpha t^{-|\alpha|}$ et que $\int\theta_Q(x)dx = |Q|$.

On calcule $\langle T(u_Q),\theta_Q\rangle$ de deux façons différentes. Puisque T est d'ordre 0, il vient $|\langle T(u_Q),\theta_Q\rangle| \le C|Q|$. D'autre part $T(u_Q) = \lambda_Q + r_Q$ sur Q et donc $\langle T\,u_Q,\theta_Q\rangle = \lambda_Q|Q| + \int_Q r_Q(x)\theta_Q(x)dx$.

On a donc $|\lambda_Q| \le C'_\varepsilon$.

Nous venons de prouver que la restriction de Tu_Q à Q est une fonction $h(x) \in L^\infty(Q)$ vérifiant $\|h\|_{L^\infty(Q)} \le C''_\varepsilon$. Nous avons, en fait, sur la boule Q de centre u,

$$(8.2) \qquad Tu_Q(x) = \int(K(u,y)-K(x,y))v_Q(y)dy + \varepsilon_Q$$

et la constante ε_Q vérifie $|\varepsilon_Q| \le C$.

Revenant à I_2, il vient $I_2 = \langle Tu_Q,\theta\phi_Q\rangle = \int\phi_Q(x)\theta(x)h(x)dx$. Nous avons démontré que $T\phi_Q(x) = \int K(x,y)(\phi_Q(y)-\phi_Q(x))u_Q(y)dy + \phi_Q(x)(Tu_Q)(x)$ et appartient à $L^\infty(\mathbb{R}^n)$ avec une norme uniformément bornée. En fait (8.2) nous donne un renseignement plus précis: sur Q, on a $|Tu_Q(x') - Tu_Q(x)| \le C|x-x'|^\varepsilon t^{-\varepsilon}$ et, de même $g(x) = \int K(x,y)(\phi_Q(y) - \phi_Q(x))u_Q(y)dy$ est définie partout (et est également höldérienne comme nous le verrons dans un instant). Retenons que $T\phi_Q(x)$ est définie partout (et non pas seulement presque-partout). En changeant les notations, nous avons établi que si $f \in \mathcal{D}(\mathbb{R}^n)$ est portée par une boule Q et si $\xi \in \mathcal{D}(\mathbb{R}^n)$ est égale à 1 sur la boule double $\tilde{Q}$, on a

$$(8.3) \qquad Tf(x) = \int K(x,y)(f(y)-f(x))\xi(y)dy + f(x)T\xi(x).$$

On est maintenant en mesure de vérifier le résultat suivant.

Proposition 2. *Soit* $f \in \mathcal{D}(\mathbb{R}^n)$ *et soit* $T \in L_\varepsilon$ *un opérateur d'ordre* 0 *vérifiant* $T(1) = 0$. *Soient* x_2 *et* x_1 *deux points de* $\mathbb{R}^n$ *et* $\xi \in \mathcal{D}(\mathbb{R}^n)$ *une fonction égale a* 1 *au voisinage de* x_1 *et de* x_2. *Posons* $\eta = 1-\xi$. *Alors on a*

$$(8.4)\quad Tf(x_2)-Tf(x_1) = \int (K(x_2,y)-K(x_1,y))(f(y)-f(x_1)\eta(y)dy$$
$$- \int K(x_1,y)(f(y)-f(x_1))\xi(y)dy + \int K(x_2,y)(f(y)-f(x_2))\xi(y)dy$$
$$+ (f(x_2)-f(x_1))(T\xi)(x_2).$$

Nous commençons par démontrer (8.4) dans le cas particulier où $\xi = 1$ sur $\tilde{Q}$, Q étant une boule assez grande pour contenir le support de f. On peut alors utiliser (8.3) et il suffit d'observer que $f(x_2)T\xi(x_2)-f(x_1)T\xi(x_1) = (f(x_2)-f(x_1))(T\xi)(x_2) +$ $+ f(x_1)(T\xi(x_2)-T\xi(x_1))$. On utilise alors (8.2) et il vient $T\xi(x_2)-T\xi(x_1) = -\int (K(x_2,y)-K(x_1,y))\eta(y)dy$. Puisque $f(y)\eta(y) = 0$, (8.4) est demontrée.

Pour passer au cas général, on remarque que la différence entre deux choix possibles de ξ conduit à une fonction (notée ξ_0), nulle au voisinage de x_1 et de x_2 et que η devient alors $\eta_0 = -\xi_0$. La somme des quatre intégrales du membre de droite de (8.4) est nulle.

Nous sommes en mesure de terminer la preuve du théorème 1. Soit s un exposant de l'intervalle $]0,\varepsilon[$. Supposons que $f \in \mathcal{D}(\mathbb{R}^n)$ vérifie $\|f\|_{\Lambda^s} \leq 1$ et cherchons à majorer $\|Tf\|_{\Lambda^s}$.

Pour cela désignons par x_1 et x_2 deux points de $\mathbb{R}^n$ et posons $d = |x_2-x_1|$. Soit Q la boule $|x-x_1| \leq 10d$ et ξ_Q une fonction de $\mathcal{D}(\mathbb{R}^n)$, dont le support est contenu dans Q, égale a 1 si $|x-x_1| \leq 5d$ et vérifiant les conditions usuelles $|\partial^\alpha \xi_Q| \leq C_\alpha\, d^{-|\alpha|}$. Les quatre morceaux de (8.4) se majorent par Cd^s, de façon évidente, ce qui donne la continuité Λ^s désirée. Il est intéressant de prolonger $T : \mathcal{D}(\mathbb{R}^n) \to \Lambda^s(\mathbb{R}^n)$ en un opérateur linéaire continu de $\Lambda^s(\mathbb{R}^n)$ dans lui-même. Pour cela on observe que $\Lambda^s(\mathbb{R}^n)$ est le dual de l'espace $H^p(\mathbb{R}^n)$ de Stein et Weiss lorsque $\frac{s}{n} = \frac{1}{p} - 1$ $(0 < p < 1)$. Il est donc naturel de munir $\Lambda^s(\mathbb{R}^n)$ de la topologie $\sigma(\Lambda^s,H^p)$ définie par cette dualité. Alors $\mathcal{D}(\mathbb{R}^n)$ est dense dans Λ^s; plus précisément il existe une constante C telle que, pour toute $f \in \Lambda^s$, il existe une suite f_k, $k \in \mathbb{N}$, de fonctions de $\mathcal{D}(\mathbb{R}^n)$ telle que $\|f_k\|_{\Lambda^s} \leq C\|f\|_{\Lambda^s}$ et $f_k \rightharpoonup f$ pour la topologie $\sigma(\Lambda^s,H^p)$.

Montrons alors que $T(f_k) = g_k$ converge pour cette même topologie vers $g \in \Lambda^s$.

On a d'une part $\|g_k\|_{\Lambda^s} \leq C' \|f\|_{\Lambda^s}$. Il suffit alors de tester la convergence faible sur des fonctions de test appartenant à $\mathcal{D}_0$, sous-espace dense dans H^p. Si $T \in L_\varepsilon$ vérifie $T(1) = 0$ et si T est d'ordre 0, alors $T^*(\psi) \in H^p$ pour toute $\psi \in \mathcal{D}_0$ et pour $(n+\varepsilon)p > n$. On a donc $\lim\limits_{k\to+\infty} \langle T(f_k),\psi\rangle = \lim\limits_{k\to+\infty} \langle f_k, T^*(\psi)\rangle = \langle f, T^*(\psi)\rangle$. La suite des g_k est bornée dans Λ^s et converge (faiblement) sur $\mathcal{D}_0 \subset H^p$, sous-espace dense dans $H^p(\mathbb{R}^n)$. Il en résulte que g_g converge faiblement vers une forme linéaire continue sur H^p, c'est-à-dire une fonction $g \in \Lambda^s$.

9. La preuve de P.G. Lemarié du théorème 5

La démonstration directe du théorème 5 due a P.G. Lemarié n'utilise en fait ni la structure de groupe de $\mathbb{R}^n$, ni la transformation de Fourier. Elle s'étend donc aux espaces de nature homogènes de Coifman et Weiss (voir la thèse de troisième cycle de Lemarié).

On utilise la condition suivante d'appartenance à B^s lorsque $0 < s < 1$.

Lemme 5. *Soit* $f \in L^2(\mathbb{R}^n)$ *une fonction telle que*

$$\iint |f(x)-f(y)|^2 |x-y|^{-n-2s}\, dx\, dy < +\infty.$$

Alors $f \in B^s$ *et* $c(s,n)\|f\|_{B^s} = \left(\iint |f(x)-f(y)|^2|x-y|^{-n-2s} dx\, dy\right)^{1/2}$.

Ce résultat est classique. Il convient d'observer que la condition $f \in L^2(\mathbb{R}^n)$ ne peut y être remplacée par $f \in L^2_{loc}(\mathbb{R}^n)$.

Nous allons démontrer l'inégalité a priori

$$\|Tf\|_{B^s} \leq C(s,\varepsilon,n)\|f\|_{B^s} \tag{9.1}$$

pour $0 < s < \varepsilon$, $T \in L_\varepsilon$ vérifiant $T(1) = 0$ et d'ordre 0 et pour $f \in \mathcal{D}(\mathbb{R}^n)$. Observons que $g = Tf \in L^\infty(\mathbb{R}^n)$ (proposition 1) et $g(x) = O(|x|^{-n})$ à l'infini. On a donc $g \in L^2(\mathbb{R}^n)$ et il suffit de calculer $I = \iint |g(x)-g(y)|^2|x-y|^{-n-2s}\, dx\, dy$.

On appelle $\xi \in \mathcal{D}(\mathbb{R}^n)$ une fonction radiale, égale à 1 sur la boule $|u| \leq 2$ et l'on pose $1-\xi(u) = \eta(u)$.

La proposition 2 donne alors $g(y) - g(x) = g_1(x,y) + g_2(x,y) + g_3(x,y) + g_4(x,y)$ où

$$g_1(x,y) = \int_{\mathbb{R}^n} \{K(y,u)-K(x,u)\}(f(u)-f(x))\eta\left(\frac{u-x}{|y-x|}\right)du$$

$$g_2(x,y) = -\int_{\mathbb{R}^n} K(x,u)(f(u)-f(x))\ \xi\left(\frac{u-x}{|y-x|}\right) du$$

$$g_3(x,y) = \int_{\mathbb{R}^n} K(y,u)(f(u)-f(y))\ \xi\left(\frac{u-x}{|y-x|}\right) du$$

et
$$g_4(x,y) = (f(y)-f(x)) \int_{\mathbb{R}^n} K(y,u)\ \xi\left(\frac{u-x}{|y-x|}\right) du.$$

Nous allons successivement majorer les quatre intégrales

$$\iint |g_j(x,y)|^2 |x-y|^{-n-2s}\, dx\, dy \qquad (j = 1,2,3 \text{ et } 4).$$

On introduit un exposant α tel que $s < \alpha < \varepsilon$ et l'on écrit

$$|g_1(x,y)| \le C \int_{\{|u-x| \ge 2|y-x|\}} |x-y|^{\varepsilon} |u-x|^{-n-\varepsilon} |f(u)-f(x)|\, du =$$

$$C|y-x|^{\varepsilon} \int_{\{|u-x|\ge 2|y-x|\}} |u-x|^{-n/2-\varepsilon+\alpha}\ |u-x|^{-n/2-\alpha} |f(u)-f(x)|\, du.$$

L'inégalité de Cauchy-Schwarz donne alors

$$|g_1(x,y)|^2 \le$$

$$\left(C^2 |x-y|^{2\varepsilon} \int_{\{|u-x|\ge 2|y-x|\}} |u-x|^{-n+2\varepsilon+2\alpha}\ du\right)\left(\int_{\{|u-x|\ge 2|y-x|\}} |u-x|^{-n-2\alpha}\ |f(u)-f(x)|^2\ du\right)$$

$$= C'|x-y|^{2\alpha} \int_{\{|u-x|\ge 2|y-x|\}} |u-x|^{-n-2\alpha}\ |f(u)-f(x)|^2\ du.$$

On calcule alors $\iint |g_1(x,y)|^2 |x-y|^{-n-2s}\, dx\, dy$ en intégrant d'abord en y; cette intégrale double se majore par $C''\iint |f(u)-f(x)|^2 |u-x|^{-n-2s}\, du\, dx$. On a ensuite

$|g_2(x,y)| \le C \int_{\{|u-x|\le 10|y-x|\}} |u-x|^{-n} |f(u)-f(x)|\, du$ si le support de ξ est contenu dans le boule $|u| \le 10$, ce que l'on peut toujours supposer.

On introduit alors un exposant β tel que $0 < \beta < s$ et l'on écrit

$$|g_2(x,y)| \le C \int_{\{|u-x|\le 10|y-x|\}} |u-x|^{-\frac{n}{2}+\beta}\ |u-x|^{-\frac{n}{2}-\beta}\ |f(u)-f(x)|\, du.$$

L'inégalité de Cauchy-Schwarz conduit a

$|g_2(x,y)|^2 \le C'|x-y|^{2\beta} \int_{\{|x-u|\le 10|y-x|\}} |u-x|^{-n-2\beta}\ |f(u)-f(x)|^2 du$ et le calcul se termine comme plus haut en intégrant d'abord en y.

On a $|g_3(x,y)| \leq C\int_{\{|u-y|\leq 11|x-y|\}} |y-u|^{-n}|f(u)-f(y)|du.$

Il suffit alors d'échanger les rôles de x et de y dans le traitement de $g_2(x,y)$ pour majorer $\iint |g_3(x,y)|^2|x-y|^{-n-2s}\,dx\,dy.$ Enfin on observe que $u \to \xi(\frac{u-x}{|y-x|})$ est l'une des fonctions ϕ_Q de la proposition 1. On a donc $|g_4(x,y)| \leq C'|f(y)-f(x)|.$

10. Continuité des opérateurs $T \in L_\varepsilon$ lorsque $\varepsilon > 1$

Désignons par $D_j = -i\frac{\partial}{\partial x_j}$ l'opérateur usuel de dérivation partielle et, si $\varepsilon > 0$, par $L_\varepsilon^{(0)} \subset L_\varepsilon$ le sous-espace des opérateurs d'ordre 0 et tels que $T(x^\alpha) = 0$ pour tout $\alpha \in \mathbb{N}^n$ tel que $|\alpha| < \varepsilon$.

La continuité des opérateurs $T \in L_\varepsilon^{(0)}$ sur Λ^s pour $0 < s < \varepsilon$ s'obtient (par récurrence sur la partie entière de ε) à l'aide de la proposition suivante

Proposition 3. *Si* $T \in L_\varepsilon^{(0)}$, *alors* $D_j\,T = \sum_1^n T_m\,D_m$ *où* $T_m \in L_{\varepsilon-1}^{(0)}$.

Naturellement, les opérateurs T_m dépendent aussi de j.

La façon la plus commode de faire ces vérifications est d'effectuer au préalable une décomposition de Littlewood-Paley de l'opérateur T.

On désigne, à cet effet, par $\phi \in S(\mathbb{R}^n)$ une fonction radiale dont la transformée de Fourier vaut 1 sur le boule $|\xi| \leq 1$ et 0 sur $|\xi| \geq 2$. On pose $\psi(\xi) = \phi(\frac{\xi}{2}) - \phi(\xi)$ et, pour tout $k \in \mathbb{Z}$, on désigne par S_k l'opérateur de convolution défini, via la transformation de Fourier, par le multiplicateur $\phi(\frac{\xi}{2^k})$ tandis que $\Delta_k = S_{k+1} - S_k$ est défini par le multiplicateur $\psi(\frac{\xi}{2^k})$.

Lemme 6. *Soit* $T : \mathcal{D}(\mathbb{R}^n) \to \mathcal{D}'(\mathbb{R}^n)$ *un opérateur linéaire continu d'ordre* 0. *Alors on a, au sens faible,*

$$(10.1) \qquad \lim_{k\to+\infty} S_k\,T\,S_k = T$$

$$(10.2) \qquad \lim_{k\to-\infty} S_k\,T\,S_k = 0.$$

Rappelons que $L_k : \mathcal{D}(\mathbb{R}^n) \to \mathcal{D}'(\mathbb{R}^n)$ tend vers $L : \mathcal{D} \to \mathcal{D}'$ au sens faible si, pour toute $f \in \mathcal{D}(\mathbb{R}^n)$ et toute $g \in \mathcal{D}(\mathbb{R}^n)$, on a $\lim_{k\to+\infty} \langle L_k f,g\rangle = \langle Lf,g\rangle$.

La preuve du lemme 6 est élémentaire et laissée au lecteur. Grâce au lemme 6, on peut écrire T sous la forme de la série télescopique

$$\sum_{-\infty}^{+\infty} \{S_{k+1}\, T\, S_{k+1} - S_k\, T\, S_k\} = \sum_{-\infty}^{+\infty} (\Delta_k\, T\, S_{k+1} + S_k\, T\, \Delta_k) =$$

$$= \sum_{-\infty}^{+\infty} L_k + M_k .$$

Nous allons étudier les propriétés des noyaux $L_k(x,y)$ et $M_k(x,y)$ des opérateurs L_k et M_k.

On a $$|L_k(x,y)| \leq C2^{kn}(1 + 2|x-y|)^{-n-\varepsilon}$$

$$\int L_k(x,y)y^{\alpha}dy = 0 \quad \text{pour} \quad |\alpha| < \varepsilon, \quad \alpha \in \mathbb{N}^n$$

et $$|\partial_x^{\alpha}\, L_k(x,y)| \leq C_{\alpha} 2^{k(n+|\alpha|)}(1+2^k|x-y|)^{-n-\varepsilon}$$

pour tout $\alpha \in \mathbb{N}^n$.

En ce qui concerne $M_k(x,y)$, l'absence de régularité en y du noyau de T ne permet pas d'obtenir une estimation comparable à celle de $L_k(x,y)$. On a

$$|M_k(x,y)| \leq C2^{kn}(1+2^k|x-y|)^{-n}$$

puis $$|\partial_x^{\alpha}\, M_k(x,y)| \leq C\; 2^{k(n+|\alpha|)}\; (1+2^k|x-y|)^{-n|\alpha|}$$

lorsque $|\alpha| < \varepsilon$ et enfin

$$|\partial_x^{\alpha}\, M_k(x,y)| \leq C_{\alpha}\; 2^{k(n+|\alpha|)}(1+2^k|x-y|)^{-n-\varepsilon}$$

pour $|\alpha| \geq \varepsilon$.

On a, par ailleurs, si $0 \leq |\beta| < |\alpha|$,

$$\int \partial_x^{\alpha}\, M_k(x,y)y^{\beta}\, dy = 0.$$

Ces vérifications reposent sur le lemme suivant

Lemme 7. *On suppose que* $T : \mathcal{D}(\mathbb{R}^n) \to \mathcal{D}'(\mathbb{R}^n)$ *est un opérateur linéaire continu, d'ordre* 0 *et appartenant à* L_{ε}.

Appelons ϕ *et* ψ *deux fonctions de* $S(\mathbb{R}^n)$ *et supposons que* $\int x^{\alpha}\psi(x)dx = 0$ *pour tous les multi-indices* $\alpha \in \mathbb{N}^n$ *tels que* $|\alpha| < \varepsilon$. *Alors il existe une constante* $C(\phi,\psi)$ *telle que, pour tout couple* R_u, R_v *de deux opérateur de translation on ait*

$$(10.3) \qquad |<T\ R_u\ \phi,\ R_v\ \psi>| \le \frac{C(\phi,\psi)}{(1+|u-v|)^{n+\varepsilon}}$$

et

$$(10.4) \qquad |<T\ R_u\ \psi,\ R_v\ \phi>| \le \frac{C(\phi,\psi)}{(1+|u-v|)^{n}} .$$

On commence par vérifier ces inégalités si ϕ et ψ appartiennent à $\mathcal{D}(\mathbb{R}^n)$. On appelle R le rayon d'une boule de centro 0 contenant les supports de ϕ et de ψ. Si $|u-v| \ge 3R$, (10.3) et (10.4) découlent immédiatement des propriétés du noyau-distribution $K(x,y)$ et d'une intégration par parties en x, en ce qui concerne (10.3). Si $|u-v| \le 3R$, on observe que $R_u\psi$ et $R_v\phi$ appartiennent à une partie bornée $\mathcal{B} \subset \mathcal{D}(\mathbb{R}^n)$ et l'on utilise le fait que T est d'ordre 0.

Un simple changement d'échelle fournit

$$(10.5) \qquad |<T\ \phi^{(u,t)},\ \psi^{(v,t)}>| \le \frac{C(\phi,\psi)t^n}{(t+|u-v|)^{n+\varepsilon}}$$

et

$$(10.6) \qquad |<T\ \psi^{(u,t)},\ \phi^{(v,t)}>| \le \frac{C(\phi,\psi)t^n}{(t+|u-v|)^{n}}$$

Le cas général où ϕ et ψ appartiennent à $\mathcal{S}(\mathbb{R}^n)$ s'obtient en écrivant $\phi(x) = \sum_{k\in\mathbb{Z}^n} \alpha_k\ \phi_k\ (x-k)$, $\psi(x) = \sum_{k\in\mathbb{Z}^n} \beta_k\ \psi_k(x-k)$ où les constantes α_k et β_k sont à décroissance rapide, où les fonctions ϕ_k et ψ_k appartiennent à une partie bornée de $\mathcal{D}(\mathbb{R}^n)$ et où $\int x^\alpha\ \psi_k(x)dx = 0$ si $|\alpha| < \varepsilon$.

Les détails sont laissés au lecteur.

Si ϕ et ψ sont paires et si P_t ou Q_t représente l'opérateur de convolution avec $\phi_t(x) = t^{-n}\phi(\frac{x}{t})$ ou avec $\psi_t(x) = t^{-n}\psi(\frac{x}{t})$, le noyau-distribution de $P_t\ T\ Q_t$ est $M_t(x,y) = <T\ R_y\ \psi_t,\ R_x\ \phi_t>$. De même, le noyau-distribution de $Q_t\ T\ P_t$ est $L_t(x,y) =$ $= <T\ R_y\ \phi_t,\ R_x\ \psi_t>$.

Pour conclure, on pose $t = 2^{-k}$, $k \in \mathbb{Z}$, et l'on tombe sur les noyaux $L_k(x,y)$ et $M_k(x,y)$ que nous voulions estimer.

Revenons à $D_j M_k = D_j\ S_k\ T\Delta_k$. Pour extraire, à droite, les opérateurs $D_m = -i\ \partial/\partial x_m$ $(1 \le m \le n)$, on utilise l'identité évidente $\psi(\xi) = \sum_1^n \psi_m(\xi)\xi_m$ où $\psi_m(\xi) = \frac{\xi_m}{|\xi|^2}\ \psi(\xi)$. On appelle $\Delta_k^{(m)}$ l'opérateur de convolution associé au multiplicateur $\psi_m(\frac{\xi}{2^k})$ et l'on a donc

$\Delta_k = 2^{-k} \sum_1^n \Delta_k^{(m)} D_m$ ce qui conduit à $D_j M_k = 2^{-k} \sum_1^n D_j S_k T \Delta_k^{(m)} D_m$.

Il reste à vérifier que chacun des n opérateurs $T_m = \sum_{-\infty}^{+\infty} 2^{-k} D_j S_k T \Delta_k^{(m)}$ appartient à $L_{\varepsilon-1}^{(0)}$.

Cette vérification s'obtient à l'aide du lemme suivant.

Lemme 8. *Supposons que les fonctions* $f_k(x,y) \in C^\infty(\mathbb{R}^n \times \mathbb{R}^n)$ *vérifiant* $|f_k(x,y)| \le 2^{kn}(1+2^k|x-y|)^{-n-1}$ *pour tout* $x \in \mathbb{R}^n$, *tout* $y \in \mathbb{R}^n$ *et tout* $k \in \mathbb{Z}$ *ainsi que* $|\partial_x^\alpha f_k(x,y)| \le$ $\le 2^{kn} 2^{k|\alpha|} (1+2^k|x-y|)^{-n-1-|\alpha|}$ *chaque fois que* $\alpha \in \mathbb{N}^n$ *et que* $|\alpha| \le \varepsilon-1$. *On suppose, si* $\varepsilon-1 < |\alpha| \le \varepsilon$

$$|\partial_x^\alpha f_k(x,y)| \le 2^{kn} 2^{k|\alpha|}(1+2^k|x-y|)^{-n-\varepsilon}$$

En outre, on fait l'hypothèse que $\int \partial_x^\alpha f_k(x,y) y^\beta \, dy = 0$ *chaque fois que* $|\beta| < |\alpha| \le \varepsilon$.

Alors le noyau-distribution $\sum_{-\infty}^{+\infty} f_k(x,y)$ *définit un opérateur appartenant à* $L_{\varepsilon-1}^{(0)}$.

La preuve du lemme 8 ne présente aucune difficulté et est laissée au lecteur. Observons que nous n'avons pas encore utilisé les conditions $T(x^\alpha) = 0$ pour $|\alpha| < \varepsilon$; la raison en est que $M_k = S_k T \Delta_k$ et que $\Delta_k(x^\alpha) = 0$ pour tout $\alpha \in \mathbb{N}^n$. En revanche ces hypothèses ($T(x^\alpha) = 0$ si $|\alpha| < \varepsilon$) jouent un rôle essentiel dans l'analyse de $L_k = \Delta_k T S_{k+1}$.

Nous allons démontrer une variante de la proposition 3 en démontrant que $\sum_{+\infty}^{+\infty} D_j L_k(-\Delta)^{-1/2} \in L_{\varepsilon-1}^{(0)}$. Pour cela nous utilisons le lemme suivant.

Lemme 9. *Soit* $\varepsilon > 1$ *un exposant et* $f \in L^1(\mathbb{R}^n)$ *une fonction telle que* $|f(x)| \le (1+|x|)^{-n-\varepsilon}$ *et que* $\int x^\alpha f(x)dx = 0$ *pour tout* $\alpha \in \mathbb{N}^n$ *tel que* $|\alpha| < \varepsilon$. *Alors il existe une et une seule fonction* $g \in L^1(\mathbb{R}^n)$ *telle que* $\hat{f}(\xi) = |\xi|\hat{g}(\xi)$ *pour tout* $\xi \in \mathbb{R}^n$ *et l'on a* $|g(x)| \le C(n,\varepsilon)\ (1+|x|)^{-n-\varepsilon+1}$ *et* $\int x^\alpha g(x)dx = 0$ *pour tout* $\alpha \in \mathbb{N}^n$ *tel que* $|\alpha| < \varepsilon-1$.

La preuve de ce résultat ne présente pas de difficulté et est laissée au lecteur. Grosso modo l'hypothèse signifie que $\hat{f}$ a un zéro d'ordre m en 0 (m désignant la partie entier de ε) et la conclusion signifie que $\hat{f}(\xi)/|\xi|$ a un zéro d'ordre $m-1$.

Le lemme 9 s'applique à la fonction $y \to -i \frac{\partial}{\partial x_j} L_k(x,y)$ après un changement d'origine et d'échelle évident. On a donc
$-i \frac{\partial}{\partial x_j} L_k(x,y) = (-\Delta_y)^{1/2} Z_k(x,y)$ où $|Z_k(x,y)| \leq$
$\leq C\ 2^{kn}(1+2^k|x-y|)^{-n-\varepsilon+1}$ et plus généralement
$|\partial_x^\alpha Z_k(x,y)| \leq C_\alpha\ 2^{k|\alpha|} 2^{kn}(1+2^k|x-y|)^{-n-\varepsilon+1}$ pour $\alpha \in \mathbb{N}^n$ tandis que

$$\int Z_k(x,y) y^\beta\, dy = 0 \quad \text{si} \quad |\beta| < \varepsilon - 1.$$

L'opérateur dont le noyau distribution est $\sum_{-\infty}^{+\infty} Z_k(x,y)$ appartient donc bien à $\mathcal{L}_{\varepsilon-1}^{(0)}$. Pour terminer on observe qu'écrire
$-i \frac{\partial}{\partial x_j} L_k(x,y) = (-\Delta_y)^{1/2} Z_k(x,y)$ au niveau des noyaus signifie
$D_j L_k = Z_k(-\Delta)^{1/2}$ au niveau des opérateurs.

La preuve du théorème 3 s'obtient alors par une récurrence sur la partie entière m_o de s.

Si $m_o = 0$, on a $0 < s < 1$ et l'on peut alors supposer que $0 < s < \varepsilon < 1$. Le théorème 3 coincide alors avec le théorème 1.

Pour passer au cas général, il suffit d'utiliser la proposition 3.

11. La nouvelle démonstration du théorème de David et Journé

Nous nous proposons de montrer que le théorème 7 découle simplement du théorème 5 et de l'existence d'opérateurs de paramultiplication entre BMO et L^2.

Si $b(x) \in \mathrm{BMO}(\mathbb{R}^n)$ et $f \in L^2(\mathbb{R}^n)$, on pose $\pi(b,f) =$
$= 4 \int_0^\infty Q_t\{(Q_t b)(P_t f)\} \frac{dt}{t}$ où P_t est le semi-groupe de Poisson et où $Q_t = -t \frac{\partial}{\partial t} P_t$. Il est facile de vérifier qu'il existe une constante $C > 0$ telle que

$$\|\pi(b,f)\|_2 \leq C\|b\|_{\mathrm{BMO}} \|f\|_2 \tag{11.1}$$

lorsqu'on dispose de la caractérisation de BMO par le fait que $|Q_t b|^2 \frac{dx\, dt}{t}$ est une mesure de Carleson.

On désigne alors L_b l'opérateur défini par $L_b(f) = \pi(b,f)$. Il est aisé de vérifier que L_b et L_b^* appartiennent à $\mathcal{L}_\varepsilon$ si $0 < \varepsilon < 1$.

Grâce a cette nouvelle opération algébrique, nous pouvons corriger l'opérateur T du théorème 7 pour obtenir un opérateur R vérifiant $R(1) = R^*(1) = 0$.

Pour faire cette correction, on pose $T(1) = b \in BMO$ et $T^*(1) = c \in BMO$ et l'on forme $R = T - L_b - L_c^*$. Puisque $L_b(1) = b$, $L_b^*(1) = 0$, $L_c(1) = c$ et $L_c^*(1) = 0$, il vient $R(1) = R^*(1) = 0$.

Par ailleurs $R \in \mathcal{L}_\varepsilon$ et $R^* \in \mathcal{L}_\varepsilon$ pour $0 < \varepsilon < 1$.

C'est alors que le théorème 5 intervient. Puisque $R \in \mathcal{L}_\varepsilon$ vérifie en outre $R(1) = 0$ et que R est d'ordre 0, R est automatiquement continu sur B^s pour $0 < s < \varepsilon$. De même R^* est continu sur B^s; c'est-à-dire que R est continu sur B^{-s}. Par interpolation, R est continu sur $L^2(\mathbb{R}^n)$.

12. La preuve du théorème 4

Deux difficultés se présentent pour rattacher le théorème 4 au théorème 3. D'une part, il faut obtenir l'estimation L^∞ manquante; d'autre part il faut introduire l'hypothèse $T(1) = 0$ ou $T(x^\alpha) = 0$, $|\alpha| \leq m_o$, manquante.

Tout d'abord, nous pouvons nous limiter au cas où $K(x,y) = 0$ dès que $|x-y| \geq 1$. Il suffit d'appeler $\lambda \in \mathcal{D}(\mathbb{R}^n)$ une fonction égale à 1 sur la boule $|x| \leq 1/2$ et nulle hors de $|x| \leq 1$ et d'écrire $K(x,y) = K(x,y)\lambda(y-x) + R(x,y)$.

De façon évidente $\int R(x,y)f(y)dy$ appartient à C^ε pour toute fonction $f \in L^\infty(\mathbb{R}^n)$. Si bien que toute la discussion porte sur $K(x,y)\lambda(y-x)$ que nous noterons $K(x,y)$ pour alléger.

Nous allons maintenant opérer une seconde réduction du problème en supposant que $f \in C^s(\mathbb{R}^n)$ est nulle si $|x| \geq R$; $R > 0$ est une constante qui sera fixée dans un instant.

Pour voir que ce cas particulier est suffisant, nous partons d'une décomposition de l'identité $1 = \sum_{k\in\mathbb{Z}^n} \phi(x-k)$ où $\phi \in \mathcal{D}(\mathbb{R}^n)$ et nous appelons R le rayon d'une boule contenant le support de ϕ.

On écrit $f = \sum_{k\in\mathbb{Z}^n} f(x)\phi(x-k) = \sum_{k\in\mathbb{Z}^n} f_k(x-k)$; alors les supports des f_k sont eux-mêmes contenus dans la boule $|x| \leq R$ et l'on a $\|f_k\|_{C^s} \leq C(f)$. Réciproquement si $g = \sum_{k\in\mathbb{Z}^n} g_k(x-k)$, si les supports des g_k sont contenus dans la boule $|x| \leq R'$ et si $\|g_k\|_{\Lambda^s} \leq C$, alors g appartient a C^s. En effet, on observe d'abord qu'une forme très rudimentaire, du lemme de Poincaré donne $\|g_k\|_{C^s} \leq C'$ et qu'ensuite g appartient à C^s parce que la somme donnant g est localement finie.

Dans le même optique, nous utiliserons le résultat suivant

Lemme 10. *Avec les notations ci-dessus, si* $f \in H^s(\mathbb{R}^n)$, *on a* $\|f_k\|_{H^s(\mathbb{R}^n)} = \varepsilon_k \in \ell^2(\mathbb{Z}^n)$. *Réciproquement, si* $g = \sum_{k \in \mathbb{Z}^n} g_k(x-k)$, *si les supports des* g_k *sont tous contenus dans la boule* $|x| \leq R$ *et si* $\|g_k\|_{B^s(\mathbb{R}^n)} = \eta_k \in \ell^2(\mathbb{Z}^n)$, *alors* $g \in H^s(\mathbb{R}^n)$.

On utilise cette fois le véritable lemme de Poincaré pour écrire $\|g_k\|_{H^s} \leq C\,\|g_k\|_{B^s}$.

Revenons à la continuité de T sur C^s ou sur H^s. En employant le lemme 10 et en se souvenant de ce que $K(x,y) = 0$ si $|y-x| \geq 1$, on se ramène à prouver que $\|R_k^{-1}\, TR_k(g_k)\|_{\Lambda^s} \leq C\|g_k\|_{C^s}$ ou bien $\|R_k^{-1}\, TR_k\, g_k\|_{B^s} \leq C\|g_k\|_{H^s}$.

Nous allons démontrer ces inégalités en utilisant les théorèmes 3 et 5 et en corrigeant l'opérateur T (et les opérateurs $R_k^{-1}\, T\, R_k$, $k \in \mathbb{Z}^n$) pour avoir $T(x^\alpha) = 0$ lorsque $|\alpha| \leq m_0$.

Si $m_0 = 0$ et si $T(1) = \mu(x)$ appartient à C^s, la correction évidente est de soustraire l'opérateur de multiplication ponctuelle par $\mu(x)$. Dans le cas de l'espace de Sobolev, on est amené à supposer que $\mu(x)$ est un multiplicateur ponctuel de H^s.

Dans le cas général, nous supposons que $T(x^\alpha) = \mu_\alpha(x)$ appartient à C^ε lorsque $|\alpha| \leq m_0$ et nous corrigerons T pour aboutir à $L(x^\alpha) = 0$ lorsque $|\alpha| \leq m_0$. A cet effet, on désigne par $\psi_\alpha \in \mathcal{D}(\mathbb{R}^n)$, $|\alpha| \leq m_0$, des fonctions telles que $\int \psi_\alpha(x) x^\beta\, dx = 0$ si $\alpha \neq \beta$, $|\beta| \leq m_0$ et $\int \psi_\alpha(x) x^\alpha\, dx = 1$ si $|\alpha| \leq m_0$.

On appelle L l'opérateur défini par le noyau

$$L(x,y) = \sum_{|\alpha| \leq m_0} m_\alpha(x)\, \psi_\alpha(x-y)$$

où les fonctions $m_\alpha(x) \in C^\varepsilon$ sont choisies de sorte que $L(x^\beta) = \mu_\beta(x)$ pour $|\beta| \leq m_0$. Ces conditions conduisent à

$\sum'_{|\alpha| \leq m_0} \frac{\beta!}{\alpha!(\beta-\alpha)!} m_\alpha(x)\, x^{\beta-\alpha} = \mu_\beta(x)$; Σ' signifie ici que la somme est étendu aux α tels que $\alpha_j \leq \beta_j$ pour $1 \leq j \leq n$. Ce système en cascade conduit de proche en proche aux choix de $m_o(x)$, puis $m_\alpha(x)$ pour $|\alpha| = 1$ etc...

L'opérateur L appartient à M^ε; les théorèmes 3 et 5 s'appliquent à T-L et cela termin r la preuve des théorèmes 4 et 6.

Bibliographie

[1] A.P. Calderón. Commutators, Singular Integrals on Lipschitz Curves and Applications, I.C.M., Helsinki (1978) tome 1, 85-96.

[2] G. David et J.L. Journé. Une caractérisation des opérateurs intégraux singuliers bornés sur $L^2(\mathbb{R}^n)$. C.R. Acad. Sc. Paris t.296 (16 mai 1983) 761-764.

[3] C. Fefferman and E.M. Stein. H^p spaces of several variables. Acta Math. 129 (1972) 137-193.

[4] P.G. Lemarié. Thèse de troisième cycle. A paraître aux publications mathématiques d'Orsay. Département de Mathématique 91405 Orsay.

[5] Y. Meyer. Théorie du potentiel dans les domaines lipschitziens d'après G.C. Verchota (Séminaire Goulaouic-Schwartz, 1982-1983)

[6] Y. Meyer. Les nouveaux opérateurs de Calderón-Zygmund. Actes du Colloque L. Schwartz, Ecole Polytechnique, Juin 1983 (à paraître dans Astérisque, S.M.F.)

[7] Y. Meyer. Continuité sur les espaces de Hölder et de Sobolev des opérateurs définis par des intégrales singulières (Séminaire Goulaouic-Schwartz 1983-1984).

Recent Progress in Fourier Analysis
I. Peral and J.-L. Rubio de Francia (Editors)

ANALYTIC FAMILIES OF BANACH SPACES AND SOME OF THEIR USES

Richard Rochberg and Guido Weiss
Washington University in St. Louis

TABLE OF CONTENTS

1. SOME EXAMPLES OF ANALYTIC FAMILIES OF BANACH SPACES

Let D be a connected domain in $\mathbf{C}$ and suppose that corresponding to each $\xi \in \partial D$ we are given a Banach space B_ξ. In some sense we would like to obtain a function $z \to B_z$, $z \in D$, whose values are Banach spaces, that is the solution of a Dirichlet problem with boundary data $\{B_\xi\}$, $\xi \in \partial D$. Our original motivation for considering such a construction comes from the theory of interpolation of operators; our aim was to extend the "complex method" of A.P. Calderón and J.L. Lions. Before discussing these notions we present some features of the one-dimensional case of our problem and some immediate higher dimensional extensions. This will give us a more precise idea of the analytic properties of these spaces and the connection they have with the Dirichlet problem.

The construction of the interior spaces $\{B_z\}$, $z \in D$, obtained from the boundary spaces $\{B_\xi\}$, $\xi \in \partial D$, in the one dimensional case is closely connected to Szegö's theorem. A version of this result when D is the unit disk $= \{z \in \mathbf{C} : |z| < 1\}$ is the following:

THEOREM 1.1. (Szegö). *Suppose* $h(\theta) \geq 0$ *on* $[-\pi,\pi)$ *and* $\log h$ *is integrable. If* a *is the class of all analytic functions on* $\overline{D}$ *and* $0 < p < \infty$, *then*

$$\inf_{f \in a, f(0)=1} \int_{-\pi}^{\pi} |f(e^{i\theta})|^p h(\theta)\, \frac{d\theta}{2\pi} = \exp \int_{-\pi}^{\pi} \{\log h(\theta)\} \frac{d\theta}{2\pi}\,. \tag{1.2}$$

Suppose we write $|1|_{e^{i\theta}} = |h(\theta)|^{1/p}$ and, thus, interpret $|\ |_{e^{i\theta}}$ to be a norm on C associated with the point $e^{i\theta} \in \partial D$. Then (2.2) becomes

$$\{\inf_{f \in a,\ f(0)=1} \int_{-\pi}^{\pi} (|f(e^{i\theta})|_{e^{i\theta}})^p \frac{d\theta}{2\pi}\}^{1/p} =$$

$$= \exp \int_{-\pi}^{\pi} \{\log |1|_{e^{i\theta}}\} \frac{d\theta}{2\pi}$$

and we may think of the left side of this expression to be the definition of the norm of $1 \in C$ corresponding to the point $0 \in D$. In this way the boundary spaces $B_{e^{i\theta}} = (\mathbf{C}, |\ |_{e^{i\theta}})$ are used to determine the space $B_0 = (\mathbf{C}, |\ |_0)$. If we apply the change of variables

$e^{i\psi} = b(e^{i\theta})$, induced by the linear fractional transformation $b(z) = (z-z_0)/(1-\bar{z}_0 z)$, we obtain the natural extension, from the point of view of conformal invariance, of this last equality to the other points $z_0 \in D$:

$$(1.3) \qquad \{\inf_{f\in a,\ f(z_0)=1} \int_{-\pi}^{\pi} (|f(e^{i\theta})|_{e^{i\theta}})^p P_{z_0}(\theta)d\theta\}^{1/p} =$$

$$= \exp \int_{-\pi}^{\pi} \{\log |1|_{e^{i\theta}}\} P_{z_0}(\theta)d\theta,$$

where $P_{z_0}(\theta)$ is the Poisson kernel. Thus, the left side of (1.3) defines the Banach space $B_{z_0} = (C,|\ |_{z_0})$ and the right side gives the explicit formula

$$(1.4) \qquad |1|_{z_0} = \exp \int_{-\pi}^{\pi} \{\log|1|_{e^{i\theta}}\} P_{z_0}(\theta)d\theta$$

for the norm of $1 \in B_{z_0}$. Observe that $|1|_{z_0}$, therefore, is defined independently of p. Furthermore, the function $z \to |1|_z$ is the absolute value of the analytic function

$$n(z) = \exp \int_{-\pi}^{\pi} \{\log |1|_{e^{i\theta}}\} h_z(e^{i\theta})d\theta,$$

where $h_z(e^{i\theta}) = \frac{1}{2\pi}(1 + ze^{-i\theta})/(1-ze^{-i\theta})$ is the Herglotz kernel. It follows, therefore, that if u is any "vector" in C^1 then $|u|_z = |u|\ |n(z)|$ and, if $u \neq 0$,

$$(1.5) \qquad \log |u|_z \quad \underline{\text{is harmonic in}} \quad D$$

More generally,

$$\log |f(z)|_z \quad \underline{\text{is subharmonic in}} \quad D$$

whenever f is analytic in D. If we write $A(z;z_0) = n(z_0)/n(z)$ for $z,z_0 \in D$ we have

$$(1.6) \qquad |A(z;z_0)u|_z = |u|_{z_0}$$

for $z \in D$. In fact, this equality extends to $z \in \partial D$ (a.e.) and it follows that the infimum on the left of (1.3) is attained when $f(z) = A(z;z_0)1$. Moreover, $|A(e^{i\theta};z_0)1|_{e^{i\theta}} = |n(z_0)| = |1|_{z_0}$ for almost all $\theta \in [-\pi,\pi)$ (the existence of the non-tangential limits $n(e^{i\theta}) = \lim_{z \rhd e^{i\theta}} n(z)$ follows from the classical H^p theory).

The most immediate extension of these observations to n dimensions involves C^n with weighted Euclidean norms: Suppose $w(e^{i\theta}) = (w_1(e^{i\theta}),\dots,w_n(e^{i\theta}))$ is a vector valued function on ∂D whose components are non-negative logarithmically integrable functions, we then have the Banach spaces $B_{e^{i\theta}} = (\mathbf{C}^n, |.|_{e^{i\theta}})$, where $|u|_{e^{i\theta}} = (\sum_{j=1}^{n} |u_j w_j(e^{i\theta})|^2)^{1/2}$ for $u = (u_1,u_2,\dots,u_n) \in C^n$. We can then define norms $|.|_z$ associated with the points $z \in D$ by introducing the analytic functions

$$n_j(z) = \exp \int_{-\pi}^{\pi} \{\log w_j(e^{i\theta})\} h_z(e^{i\theta})d\theta,$$

$j=1,2,\dots,n$, and define

$$|u|_z = (\sum_{j=1}^{n} |u_j n_j(z)|^2)^{1/2}.$$

It is no longer true that $\log|u|_z$ is harmonic; however

$$(1.7) \qquad \log|F(z)|_z \quad \underline{\text{is subharmonic}}$$

whenever F is a $\mathbf{C}^n$-valued analytic function on D. Also, it can be shown that for each $u \in \mathbf{C}^n$ and $z_0 \in D$

$$(1.8) \qquad \inf_{F \in \mathcal{a}^{(n)}, F(z_0)=u} \{\int_{-\pi}^{\pi} (|F(e^{i\theta})|_{e^{i\theta}})^p P_{z_0}(\theta)d\theta\}^{1/p} = |u|_{z_0},$$

where $\mathcal{a}^{(n)}$ is the class of all C^n-valued analytic functions on $\overline{D}$. In particular, the expression on the left of (1.8) is independent of p, $0 < p < \infty$. Moreover, equality (1.6) is valid if we put

$$A(z;z_0)u \equiv (u_1[n_1(z_0)/n_1(z)],\dots,u_n[n_n(z_0)/n_n(z)])$$

for $z,z_0 \in D$ and $F(z) = A(z;z_0)u$ is the extremal function for the expression on the left side in (1.8).

If we let $n(z)$ denote the diagonal matrix valued function with entry $n_j(z)$ in its jth row and column, we can write

$$(1.9) \qquad |u|_z = \|n(z)u\|$$

where $\|.\|$ denotes the Euclidean norm on $\mathbf{C}^n$, for each $u \in \mathbf{C}^n$. This leads us to a further extension of these notions to non-diagonal $n(z)$ that is connected with the theorem of Wiener-Masani:

THEOREM 1.10. (Wiener-Masani). *Suppose* $W(e^{i\theta})$ *is a positive definite* $n \times n$ *matrix valued function on* $[-\pi,\pi)$ *having integrable coefficients and such that* $\log \det W(e^{i\theta})$ *is integrable, then there exists an analytic invertible matrix-valued function* $n(z)$ *with coefficients in the Hardy space* $H^2(D)$ *such that*

$$W(e^{i\theta}) = n(e^{i\theta})^* n(e^{i\theta}) \tag{1.11}$$

almost everywhere, where $n(e^{i\theta})^*$ *is the adjoint of* $n(e^{i\theta})$. (1)

There are more general versions of this result, where the integrability of the coefficients of $W(e^{i\theta})$ can be weakened considerably (this will be apparent once we introduce the general theory). We pospone the discussion of such extensions and will restrict our attention to describing the features of this theorem that allow us to construct analytic families of Banach spaces. The positive definite matrix $W(e^{i\theta})$ defines the family of Banach spaces $B_{e^{i\theta}} = (\mathbb{C}^n, |\cdot|_{e^{i\theta}})$ associated with the points $e^{i\theta} \in \partial D$ via the norm

$$|u|_{e^{i\theta}} = \sqrt{W(e^{i\theta})u.\bar{u}} \tag{1.12}$$

for $u \in \mathbb{C}^n$. The Wiener-Masani theorem can then be used to define the norms $|\cdot|_z$ by equality (1.9). It can be shown that equality (1.8) is also valid and the operator $A(z;z_0) = n(z)^{-1}n(z_0) : \mathbb{C}^n \to \mathbb{C}^n$ satisfies the extensions of (1.6):

$$|A(z;z_0)u|_z = |u|_{z_0} \tag{1.13}$$

for all $u \in C^n$ and $z, z_0 \in D$. Moreover, $F(z) = A(z;z_0)u$ is the extremal function for the expression on the left side of (1.8).

There are several straighforward extensions of these ideas. For example, instead of the Euclidean norm in (1.9) we could have used the ℓ_q-norm on $\mathbb{C}^n$, $1 \leq q \leq \infty$: $\|u\|_q = (\sum_{j=1}^{n} |u_j|^q)^{1/q}$ (with the usual convention, $\|u\|_\infty = \max_j\{|u_j|\}$, when $q = \infty$). Even more

---

(1) Szegö's theorem (theorem (1.1)) is closely related to "another theorem of Szegö': *A non-negative function* $w(e^{i\theta})$ *on* $[-\pi,\pi)$ *is the (non-tangential) boundary value of* $|n(e^{i\theta})|^2$, *with* $n(e^{i\theta})$ *the boundary value of a function in* $H^2(D)$, *if and only if* w *and* $\log w$ *are integrable*. The Wiener-Masani theorem is clearly an n-dimensional extension of this result of Szegö.

generally, all this applies to general weighted L^q spaces on a measure space (M,μ). Weights $w(t;e^{i\theta})$, $t \in M$, can be assigned to the boundary points $e^{i\theta}$; the results we announced are valid if we assume the integrability on ∂D of $\log w(t;e^{i\theta})$ for each (or almost every) $t \in M$. The norms associated with the points $z \in D$ are the weighted L^q-norms

$$|f|_z = \left(\int_M |f(t)n(t;z)|^q \, d\mu(t)\right)^{1/q},$$

where

$$n(t;z) = \exp \int_{-\pi}^{\pi} \{\log w(t;e^{i\theta})\} h_z(e^{i\theta}) d\theta .$$

All these examples arise from a family of norms $\{|\cdot|_z\}$, $z \in D$, given by an analytic, invertible linear operator $T(z)$ on $\mathbf{C}^n$ and a (fixed) norm on $\mathbf{C}^n$. More precisely, if $|v|$, $v \in C^n$, in this fixed norm then

(1.14) $$|u|_z = |T(z)u|$$

for all $u \in \mathbf{C}^n$ (or, more generally, as indicated by the example involving $L^q(M,\mu)$, we can consider certain infinite dimensional function spaces instead of $\mathbf{C}^n$). The function $A(z;z_0) = T(z)^{-1}T(z_0)$ determines our analytic family completely via equality (1.6). In all these cases the function $A(.;z_0)$ and $A(z;.)$ are both linear and analytic, This linearity and the analyticity of the latter function, however, are not true for the general class of families $\{B_z\}$ we want to consider. In order to understand more fully the general case let us discuss the spaces associated with the Riesz-Thorin theorem.

The classical Riesz-Thorin theorem on interpolation of operators involves the following simple construction of an analytic family of L^p spaces ("intermediate spaces"): Let D be the strip $\{z = x + iy \in \mathbf{C} : 0 < x < 1\}$. To the boundary points $\xi = i\eta$, $-\infty < \eta < \infty$, we assign the space $B_\xi = L^{p_0} = L^{p_0}(M,\mu)$ and to the boundary points $\xi = 1 + i\eta$, $-\infty < \eta < \infty$, we assign the space $B_\xi = L^{p_1} = L^{p_1}(M,\mu)$, where (M,μ) is a measure space and $1 \leqq p_0$, $p_1 \leqq \infty$. The spaces B_z are, then, equal to $L^{p(z)} = L^{p(z)}(M,\mu)$, where, for $z = x + iy \in D$,

(1.15) $$\frac{1}{p(z)} = \frac{1 - x}{p_0} + \frac{x}{p_1} .$$

This function $1/p(z)$ is obviously harmonic in D. Thus, if $z_0 \in D$ there exists a unique analytic function $H(z)$ in D satisfying

$$\text{(1.16)} \qquad \operatorname{Re} H(z) = 1/p(z); \quad \operatorname{Im} H(z_0) = 0.$$

The basic idea in the proof of the Riesz-Thorin theorem is the introduction of the operator $A(z;z_0) : L^{p(z_0)} \to L^{p(z)}$ defined by

$$\text{(1.17)} \qquad A(z;z_0)f = \|f\|_{p(z_0)} \left| \frac{\|f\|}{\|f\|_{p(z_0)}} \right|^{p(z_0)H(z)} \frac{f}{|f|}.$$

Since, in this case, $|.|_z = \|.\|_{p(z)}$, an immediate calculation gives us relation (1.13) for these spaces:

$$\text{(1.18)} \qquad |A(z;z_0)f|_z = \|A(z;z_0)f\|_{p(z)} = \|f\|_{p(z_0)} = |f|_{z_0}.$$

The operator $A(.;z_0)$ is analytic; however, neither this operator nor $A(z;.)$ is linear and the latter is not analytic. Nevertheless, these spaces B_z enjoy many of the properties shared by those arising from an analytic linear operator $T(z)$ via equality (1.14). In particular, (1.18) and

$$\text{(1.19)} \qquad \log |F(z)|_z \quad \underline{\text{is subharmonic}}$$

whenever $F(z)$ is a holomorphic function with values in $L^{p_0} \cap L^{p_1}$, are the basic tools needed for the proof of the Riesz-Thorin theorem[(2)]. Indeed, suppose T is a linear operator on L^{p_j} with operator norm M_j, $j = 0,1$. Let $z_0 = t$, $0 < t < 1$; thus

$$\frac{1}{p(z_0)} = \frac{1}{p} = \frac{1-t}{p_0} + \frac{t}{p_1}.$$

If we put $F(z) = T(A(z;z_0)f)$, where $f \in L^{p_0} \cap L^{p_1}$ (say, f is simple), then the subharmonicity of $\log |F(z)|_z$ and <u>the three lines theorem</u> (which is a technical extension of the maximal principle) give us the inequality

$$\text{(1.20)} \qquad \log |F(t)|_t \leqq (1-t)\log M_0 \|f\|_p + t \log M_1 \|f\|_p$$

since, by hypothesis,

(2) We shall show how the subharmonicity (1.19) follows from the general theory we describe below. It is not immediately obvious in this particular situation.

$$\log |F(iy)|_{iy} = \log |T(A(iy;t)f)|_{iy} = \log \|T(A(iy,t)f\|_{p_o} \leqq$$

$$\leqq \log M_o \|A(iy;f)\|_{p_o} = \log M_o \|f\|_p$$

and, similarly,

$$\log |T(A(1+iy;t)f|_{1+iy} \leqq \log M_1 \|f\|_p .$$

Since $F(t) = Tf$, exponentiating both sides of (1.20) we obtain

$$\|Tf\|_p \leqq M_o^{1-t} M_1^t \|f\|_p . \tag{1.21}$$

But this gives us the conclusion of the Riesz-Thorin theorem: T <u>is a bounded operator on</u> L^p <u>with operator norm not exceeding</u> $M_o^{1-t} M_1^t$.(3)

2. THE CONSTRUCTION OF ANALYTIC FAMILIES OF BANACH SPACES

We shall now give a fairly detailed description of the methods used to obtain general families $\{B_z\}$ from appropriate boundary data $\{B_\xi\}$. Let us assume that D is a bounded domain whose boundary ∂D is a simple closed analytic curve (the theory can be developed for less smooth boundaries). The Hardy space $H^p(D)$, $0 < p \leqq \infty$, can be defined to be the collection of all those analytic f on D such that $|f|^p$ has a harmonic majorant on D. It is well known that the classical theory on the disk extends to this case (we have the existence of non-tangential boundary values, the Blaschke product factorization, etc.). Associated with D there exists a Herglotz kernel $h_z(\xi)$. Suppose that for each $\xi \in \partial D$ we have a norm on $\mathbf{C}^n$ such that $\xi \to |u|_\xi$ is a measurable function for each $u \in \mathbf{C}^n$ and

$$k_1(\xi) \|u\| \leqq |u|_\xi \leqq k_2(\xi) \|u\| \tag{2.1}$$

(3) Actually $F(z)$ is analytic in $\bar{D}$; moreover, since f is simple, F is bounded on $\bar{D}$. Thus, the three lines theorem is clearly applicable. We also should point out that this argument extends to the more general setting in which T maps L^{p_o} (or L^{p_1}) into <u>another</u> Lebesgue space L^{q_o} (or L^{q_1}).

for all $u \in \mathbb{C}^n$, where $\log k_j(\xi)$, $j=1,2$, is integrable on ∂D. The $k_j(\xi)$ can, at first reading, be thought of as constant. Let

$$W_j(z) = \exp \int_{\partial D} h_z(\xi) \log k_j(\xi)\, ds(\xi),$$

where $j=1,2$ and $ds(\xi)$ denotes the element of arc length on ∂D. The functions $\log W_j$ belong to $H^p(D)$ for $0 < p < 1$ and, in particular,

$$\text{(2)} \qquad \lim_{z \rhd \xi} |W_j(z)| = k_j(\xi)$$

a.e. on ∂D, where $z \rhd \xi$ denotes the general non-tangential approach of $z \in D$ to the boundary point ξ. We now introduce the spaces $H_j^p = H_j^p(D;\mathbb{C}^n)$ for all $\mathbb{C}^n$-valued analytic functions $F = (f_1,\ldots,f_n)$ on D such that $W_j f_k \in H^p(D)$ for $k=1,2,\ldots,n$. $j=1,2$. Since the non-tangential limits (2.2) are positive a.e. it follows that

$$\lim_{z \rhd \xi} F(z) = F(\xi)$$

exists a.e. on ∂D. We can, therefore, introduce, for $0 < p \leqq \infty$, the spaces

$$H_\#^p = \{F \in H_1^p : ||F||_{z_o,p} = (\int_{\partial D} (|F(\xi)|_\xi)^p p_{z_o}(\xi) ds(\xi))^{1/p} < \infty\},$$

where $p_{z_o}(\xi) = \mathrm{Re}\{h_{z_o}(\xi)\}$ is the Poisson kernel(4). This space $H_\#^p$ is basic for the introduction of the norms (on $\mathbb{C}^n$) of the "intermediate spaces" B_{z_o}, $z_o \in D$: If $v \in \mathbb{C}^n$ and $E = E_{z_o,p}^v = \{F \in H_\#^p : F(z_o) = v\}$ we let

$$|v|_{z_o,p} = \inf_{F \in E} \{||F||_{z_o,p}\}.$$

(4) The spaces H_j^p are introduced for technical reasons. The condition $||F||_{z_o,p} < \infty$ alone is not sufficient for guaranteing that the size of the analytic function on the boundary (even when the boundary values exist) determines the size of F in D. For example, $F(z) = \exp\{\frac{z+1}{z-1}\}$, $|z| \leqq 1$, has modulus 1 of the unit disk but is not in H^∞ (nor any other H^p). The restriction $f \in H_1^p$ gives us the needed control for developing our theory. Also note that the index z_o is not needed in $||.||_{z_o,\infty}$ since $P_{z_o}(\xi)ds(\xi)$ and $ds(\xi)$ are absolutely continuous with respect to each other.

The first principal result in our theory is

THEOREM 1. *Suppose* $v \in \mathbb{C}^n$, $z_0 \in D$ *and* $0 < p \leqq \infty$. *Then* $|v|_{z_0,p}$ *is independent of* p; *in fact,*

$$|v|_{z_0,p} = \inf \{ \| G \|_\infty : G \in E, \ |F(\xi)|_\xi = \text{const. a.e.} \}.$$

Furthermore, there exists an extremal function $F = F_{z_0,v} \in H^\infty_\# \subset H^p_\#$ *such that* $F(z_0) = v$ *and* $|v|_{z_0} = |v|_{z_0,p} = |F(\xi)|_\xi$ *a.e.*

The analytic family $\{B_z\}$, $z \in D$, where B_z is $\mathbb{C}^n$ normed by $|.|_z$ is, then, the one that is determined by the boundary data $\{B_\xi\} = \{(C^n, |.|_\xi)\}$, $\xi \in \partial D$. We shall denote this relation of B_z, $z \in D$, to the boundary spaces $\{B_\xi\}$, $\xi \in \partial D$, by writing $B_z = [B_\xi]_z$. The spaces B_z introduced in the first section are all examples of such families.

The second principal result in our theory is the "duality theorem". If N is a norm on C^n, $N^*(w) = \sup\{|<v,w>| : N(v) = 1\}$, where $<v,w> = \sum_{j=1}^{n} v_j w_j$, defines the <u>dual norm</u> to N, N^*. The dual $(\mathbb{C}^n, N)^*$ of the Banach $(\mathbb{C}^n, N)$ is then represented by the linear functionals $L_w : v \to <v,w>$ and the norm of L_w is $N^*(w)$. Hence, we write $(\mathbb{C}^n, N)^* = (\mathbb{C}^n, N^*)$. Moreover, from (2.1) we obtain the fact that the dual boundary spaces $B^*_\xi = (\mathbb{C}^n, |.|^*_\xi)$ satisfy

$$\|v\|(1/k_2(\xi)) \leqq |v|^*_\xi \leqq \|v\|(1/k_1(\xi)) \tag{2.3}$$

for all $v \in \mathbb{C}^n$. Thus, since the logarithmic integrability of k_1 and k_2 is the only assumption needed to obtain theorem I, we see that this theorem can be applied to the family of duals $B^*_\xi = (\mathbb{C}^n, |.|^*_\xi)$. It is natural to inquire into the relationship between $[B^*_\xi]$ and $[B_\xi]^*_z$:

THEOREM 2. (The Duality Theorem). $[B^*_\xi]_z = [B_\xi]^*_z$ *for all* $z \in D$.

From theorem 1 and 2 one can derive fairly easily many properties of such families of Banach spaces that extend the relations we described in the special cases presented in the first section. In particular, we shall show how an operator $A(z;z_0) : \mathbb{C}^n \to \mathbb{C}^n$ satisfying (1.6) (or (1.13)) completely characterizes these families.

It follows immediately from theorem 1 that

$|v|_{z_0} \leqq |W_2(z_0)|\ \|v\|$ for each $z_0 \in D$ and $v \in \mathbf{C}^n$. Using (2.3) and applying this inequality to the duals we obtain

$$(2.4) \qquad |W_1(z_0)|\ \|v\| \leqq |v|_{z_0} \leqq |W_2(z_0)|\ \|v\|$$

for all $z_0 \in D$ and $v \in \mathbf{C}^n$. Suppose that γ is a simple closed analytic curve within D. Inequalities (2.4) applied to the points $z_0 = \xi \in \gamma$ guarantee that we can carry out the construction in theorem 1 to obtain the intermediate spaces $\{[B_\xi]_z : \xi \in \gamma,\ z \in \Omega\}$, where $\Omega \subset D$ is the domain bounded by γ. It is clear that the norm of B_z is greater than or equal to the norm of $[B_\xi]_z$, $\xi \in \gamma$, whenever $z \in \Omega$. Since the same relation must also be true for the duals, we obtain

<u>Theorem 2.5. (The Iteration Theorem).</u> *If* $\Omega \subset D$ *is a domain of the type just described, then* $[B_\xi]_z = B_z$ *for all* $z \in \Omega$ *(where* ξ *denotes the general point of* $\partial\Omega = \gamma$*).*

This result can be regarded as "evidence" of the fact that the family $\{B_z\}$, $z \in D$, represents the solution of the Dirichlet problem on D with boundary data $\{B_\xi\}$, $\xi \in \partial D$. An "analytic" aspect of these families is the following extension of (1.19):

<u>THEOREM 2.6.</u> *If* F *is a* $\mathbf{C}^n$-valued *analytic function on* D *then* $\log |F(z)|_z$ *is subharmonic.*

This last result is an immediate consequence of the iteration theorem. In fact, if the closed disk about z_0 of radius $r > 0$ is contained in D and we let $w(\theta) = z_0 + re^{i\theta^o}$, then it follows from (2.5) that, if $p > 0$,

$$|F(z_0)|_{z_0} \leqq \left(\int_0^{2\pi} \{|F(w(\theta))|_{w(\theta)}\}^p \frac{d\theta}{2\pi}\right)^{1/p}$$

whenever F is analytic $\mathbf{C}^n$-valued function. Letting p tend to 0 we obtain

$$(2.7) \qquad |F(z_0)|_{z_0} \leqq \exp\left(\int_0^{2\pi} \{\log |F(w(\theta)|_{w(\theta)}\} \frac{d\theta}{2\pi}\right)$$

and theorem (2.6) is established.

A corollary of (2.6) is the following characterization of the extremal function in theorem 1:

THEOREM 2.8. *If F is an extremal function corresponding to $z_0 \in D$ and $v \in \mathbb{C}^n$ then $|F(z)|_z = |v|_{z_o}$ for all $z \in D$. Conversely, if and analytic $\mathbb{C}^n$-valued F satisfies $|F(z)|_z$ = const. for all $a \in D$, then it is an extremal function for each $z_0 \in D$ and vector $F(z_0)$.*

The extremal function $F_{z_0,v}$ of theorem 1 is unique in most situations: this is the case when the unit spheres of the spaces B_ξ are strictly convex for ξ in a subset of ∂D of positive measure. When this is so we introduce the notation

$$A(z;z_0)v \equiv F_{z_0,v}(z)$$

for $z, z_0 \in D$ and $v \in \mathbb{C}^n$. Thus, for each $z_0 \in D$ we have a mapping $A(z;z_0) : \mathbb{C}^n \to \mathbb{C}^n$ such that $A(.;z_0)v$ is the extremal function associated with z_0 and v. The following is a list of the basic properties of this mapping:

(i) $A(z;z_0)v$ is an analytic function of $z \in D$ for each $z_0 \in D$, $v \in \mathbb{C}^n$;

(ii) $A(z;z) = I$ = identity operator, for all $z \in D$;

(iii) (The propagator equation) $A(z;z_0)A(z_0;z_1) = A(z;z_1)$ for all $z, z_0, z_1 \in D$;

(iv) $A(z;z_0)$ maps $\mathbb{C}^n$ one-to-one continuously onto $\mathbb{C}^n$ and $A(z_0;z)$ is both the left and right inverse of $A(z;z_0)$;

(v) $A(z;z_0)(\lambda v) = \lambda A(z;z_0)v$ for all $\lambda \in \mathbb{C}$, $v \in \mathbb{C}^n$;

(vi) Let $T(z) = A(z_0;z)$ for z_0 a fixed point of D and put $|u| \equiv |u|_{z_0}$ for each $u \in \mathbb{C}^n$; then $T(z)$ is a sublinear operator (with respect to the norm $|.|$) such that

$$|u|_z = |T(z)u|$$

for all $u \in \mathbb{C}^n$ and $z \in D$.

From these properties we see to what extent the general case is similar to the especial cases described in the first section. In particular, the norms $|\ |_z$ are obtained from a fixed norm and an operator T, as in (1.14) (see (vi)); however, T is neither analytic nor linear in the general case. These six properties are particularly useful for developing our theory further (as we shall see in the sequel); in particular, the mapping $A(z;z_0)$, together with its analog, $A_*(z;z_0)$, associated with the family of duals $\{B_z^*\}$, can be used to show

$$(2.9) \qquad \lim_{z \rhd \xi} |u|_z = |u|_\xi \quad \text{a.e. for} \quad \xi \in \partial D$$

for all $u \in \mathbf{C}^n$.

In many applications of this theory of analytic families of Banach spaces (such as their use for interpolation of operators) it suffices to consider the finite dimensional cases we have described. This is the case when estimates that are independent of the dimension easily imply general inequalities. There are situations, however, where one does need the theory we are considering that is associated with infinite-dimensional Banach spaces. This theory is technically more complicated than the finite dimensional case we just described. We shall not discuss it here in any detail; however, it does enjoy most of the basic features we presented here in the case of C^n with varying norms. In many situations one can exhibit an appropriate operator $A(z;z_0)$ which satisfies properties (i) - (vi). We did this in the case of arbitrary L^p-spaces in formula (1.17). In general, the reader should not find major difficulties in following subsequent discussions, even when infinite-dimensional analytic families of Banach spaces are involved.

Finally, we shall make some observations about the nature of the boundary valued problem that has been solved by theorems 1 and 2. One way to describe the norms of the spaces $\{B_\xi\}$, $\xi \in \partial D$, is with the so-called <u>duality</u> maps J_ξ. These are the maps $J_\xi : \mathbf{C}^n \to \mathbf{C}^n$ for which

$$|u|_\xi^2 = (|J_\xi u|_\xi^*)^2 = \langle u, J_\xi u\rangle$$

(with appropriate convexity assumptions on $|.|_\xi$, this equality determines J_ξ completely). If $|.|_\xi = |.|$ (the Euclidean norm) then $J_\xi u = \bar{u}$. If $|.|_\xi$ is given by (1.12) then $J_\xi u = \overline{W(\xi)u}$. If $|.|_\xi$ is not a Hilbert space norm then J_ξ is not conjugate linear (though it is conjugate homogeneous). The occurence of complex conjugation in these examples suggests that it might be difficult to find holomorphic vector valued functions $A(z)$, $z \in D$, so that for $\xi \in \partial D$

(2.10) <u>both</u> $A(\xi)$ <u>and</u> $J_\xi(A(\xi))$ <u>are boundary values of holomorphic functions</u>.

The functions $A(z) = A(z;z_0)v$, $v \in \mathbb{C}^n$, $z_0 \in D$, satisfy (2.10) and are the only functions which do. In fact, solving the "novel" boundary value problem (2.10), for all possible choices of $A(z_0)$, is equialent to the construction of the spaces $\{B_z\}$, z in D, via theorem 1. Note that if $A_*(z;z_0)$ is the operator satisfying (i) - (vi) for the dual family $\{B_z^*\}$ then

$$(2.11) \qquad J_z A(z;z_0) = A_*(z;z_0) J_{z_0} .$$

This equivalence is established by using (2.11) as well as theorems 1 and 2. To go from the solution of (2.10) to the definition of the spaces $\{B_z\}$ we define J_z in D by requiring that $J_z A(z)$ be the analytic function with boundary values $J_\xi A(\xi)$. This specifies, J_z and, hence, $|.|_z$. This interpretation of the construction of analytic families of Banach spaces as solving a Dirichlet problem for J_ξ can be used as a starting point for obtaining the differential equation (5.5) satisfied by the norm $|.|_z$ that is discussed later in Section 5.

3. ANALYTIC FAMILIES OF BANACH SPACES AND THE INTERPOLATION OF OPERATORS

As we said at the beginning of this article, our original motivation for this study was to extend the complex method of interpolation developed by Calderón and Lions. One familiar with this method can see that some of the ideas developed in the first two sections are natural extensions of those introduced by these two authors. We already described how the Riesz-Thorin theorem is associated with certain classes of analytic families of Banach spaces. We shall now discuss the general problem of interpolation of operators associated with these families.

Suppose we are given two of our families of Banach spaces, $\{B_z\}$ and $\{C_z\}$, defined for z in a domain D. Let $\{T_z\}$ be a family of (linear) operators mappings B_z into C_z with operator norm $N(z)$. We assume that the mapping $z \to T_z$ is analytic (in the $\mathbb{C}^n$-case this means $z \to T_z v$ is analytic for each $v \in \mathbb{C}^n$; in the infinite dimensional case we have to assume that all the spaces $\{B_z\}$, and $\{C_z\}$, contain common dense subspaces and analyticity is defined in terms of an appropriate topology on these subspaces). We then have

THEOREM 3.1. (The Interpolation Theorem). $\log N(z)$ *is a subharmonic function on* D.

Before indicating the simple argument that establishes this theorem let us make some observations about it. First, it is immediate that the Riesz-Thorin theorem is a very special case of (3.1). Indeed, given $1 \leqq p_0, p_1 \leqq \infty$ and $1 \leqq q_0, q_1 \leqq \infty$, let D be the unit disk and $1/p(z)$, $1/q(z)$ the solutions of the Dirichlet problem with boundary data

$$\frac{1}{p(e^{i\theta})} = \begin{cases} \frac{1}{p_0} & \text{if } -\pi \leqq \theta \leqq \pi(1-2t) \\ \frac{1}{p_1} & \text{if } \pi(1-2t) < \theta < \pi \end{cases}, \qquad \frac{1}{q(e^{i\theta})} = \begin{cases} \frac{1}{q_0} & \text{if } -\pi \leqq \theta \leqq \pi(1-2t) \\ \frac{1}{q_1} & \text{if } \pi(1-2t) < \theta < \pi \end{cases},$$

where $0 < t < 1$. Then, the mean-value inequality

$$\log N(0) \leqq \frac{1}{2\pi} \int_0^{2\pi} \log N(e^{i\theta})d\theta = (1-t)\log M_0 + t \log M_1$$

must be satisfied, since $\log N(z)$ is subharmonic, whenever $T_z \equiv T$ is an operator mapping L^{p_j} with norm M_j, $j=0,1$. But this means that

$$\|Tf\|_q = M_0^{1-t} \, M_1^t \|f\|_p \tag{3.2}$$

for all $f \in L^p$, where $\frac{1}{p} = \frac{1}{p(0)} = \frac{1-t}{p_0} + \frac{t}{p_1}$ and $\frac{1}{q} = \frac{1}{q(0)} = \frac{1-t}{q_0} + \frac{t}{q_1}$, but this is the precise conclusion of the Riesz-Thorin theorem (in fact, (3.2) extends (1.21)).

Essentially the same considerations show that E.M. Stein's theorem on interpolation of analytic families of operators is as immediate consequence of theorem (3.1). This can be seen by using the strip $\overline{D}$ $\{z = x+iy : 0 \leqq x \leqq 1\}$ as the domain on which the spaces and $\{T_z\}$ is defined; Stein's condition of admissible growth is precisely the one needed to construct the least harmonic majorant of the subharmonic function $\log N(z)$, which is used to obtain the bounds for the norm of the operator T_t, $0 \leqq t \leqq 1$.

The usual problem posed in the theory of interpolation of operators is the following one: Suppose T is defined on two Banach spaces B_0 and B_1 and maps these continuously into two Banach spaces C_0 and C_1, can one then construct intermediate spaces $B_t = [B_0, B_1]_t$ and $C_t = [C_0, C_1]_t$, for $0 \leqq t \leqq 1$, in such a way that $T : B_t \to C_t$ is a bounded operator? One of the novelties introduced in our theory is that, instead of the pairs of "end point" spaces (B_0, B_1) and (C_0, C_1) we are now dealing with a continuum of Banach spaces on the boundary of D for both the domain and the range of our operators. It is important to point out that this last situation is most natural in the application of our theory. Many operators $\{T_z\}$ arising in analysis vary analytically in a domain D and map spaces that vary with z into other spaces that vary with $z \in D$; this dependence on z is precisely described by our general theory. For example, the Riesz potential operators

$$(I_z f)(x) = \frac{1}{\gamma(z)} \int_{\mathbb{R}^n} |x-y|^{-n+z} f(y)dy$$

map varying families of Lebesgue (or Sobolev) spaces into (other) Lebesgue (or Sobolev) spaces, where z is restricted to a strip $0 < Re(z) < n$, and the indices p and q of the domain and range spaces are also related by $z(1/q - 1/p = R(z/n))$. Similarly, the Bessel potentials exhibit such features. Moreover, these same operators act naturally on families of Lipschitz spaces. The Fourier transform can be studied by considering the family of operators $T_z : H_n \to z^n H_n$, where the H_n's are the Hermite polynomials and z belongs to the closed unit disk. It was in this context that Beckner obtained the best constants in the Hausdorff-Young inequality for the Fourier transform.

Once it is established that boundedness of an operator at the "end-points" implies boundedness on the intermediate spaces, the next important step in the theory of interpolation is to identify these intermediate spaces. The well known fact that Lebesgue spaces have intermediate spaces that are also Lebesgue spaces extends to a large class of spaces that are commonly used in analysis: The intermediate spaces of Lorentz spaces are Lorentz spaces, of Sobolov spaces are Sobolov spaces, of Besov-Lipschitz space are Besov-Lipschitz spaces, etc. A.P. Calderón developed a theory of interpolation of Banach lattices that enables one to obtain many of these identifications. This theory of Calderón's can be extended to the analytic families of Banach spaces we have presented. Thus, if the boundary spaces belong to certain classes (such as the Lorentz spaces) so do the interior spaces.

The proof of (3.1) is very easy. In fact, let $z_o \in D$ and suppose that $\{z : |z-z_o| \leqq \rho\} \subset D$ and let us write $w(\theta) = z_o + e^{i\theta}$. For simplicity, let us suppose $\{B_z\} = \{C_z\}$ and let $A_*(z;z_o)$ denote the operator we introduced in the last section that satisfies (i) - (vi) for the dual family $\{B_z^*\}$. We now select $u \in B_z$ and $v \in B_z^*$ such that $|u|_{z_o} = 1 = |v|_{z_o}^*$. Thus, from (1.13) (or property (vi))

$$(3.3) \qquad |A(z;z_o)u|_z = |u|_{z_o} = 1 = |v|_{z_o}^* = |A_*(z;z_o)v|_z^*$$

for all $z \in D$. Since the function $\Phi(z) = \langle T_z A(z;z_o)u, A_*(z;z_o)v\rangle$ is analytic in z we must have

$$\log | T_{z_o} u,v | = \log |\Phi(z_o)| \leqq \int_{-\pi}^{\pi} \log |\Phi(w(\theta))| \frac{d\theta}{2\pi}$$

$$= \int_{-\pi}^{\pi} \log |\langle T_{w(\theta)} A(w(\theta);\ z_o)u,\ A_*(w(\theta);\ z_o)v\rangle \frac{d\theta}{2\pi}$$

But, because of (3.3),

$$|\langle T_{w(\theta)} A(w(\theta);z_o)u,\ A_*(w(\theta;z_o)v\rangle| \leqq$$

$$\leqq (w(\theta))|A(w(\theta);z_o)u|_{w(\theta)} |A_*(w(\theta);z_o)w|_{w(\theta)} = N(w(\theta)).$$

Consequently,

$$\log |\langle T_{z_o} u,v\rangle| \leqq \int_{-\pi}^{\pi} \log N(w(\theta)) \frac{d\theta}{2\pi}.$$

If we now take the supremum of the left side as u,v range over the surfaces of the unit spheres of B_{z_o} and $B_{z_o}^*$ we obtain

$$(3.4) \qquad \log \ (z_o) \leqq \int_{-\pi}^{\pi} \log N(w(\theta)) \frac{d\theta}{2\pi}.$$

This proves theorem (3.1).

4. DERIVATIVES OF ANALYTIC FAMILIES OF BANACH SPACES

The maximum principle has played a prominent role in the development of the theory we presented. We shall now show that the properties of derivatives of holomorphic functions can be exploited to obtain new results associated with the interpolation of operators and analytic families of Banach spaces. We shall see that operators that act boundedly on such parametrized families of spaces have additional features that reflect the interaction between the operator and the changing structure of such spaces.

Suppose we have an operator T on the family of boundary spaces B_ξ, $\xi \in \partial D$. If, say, the norms $N(\xi)$, $\xi \in \partial D$, of T acting on B_ξ, are bounded by 1, then, the maximum principle applied to the function $\Phi(z)$, introduced at the end of the last section, implies that the norms $N(z)$, of T acting on B_z, are bounded by 1 for all $z \in D$. But, if $z_0 \in D$ we can also estimate the size of $\Phi'(z_0)$ by means of the Cauchy formula

$$(4.1) \qquad \Phi'(z_0) = \frac{1}{2\pi i}\int_\gamma \Phi(z)(z_0-z)^{-2}\,dz$$

where γ is the circle of radius $\rho = \text{dist}(z_0, \partial D)$ about z_0. In fact, if $|\Phi(z)|$ is bounded by 1 in D we see from (4.1) that $|\Phi'(z_0)| \leqq \rho^{-1} = 1/\text{dist}(z_0, \partial D)$. It we let $A'(z_0) = \frac{\partial}{\partial z} A(z;z_0)|_{z_0}$ and $A_*'(z_0) = \frac{\partial}{\partial z} A_*(z;z_0)$, then we see that

$$(4.2) \qquad \Phi'(z_0) = \langle TA'(z_0)u,v\rangle + \langle Tu,A_*'(z_0)v\rangle .$$

We can perform the same type of analysis for the function $\Psi(z) = \langle A(z;z_0)Tu, A_*(z;z_0)v\rangle$. First observe that
$|\Psi(z)| \leqq |A(z;z_0)Tu|_z\ |A_*(z;z_0)v|_z^* = |Tu|_{z_0} |v|_{z_0}^* \leqq N(z_0)|u|_{z_0} |v|_{z_0}^* \leqq 1$.
Using this estimate and the Cauchy formula (4.1) for $\Psi'(z_0)$ instead of $\Phi'(z_0)$ we obtain $|\Psi'(z_0)| \leqq \rho^{-1}$. Since

$$(4.3) \qquad \Psi'(z_0) = \langle A'(z_0)Tu,v\rangle + \langle Tu,A_*'(z_0)v\rangle$$

we can subtract this last expression from (4.2), and use these estimates for $|\Phi'(z_0)|$ and $|\Psi'(z_0)|$, to obtain

$$|\langle TA'(z_0)u-A'(z_0)Tu,v\rangle| \leqq 2/\text{dist}(z_0, \partial D).$$

If we now take the supremum of the expressions on the left over all $v \in B_{z_0}^*$ satisfying $|v|_{z_0}^* = 1$ we obtain

$$|(TA'(z_0) - A'(z_0)T)u|_{z_0} \leqq 2/\text{dist}(z_0, D)$$

whenever $|u|_{z_0} = 1$. Writing $[T,A'(z_0)] = TA'(z_0) - A'(z_0)T$ and using the homogeneity of the operators T and $A'(z_0)$ we see that

$$(4.4)\qquad |[T,A'(z_0)]u|_{z_0} \leqq \frac{2}{\mathrm{dist}(z_0,\partial D)}|u|_{z_0}.$$

That is, the commutator $[T,A'(z_0)]$ is a bounded operator on B_{z_0} with norm not exceeding $2/\mathrm{dist}(z_0,\partial D)$.

It is natural to ask what form this commutator takes in specific cases. In the L^p-cases we can compute the derivative $A'(z_0)$ explicitly and (4.4) then gives us the boundedness of $[T,L]$, where L is the operator defined by

$$Lf = f \log |f|$$

for f measurable. The precise inequality is

$$(4.5)\qquad \|[T,L]f\|_{p(z_0)} \leqq c(z_0)\|f\|_{p(z_0)},$$

where $c(z_0) = cp(z_0)|H'(z_0)|/\mathrm{dist}(z_0,\partial D)$. If the spaces B_z are weighted L^p spaces on $\mathbb{R}^n$, with p fixed and weights $\omega(z;x) = e^{b(z;x)}$, $z \in D$, $x \in \mathbb{R}^n$, then the operator $A'(z_0)$ is simply multiplication by $\frac{d}{dz}b(z;x)|_{z=z_0}$. An important special case arises when $b(z;x) = zb(x)$ with $b \in BMO$ having small BMO-norm $\|b\|_*$ (so that $\omega(x) = e^{b(x)}$ and $\omega(x)^{-1}$ belong to the Muckenhoupt class A_p). Then, if T is a <u>Calderón-Zygmund singular integral operator</u> we obtain the L^p-boundedness of the commutator $T(bf) - bTf$. By homogeneity we then obtain the inequality

$$(4.6)\qquad \|T(bf) - bTf\|_p \leqq c\|f\|_p\|b\|_*$$

for all $f \in L^p(\mathbb{R}^n)$ and $b \in BMO$. It can be shown that an inequality of the type (4.5) is <u>not</u> true for $p(z_0) = 2$ when T is the disk multiplier on $L^2(\mathbb{R}^2)$ (which is known to be bounded on $L^p(\mathbb{R}^2)$ if and only if $p = 2$). Thus, these results are characteristic of operators that are bounded on a class of spaces that lie in a "neighborhood of B_{z_0}".

All these results can be carried out for operators T mapping analytic families $\{B_z\}$ into <u>different</u> analytic families $\{C_z\}$. A typical example of a result obtained this way is furnished by the Riesz potentials $T_\gamma f = f*|x|^{\gamma-n}$:

$$(4.7)\qquad \|T_\gamma(bf) - b(T_\gamma f)\|_{p_2} \leqq M\|f\|_{p_1}\|b\|_*$$

when $0 < \gamma < n$, $1 < p_1 < \frac{n}{\gamma}$, $\frac{1}{p_2} = \frac{1}{p_1} - \frac{\gamma}{n}$ and $b \in BMO$.

Similar analysis, starting from the fact that smooth changes of variables preserve the Sobolev spaces, show that the commutator of a smooth change of variables with the logarithm of the differentiation operator (defined via the Fourier transform) is bounded on $L^2(R)$.

One can make a systematic study of such inequalities by introducing a class of spaces that allows us to study the operators $A'(z_0)$. It is clear that the space B_{z_0} is naturally isomorphic to the quotient space $H^\infty_\#/(z-z_0)H^\infty_\#$ and the norm on B_{z_0} defined by Theorem 1 corresponds precisely to the coset norm on this quotient space. The study of $A'(z_0)$ requires specifying the value of the derivative of an extremal function at z_0, as well as its value at this point. Thus, we are led to the consideration of the spaces $B^{(2)}_{z_o} = H^\infty_\#/(z-z_0)^2H^\infty_\#$, $z_0 \in D$, normed by the coset norm on this quotient space. Each coset is uniquely determined by an ordered pair $(u,v) \in C^n \times C^n$, where $F(z_0) = u$ and $F'(z_0)(1 - |z_0|^2) = v$. It is not hard to see that the coset norm is given by

$$(4.8) \qquad |(u,v)|^{(2)}_{z_o} = \inf\{\|F\|_\infty : F \in H^\infty_\#,\ F(z_0) = u,\ F'(z_0)(1 - |z_0|^2) = v\}.$$

One can then study the family $\{B^{(2)}_z\}$ much in the same spirit of the development of the properties of $\{B_z\}$ we described in Section 2. In particular, a duality result can be shown that tells us that the duals of the space $B^{(2)}_z$ for $z \in D$ can be obtained from the duals of the boundary spaces, B^*_ξ, $\xi \in \partial D$, in analogy with theorem 2. Moreover, this general theory allows us to include analytic families of operators $\{T_z\}$ acting on the spaces $B^{(2)}_z$.

5. Partial Differential Equations and Differential Geometric Features Associated With Analytic Families of Banach Spaces

We have, by now, encountered several situations that illustrated the fundamental role played by the operator $A(z;z_0)$ in our theory. In the first section we first considered the 1-dimensional case where $\log |u|_z = \log |u| + \log |n(z)|$ is a harmonic function for each $u \in C^1$. In the n-dimensional Hilbert space case it is not true that $\log |u|_z = \log \|n(z)u\|$ is harmonic; however, as indicated by the Wiener-Masani theorem (1.10), $n(z)$ is an analytic matrix-valued function. It is not hard to see that this fact can be used to obtain a partial differential equation, that extends the Laplace equation, which characterizes these families of inner product spaces.

As we have seen, in the n-dimensional inner product case, the norms of the spaces B_z are given by

$$(5.1)\qquad |u|_z = \|n(z)u\| = \sqrt{W(z)u.\bar{u}}\quad ,$$

where $W(z) = n(z)^* n(z)$ with $n(z)$ analytic and invertible. Let $\frac{d}{dz} = \frac{d}{dx} - i\,\frac{d}{dy}$ and $\frac{d}{d\bar{z}} = \frac{d}{dx} + i\,\frac{d}{dy}$. Then, since $n(z)^*$ is conjugate analytic, $\frac{dW(z)}{dz} = n(z)^*\,\frac{dn(z)}{dz}$. Hence

$$W(z)^{-1}\,\frac{dW(z)}{dz} = n(z)^{-1} n(z)^{*-1} n(z)^*\,\frac{dn(z)}{dz} = n(z)\,\frac{dn(z)}{dz}.$$

But the last matrix-valued function is analytic; thus,

$$(5.2)\qquad \frac{d}{d\bar{z}}\,\{W(z)^{-1}\,\frac{d}{dz}\,W(z)\} = 0 \quad \text{or, equivalently,}$$

$$\Delta W - (\frac{d}{d\bar{z}}\,W)W^{-1}\,(\frac{d}{dz}\,W) = 0$$

for all $z \in D$. It the family of matrices $\{n(z) : z \in D\}$ is a commuting, normal family, then there exists a common spectral decomposition, for the members $n(z)$ and, in particular, we can find an analytic logarithm $\log W(z)$, for $z \in D$. It follows that

$$\frac{d}{dz}\log W(z) = W(z)^{-1}\,\frac{d}{dz}\,W(z).$$

When this is the case, therefore, (5.2) becomes

$$\Delta \log W(z) = \frac{d}{d\bar{z}}\,\frac{d}{dz}\log W(z) = \frac{d}{d\bar{z}}W(z)^{-1}\,\frac{d}{dz}\,W(z) = 0$$

for all $z \in D$. Thus, we can consider (5.2) to be an extension of the Laplace equation that is satisfied by the matrix-valued function $W(z)$.

As is the case in the Laplace equation, the solutions of (5.2) satisfy a maximum principle. This maximum principle asserts that for the norms defined by (5.1) we have

$$(5.3)\qquad \log |F(z)|_z \ \ \underline{\text{is subharmonic}}$$

whenever F is an analytic $\mathbb{C}^n$-valued function on D. Since $\overline{W}^{-1}$ satisfies (5.2) when W does, we also have the same maximum principle for the norms defined, as in (5.1), in terms of the positive definite operators $\overline{W}^{-1}$ (instead of $W(z)$). A duality argument then shows that $\{B_z\} = \{(C^n, \sqrt{W(z)u.\bar{u}})\}$ is the analytic family of Banach spaces determined, via Theorem 1, by the boundary norms $|u|_\xi = \sqrt{W(\xi)u.\bar{u}}$, $\xi \in \partial D$.

There is a geometric interpretation of these notions. We can regard $B = \bigcup_{z \in D} \{B_z\}$ as a complex vector bundle with base manifold D. For the general analytic family $\{B_z\} = \{(C^n, |\ |_z)\}$ put

$$(5.4) \qquad G(z,u) = |u|_z^2$$

for $z \in D$, $u \in \mathbf{C}^n$. Ignoring questions of smoothness, it can be shown that the condition

$$(5.5) \qquad \Delta G - \left(\frac{d}{dz} \nabla_u G\right)^* \left[\left(\frac{\partial}{\partial u_i} \frac{\partial}{\partial \bar{u}_j} G\right)\right]^{-1} \left(\frac{d}{dz} \nabla_u G\right) = 0$$

is equivalent to the property that the norms (5.4) are those of an analytic family of Banach spaces. The expression on the left in (5.5) is related to the <u>curvature</u> of B. If the curvature is non-positive then the bundle B satisfies property (5.3) and can be thought of as being a <u>subinterpolation</u> family (or <u>logarithmically subharmonic</u>). The basic construction of analytic families of Banach spaces gives a method for solving a boundary value problem for the equation (5.5). The iteration theorem (2.5) reflects the local nature of this problem.

6. Some Relations With the Real Method of Interpolation

Even though the subject we have been developing is motivated by and extends the complex method of interpolation, it does have some features that are analogous to the real method of interpolation. In order to explain this in some detail we need to enter into a brief discussion of the real method. Our notation is not quite the standard one but, as we shall see, it is adapted for better comparisons with the complex method as we have described it.

There are two basic functionals, the K-functional and the J-functional, that are central to the techniques associated with the real method. Suppose we are given two Banach spaces B_{-1}, B_1 with norms $|\ |_{-1}$ and $|\ |_1$, respectively. The K-<u>functional</u> (more precisely, the K_∞-functional) is the function of B_{-1}, B_1, $a \in B_{-1} + B_1$ and $s > 0$ defined by

$$K(s;a,B_1,B_{-1}) = \inf\{\max [s|a_{-1}|_{-1}, |a_1|_1] : a = a_1 + a_{-1}, \quad a_{\pm 1} \in B_{\pm 1}\};$$

the J-functional is the following function that depends on the same variables (but a is restricted to $B_{-1} \cap B_1$)

$$J(s;a,B_1,B_{-1}) = \max\{s|a|_{-1}, |a|_1\}.$$

These two functionals can be used to obtain the norms of intermediate spaces. For example

$$\|a\|_{\theta,q} \equiv \{\int_0^\infty (s^{-\theta}K(s;a,B_1,B_{-1}))^q \frac{ds}{s}\}^{1/q}, \tag{6.1}$$

where $0 < \theta < 1$, $1 < q < \infty$, defines an intermediate space "between" B_{-1} and B_1. A similar expression (involving certain integral means that equal a) in which the J-functional is used, instead of the K-functional, yields norms that can be shown to be equivalent to the ones in (6.1).

We claim that methods analogous to those described in Sections 2 and 4 lead directly to those functionals. Moreover, one can obtain results on commutators of the type announced in the fourth section by following this analogy further. We shall give precise formulations of these methods and their applications in the real n-dimensional case.

Let $I = [-1,1]$ and $B_{\pm 1}$ be the space $\mathbf{R}^n$ normed by $|\ |_{\pm 1}$. The space $(B_{-1}, |\ |_{-1})$ is associated with the boundary point -1 of I and $(B_1, |\ |_1)$ is associated with $1 \in \partial I$. The role played in §2 by holomorphic $\mathbf{C}^n$-valued functions on D is now assumed by the members of the class $\mathcal{a}$ of all affine maps $F : I \to \mathbf{R}^n$. For each t, $-1 < t < 1$, we then consider the space B_t which is $\mathbf{R}^n$ normed by

$$|a|_t = \inf\{\max_{i\in\partial I} |F(i)|_i : F \in \mathcal{a},\ F(t) = a\}.$$

It is not hard to show that the unit ball, S_t, of B_t is the slice at t of the convex hull of $\{(-1, S_{-1}) \cup (1, S_1)\}$. (This is analogous to the fact that the construction of Section 2 generates the holomorphically convex hull of the unit balls of the boundary spaces B_ξ, $\xi \in D$). A simple calculation shows that

$$S_t = \{\frac{1-t}{2} a_{-1} + \frac{1+t}{2} a_1 : a_{-1} \in S_{-1}, a_1 \in S_1\}. \tag{6.2}$$

Moreover, we have

(6.3) $$|a|_t = \inf\{\lambda > 0 : \lambda^{-1}a \in S_t\}.$$

We claim that

Theorem (6.4). *If* $s = s(t) = \frac{1+t}{1-t}$, $-1 < t < 1$, *then*

(6.5) $$K(s;a,B_1,B_{-1}) = \frac{1+t}{2}|a|_t.$$

To see this we write $F(x) = \frac{1-x}{2}b_{-1} + \frac{1+x}{2}b_1$, $-1 \leqq x \leqq 1$, for $F \in \mathcal{a}$. Thus, the condition $F(t) = a$ is equivalent to $a = a_{-1}+a_1$, where $a_{-1} = \frac{1-t}{2}b_{-1}$ and $a_1 = \frac{1+t}{2}b_1$. Hence,

$$\max\{s|a_{-1}|_{-1}, |a_1|_1\} = \max\{s\frac{1-t}{2}|b_{-1}|_{-1}, \frac{1+t}{2}|b_1|_1\} =$$

$$= \frac{1+t}{2}\max\{|F(-1)|_{-1}, |F(1)|_1\}$$

and equality (6.5) follows immediately from the definitions of K and $|\ |_t$.

There are many analogies between the theory of the spaces B_t, $t \in I$, and the theory of families B_z, $z \in D$, introduced in Section 2. For example, the analog of the fundamental subharmonicity result (1.7) is the fact that $|F(t)|_t$ is convex on I whenever F is affine on I.

If we introduce the natural analogs of the spaces $B^{(2)}_{z_o}$ studied in the fourth section, we are let to the norms

(6.6) $$|(a,b)|^{(2)}_t = \inf\{\max_{i\in\partial I}|F(i)|_i : F \in \mathcal{a},\ F(t) = a,\ (1-t^2)F'(t) = b\},$$

where $(a,b) \in \mathbb{R}^n \times \mathbb{R}^n$. This definition is virtually a copy of (4.8); however, the two valued of $F(t)$ and $F'(t)$, determined by (a,b), completely identify the affine function F. In fact, it is immediate that

$$F(x) = a + \frac{x-t}{1-t^2}b.$$

Thus,

(6.7) $$|(a,b)|^{(2)}_t = \max\{|a - \frac{b}{1-t}|_{-1}, |a + \frac{b}{1+t}|_1\}.$$

Using, again, the transformation $s = \frac{1+t}{1-t}$, $-1 < t < 1$, and putting

$a = 0$ in (6.7) we obtain

$$|(0,b)|_t^{(2)} = \frac{1}{2}\frac{s+1}{2} \max \{s|b|_{-1}, |b|_1\}.$$

That is,

<u>Theorem (6.8)</u>. *If* $s = s(t) = \frac{1+t}{1-t}$, $-1 < t < 1$, *then*

$$J(s;b,B_1,B_{-1}) = (1+t)|(0,b)|_t^{(2)}.$$

It is now possible to carry out a study modeled on the analysis presented in Section 2. Although the details are quite different (in part due to the fact that they involve technical aspectes of real interpolation theory) the conclusions are very similar. Commutator estimates, and other results, can be obtained which reflect the fact that operators bounded on families of spaces (in this case, real interpolation families) have additional characteristic properties on the individual spaces of the family.

The analog of the operator $A'(z_0)$ of Section 4 is the operator, which we denote by $A'(t_0)$, defined by

$$A'(t_0)a = (1 - t_0^2)F'(t_0),$$

where F is that function in a for which equality is attained in the definition of $|a|_t$ (the method described below is valid even when such an F is not unique). But, if $G \in a$,

$$G'(t_0) = \frac{1}{2}|G(1) - G(-1)| \tag{6.9}$$

(this is an elementary affine analog of (4.1)). Equality (6.9) can be used to show that if T is a bounded map of B_t to itself for $t = \underline{+}1$, then, for $t \in I$, $a \in B_t$,

$$J(t;[T,A'(t)]a,B_1,B_{-1}) \leqq cK(t;a,B_1,B_{-1}), \tag{6.10}$$

where c is a constant depending only of the operator norm of T. Although (6.10) is analogous to (4.4) the analogy is imperfect in two ways. First of all, (4.4) involves the same norm on both sides of the inequality, while (6.10) does not (this is related to the fact that the complex interpolation theory is self-dual and the real theory is not, the K and J functionals being dual to each other). Second, the spaces of interest in the complex theory are the B_z, but the spaces of actual interest (from the point of view of interpolation of operators) in the real theory are not the B_t, but are the more complicated spaces defined by expressions such as (6.1). When (6.10) is averaged so as to produce such expressions, technical aspects of real interpolation theory (th "equivalence of the K and J methods") come to our aid and we obtain

$$\|[T,A']a\|_{\theta,q} \leqq c\|a\|_{\theta,q}, \tag{6.11}$$

where $|T,A'|$ is, by definition, the improper integral $\int_0^\infty |T,A'(t)|\frac{dt}{t}$.

This last inequality can be regarded as an analog of (4.4). In some cases (6.11) and (4.4) are the same. For example, (4.6) is an instance of (6.11). In other cases, the results obtained are similar but not the same. For example, the analog of (4.5) is

$$\|[T,\Lambda]f\|_p \leqq c\|f\|_p,$$

where $(\Lambda f)(x) = f(x)\log|\{t : |f(t)| > |f(x)|\}|$.

7. Bibliographical Background and Further Results

Four books that present comprehensive treatments of the theory of interpolation of operators are [1], [2], [13] and [21]. For a treatment of Szego's theorem that includes the L^p-norms, $0 < p < \infty$, see [8], page 136. The Wiener-Masani theorem as formulated in theorem (1.10) can be found in [22]. The H^p-theorem on general bounded, simply-connected domains in $\mathbb{C}$ is developed in [19]. The details of the method described in Section 2 are given, for the finite dimensional case, in [5]. The infinite dimensional case is developed in [6]. A short expository article on these subjects, containing additional information, is [7]. For an independent study in this direction see [12]. The identification of the intermediate spaces, a work that was begun by A.P. Calderón, has been extended by E. Hernandez to the setting of the second section (see [3] and [10]).

The material presented in the fourth section, as well as many extensions and examples, can be found in [17]. For an earlier work in this direction see [20]. Inequality (4.7) was obtained first by Chanillo [4] by other methods. Details on the differential geometry associated with these analytic families of Banach spaces are presented in [14]. There, one can find further discussion on partial differential equations such as (5.5) and their connection with curvature. In the last section of this just mentioned article, one can also find some, but not all, of the real analogs discussed in §6. The material presented here can be considered as a "preliminary announcement" of work in progress.

There are other topics that have been developed in connection with these analytic families of Banach spaces. In [15] one can find results in the theory of functions that can be extended to functions on domains $D \subset \mathbb{C}$ having valued that are Banach spaces and vary "analytically" in the sense we described here. Schwarz's lemma, Liouville's theorem, analytic continuation, the Symmetry principle (and, of course, Szegö's theorem) are notions and results that can be extended to our setting. Moreover, the theory of invariant subspaces, as developed in Helson's book [9], has meaning in this context. Work along this direction can be found in [17] and [11]. We do not develop these topics further in this article since, in addition to the references just cited, we have included them in another expository article that will appear shortly before this one will [18].

References

[1] Bergh, J. and Löfström, J., Interpolation Spaces, an Introduction, Springer-Verlag, Berlin, Heidelberg, New York (1976) pp. 1-207.

[2] Butzer, Paul L. and Berens Hubert, Semi-Groups of Operators and Approximation, Springer-Verlag, Berlin, Heidelberg, New York (1967) pp. 1-318.

[3] Calderón, A.P., Intermediate Spaces and Interpolation, the Complex Method, Studia Math. 24 (1964) pp. 113-190.

[4] Chanillo, S., A Note on Commutators, Ind. U. Math. J., Vol 31 (1982) pp. 7-17.

[5] Coifman, R., Cwikel, M., Rochberg, R., Sagher, Y., and Weiss, Guido, The Complex Method for Interpolation of Operators Acting on Families of Banach Spaces, Lecture Notes in Mathematics 779, Springer-Verlag, Berlin, Heidelberg, New York (1980) pp. 123-153.

[6] __________, A Theory of Complex Interpolation for Families of Banach Spaces, Advances in Math. 33 (1982) pp. 203-229.

[7] __________, Complex Interpolation for Families of Banach Spaces, Proc. of Symp. in Pure Math., Vol. 35, Part 2, Am. Math. Soc. (1979) pp. 269-282.

[8] Gamelin, T.W., Uniform Algebras, Prentice-Hall, Englewood Cliffs, N.J. (1969) pp. 1-257.

[9] Helson, H., Invariant Subspaces, Acad. Press, New York (1964).

[10] Hernandez, E., Intermediate Spaces and the Complex Method of Interpolation for Families of Banach Spaces, Preprint (1982).

[11] __________, Lax's Theorem, a Generalized Wiener-Masani Theorem and Interpolation of Subspaces, Preprint (1982).

[12] Krein, S.G. and Nikolova, L.I., Holomorphic Function in a Family of Banach Spaces, Interpolation, Dokl. Akad. Nauk USSR 250 (1980) pp. 547-550.

[13] Krein, S.G., Petunin, Jr.I., Semenov, E.M., Interpolation of Linear Operators, Trans. of Mathematical Monographs, Vol 54, Am. Math. Soc., Providence (1982) pp. 1-375.

[14] Rochberg, R., Interpolation of Banach Spaces and Negatively Curved Vector Bundles, Pac. J. of Math., Vol. 109, No. 2 (1983).

[15] __________, Function Theoretic Results for Complex Interpolation Families of Banach Spaces, to appear in the Trans. of the A.M.S. (1984).

[16] Rochberg, R. and Weiss, Guido, Complex Interpolation of Subspaces of Banach Spaces, Supp. Rend. Circ. Mat. Palermo 1 (1981) pp. 179-186.

[17] __________, Derivatives of Analytic Families of Banach Spaces, Ann. of Math., Vol 118 (1983) pp. 315-347.

[18] __________, Some Topics in Complex Interpolation Theory, to appear in the Proceedings of the bimester in Harmonic Analysis held at the Universities of Milano and Torino, May-June 1982.

[19] Rudin, W., Analytic Functions of Class H_p, Trans. of the A.M.S. 78 (1955) pp. 46-66.

[20] Schechter, M., Complex Interpolation, Compositio Math., Vol. 18 (1967) pp. 117-147.

[21] Triebel, Hans, Interpolation Theory, Function Spaces, Differential Operators, North-Holland, New York, London (1978) pp. 1-528.

[22] Wiener, N. and Akutowicz, E.J., A Factorization of Positive Hermitian Matrices, J. Math. and Mech. 8 (1959) pp. 111-120.

Recent Progress in Fourier Analysis
I. Peral and J.-L. Rubio de Francia (Editors)

SOME MAXIMAL INEQUALITIES

José-L. Rubio de Francia
Universidad Autónoma de Madrid

The title of this talk is borrowed from a celebrated paper by C. Fefferman and E.M. Stein [5] in which they establish the vector valued inequalities

$$\text{(A)} \qquad \|(\sum_j (Mf_j)^q)^{1/q}\|_p \leq C_{p,q} \|(\sum_j |f_j|^q)^{1/q}\|_p \qquad (1 < p,q < \infty)$$

$$\text{(B)} \qquad \|(\sum_j (Mf_j)^q)^{1/q}\|_{WL^1} \leq C_q \|(\sum_j |f_j|^q)^{1/q}\|_1 \qquad (1 < q < \infty)$$

where $WL^1 = \text{weak-}L^1(\mathbb{R}^n)$, and $Mf = f^*$ is the Hardy-Littlewood maximal operator. I wish to describe here different generalizations and approaches to inequalities like (A) and (B). Each section explains a different method applied to some specific operator, including:

I.- The maximal operator associated to an approximation of the identity $(\Phi_t)_{t>0}$, where $\Phi_t(x) = t^{-n}\,\Phi(\frac{x}{t})$ and $\Phi \in L^1(\mathbb{R}^n)$ satisfies Zo's condition (see [17]).

II.- The dyadic version of Stein's maximal spherical means, which is known to be a bounded operator in $L^p(\mathbb{R}^n)$ for all $n \geq 2$ and $p > 1$ (see [2]), as well as maximal functions and Hilbert transform along curves, [16].

III.- The maximal operator corresponding to rectangles in $\mathbb{R}^2$ in a lacunary set of directions, [11].

IV.- The maximal Bochner-Riesz operators in $\mathbb{R}^2$ which, for arbitrary small index, are bounded in L^p, $2 \leq p \leq 4$ (see [3]). The results concerning (I) and (II) are joint work with F.J. Ruiz and J.L. Torrea.

§I. Maximal operators as Vector Valued Singular Integrals

Let me start by recalling the following result from [1]: Suppose A and B are Banach spaces, and T is a linear operator mapping (measurable) functions $f : \mathbb{R}^n \to A$ into functions $Tf : \mathbb{R}^n \to B$, which is given by

$$Tf(x) = \int K(x-y)f(y)dy \qquad (x \notin \operatorname{supp}(f))$$

for all $f \in L^\infty_A$ with compact support, where $K(x)$ takes values in $L(A,B)$ = {bounded linear from A to B}, and $\|K(x)\| \in L^1_{loc}(\mathbb{R}^n - \{0\})$. Then, we have

Theorem 0. *If* T *is bounded from* L^r_A *to* L^r_B *for some* $1 < r \leq \infty$ *and* K *satisfies*

$$(1) \qquad \int_{|x|>2|y|} \|K(x-y)-K(x)\| \, dx \leq C \qquad (y \in \mathbb{R}^n)$$

then

$$\|Tf\|_{L^p_B} \leq C_p \|f\|_{L^p_A}, \qquad 1 < p < \infty$$

$$\|Tf\|_{WL^1_B} \leq C \|f\|_{L^1_A}$$

The proof consists in a rather straightforward repetition of the classical Calderón-Zygmund argument. The important point in [1] is that a large part of the Littlewood-Paley g-functions fall under the scope of Theorem 0 by taking A = complex numbers, B = Hilbert space, and $r = 2$. However, one can also take $B = \ell^\infty$, and then, some maximal operators also fall under the scope of Theorem 0. This is so in particular for

$$M_\Phi f(x) = \sup_{t > 0} |\Phi_t * f(x)|$$

provided $\Phi \in L^1(\mathbb{R}^n)$ satisfies

$$(2) \qquad \int_{|x|>2|y|} \sup_{t>0} |\Phi_t(x-y)-\Phi_t(x)|dx \leq C \qquad (y \in \mathbb{R}^n)$$

Thus, a particular case of Theorem 0 is F. Zo's theorem ([17]):

"If Φ satisfies (2), then M_Φ is bounded in $L^p(\mathbb{R}^n)$, $1 < p \leq \infty$ and of weak type (1,1)".

Moreover, there is an extension of Theorem 0 which follows immediately from its very statement: Given $1 < q < \infty$, we consider the Banach spaces $\ell^q(A)$ and $\ell^q(B)$, and the operator

$$\tilde{T}(f_1(x), f_2(x), \ldots, f_j(x), \ldots) = (Tf_1(x), Tf_2(x), \ldots, Tf_j(x), \ldots)$$

which maps $\ell^q(A)$-valued functions into $\ell^q(B)$-valued ones, and is trivially of strong type (q,q). Since $\tilde{T}$ is given by the kernel

$$\tilde{K}(x) = \mathrm{Id}_{\ell^q} \otimes K(x) \in L(\ell^q(A), \ell^q(B))$$

which satisfies (1) (with the same constant C) we obtain that, under the hypothesis of Theorem 0, it also follows that

$$\|(\sum_j \|Tf_j\|_B^q)^{1/q}\|_p \leq C_{p,q} \|(\sum_j \|f_j\|_A^q)^{1/q}\|_p$$

$$\|(\sum_j |Tf_j|_B^q)^{1/q}\|_{WL^1} \leq C_q \|(\sum_j |f_j|_A^q)^{1/q}\|_1$$

for all $1 < p,q < \infty$. Now, it suffices to apply this remark to the maximal operator considered above in order to have the desired vector valued inequalities:

Theorem 1. *Suppose* $\Phi \in L^1(\mathbb{R}^n)$ *satisfies Zo's condition (2). Then, the inequalities* (A) *and* (B) *are verified by the operator*

$$Mf(x) = M_\Phi f(x) = \sup_{t>o} |\Phi_t * f(x)|$$

Observe that this theorem contains the inequalities of Fefferman and Stein, since the Hardy-Littlewood maximal operator is dominated by $M_\Phi f(x)$ if $\Phi(x)$ is a positive Schwartz function such that $\Phi(x) \geq 1$ when $|x| \leq 1$. The original proof given in [5] of the vector valued inequalities for the Hardy-Littlewood maximal function, f^*, is based on the inequality

$$(3) \qquad \int f^*(x)^p\, u(x)dx \leq C_p \int |f(x)|^p\, u^*(x)dx \qquad (1 < p < \infty)$$

As we have just seen, no weighted inequality like this is really necessary to derive inequalities (A) and (B), but the fact that Fefferman and Stein did obtain (3) was all the most fortunate, since this inequality is at the source of the beautiful theory of A_p weights. This method is not available for a general maximal function $M_\Phi f(x)$ due to our lack of information about the weighted inequalities satisfied by such an operator. It is now possible, however, to reverse in some sense the original path, obtaining from the vector valued inequalities in Theorem 1 some knowledge about the weights associated to M_Φ. In fact, using the arguments of [12] and [14], we get

Corollary. *Let* $\Phi \in L^1(\mathbb{R}^n)$ *satisfy (2). Given* $p,q > 1$, *to every*

$u \in L^q_+(\mathbf{R}^n)$ *we can associate* $w \in L^q_+(\mathbf{R}^n)$ *such that*

$$u(x) \leq w(x), \qquad \|w\|_q \leq 2\|u\|_q \quad \textit{and}$$

$$\int (M_\Phi f(x))^p\, w(x)dx \leq C_{p,q} \int |f(x)|^p\, w(x)dx$$

Thus, even though we lack a characterization of the weights $w(x)$ for which M is bounded in $L^p(w)$, we do know that they behave in some sense like A_p weights. We point out that the corollary follows from (and it is actually equivalent to) the strong type inequalities (A) with $q < p$. More information can be obtained from the inequalities (B) and (A) with $p < q$, but I shall skip the corresponding statements. On the other hand, if the function Φ is radial, $\Phi(x) = \Phi_0(|x|)$, it is obvious that M_Φ is an operator invariant under dilations and rotations, and [13, Theorem 3] can be applied to obtain

Corollary. *Suppose* $\Phi(x)$ *has a radial majorant in* $L^1(\mathbf{R}^n)$ *which satisfies Zo's condition (2). Then,* M_Φ *is bounded in* $L^p(|x|^a\, dx)$ *for every* $1 < p < \infty$ *and* $-n < a < n(p-1)$.

Finally, let me mention that Theorem 1 also applies to the non-isotropic maximal operator in $\mathbf{R}^n$

$$M_\Omega f(x) = \sup_{t>o} t^{-n} \int_{|y|\leq t} |f(x-y)|\, \Omega(y)dy$$

where Ω is positive, homogeneous of degree 0 and satisfying the Dini condition

$$\int_{|y|=1} \Omega(y)\, d\sigma(y) + \iint_{|y|=|z|=1} |\Omega(y)-\Omega(z)|\, |y-z|^{1-n}\, d\sigma(y)d\sigma(z) < \infty$$

Under this hypothesis, it was proved in [6] that M_Ω is bounded in $L^p(\mathbf{R}^n)$, $1 < p \leq \infty$ and of weak type (1,1). Now, denoting $\Phi(x) = \Omega(x)\, \chi_{\{|x|<1\}}(x)$, it follows from the Dini condition and some computation that

$$\sum_{j=-\infty}^{\infty} \int_{|x|>2|y|} |\Phi_{2^j}(x-y) - \Phi_{2^j}(x)|dx \leq C \qquad (y \in \mathbf{R}^n)$$

Then, the conclusions of Theorem 1 apply to M_Ω, because

$$M_\Omega f(x) = \sup_{t>o} |\Phi_t * f(x)| \leq 2^n \sup_{j\in\mathbf{Z}} |\Phi_{2^j} * f(x)|$$

§II. Maximal Inequalities and Analytic Interpolation

In some cases, the maximal operator M under consideration does not satisfy Zo's condition, but it can be extended to an analytic family M^α (with $M = M^0$) such that:

i) For $Re(\alpha) > 0$, M^α falls under the scope of Theorem 1

ii) For $-a < Re(\alpha) < 0$, M^α is bounded in L^2

In such cases, analytic interpolation can be used to obtain the strong type vector valued inequalities (A). We shall illustrate this in the case of the operator

$$(4) \qquad Nf(x) = \sup_{k\in\mathbf{Z}} \left|\int_{|y|=1} f(x-2^k y)\, d\sigma(y)\right|$$

which is the dyadic version of Stein's maximal spherical means ([2], [16]). The analytic family of operators which is well suited for this problem is

$$N^\alpha f(x) = \sup_{k\in\mathbf{Z}} |\{\hat{f}(\xi)\, m_\alpha(2^{-k}\,\xi)\}^\vee\, (x)|$$

where, for $Re(\alpha) > \frac{1-n}{2}$, the multiplier m_α is defined by

$$m_\alpha(\xi) = \pi^{1-\alpha}\, |\xi|^{1-\alpha-n/2}\, J_{\alpha+n/2-1}\, (2\pi|\xi|)$$

(J_ℓ denoting Bessel functions). Since $m_0(\xi)$ is the Fourier--Stieltjes transform of the singular measure $d\sigma$ concentrated in $\{|y| = 1\}$ and defined as Lebesgue measure on the unit sphere, it is clear that $N = N^0$. On the other hand, Plancherel's theorem and the Hardy-Littlewood maximal function can be used to show that N^α is bounded in L^2 for $Re(\alpha) > \frac{1-n}{2}$. Finally, if $Re(\alpha) > 0$, $m_\alpha(\xi)$ is the Fourier transform of the integrable function $\Phi^\alpha(x) = \Gamma(\alpha)^{-1}\ (1 - |x|^2)_+^{\alpha-1}$, so that

$$N^\alpha f(x) = \sup_{k\in\mathbf{Z}} |\Phi^\alpha_{2^k} * f(x)| \qquad (Re(\alpha) > 0)$$

Now, it is not true that Φ^α satisfies Zo's condition, but it does satisfy the analogous condition for dyadic dilations, namely

$$\int_{|x|>2|y|} \sup_{k\in\mathbf{Z}} |\Phi^\alpha_{2^k}(x-y) - \Phi^\alpha_{2^k}(x)|\, dx \leq C(\alpha) \qquad (Re(\alpha) > 0)$$

(to prove this, majorize $\int \sup_k$ by $\sum_k \int$, and use the fact that the L^1-modulus of continuity of Φ^α satisfy: $\omega_1(\Phi^\alpha;t) \leq C_\alpha t^{Re(\alpha)}$). Thus, Theorem 1 applies to the effect that (the vector valued extension of) N^α is bounded in $L^p(\ell^r)$, $1 < p,r < \infty$, for all $Re(\alpha) > 0$, while N^α is bounded in $L^2(\ell^2)$ for all $Re(\alpha) > \frac{1-n}{2}$. Analytic interpolation then gives:

Theorem 2. *In* $\mathbb{R}^n$, $n \geq 2$, *the strong type vector valued inequalities* (A) *are verified by the maximal operator* Nf *defined in (4).*

Without going into the details, let me simply mention that the same method applies to the operators

$$M_\gamma f(x) = \sup_{h>0} \frac{1}{h} \left|\int_0^h f(x-\gamma(t))dt\right|$$

$$H_\gamma f(x) = \text{p.v.} \int_{-\infty}^{\infty} f(x-\gamma(t)) \frac{dt}{t}$$

(maximal function and Hilbert transform along γ) provided that the curve $\gamma(t)$ in $\mathbb{R}^n$, $n \geq 2$, is "well curved". The main difference now is that Theorem 0 and 1 are applied to approximations of the identity defined in terms of non-isotropic dilations. The analytic families of operators to be considered in each case are described in [16]. Thus, we conclude that Theorem 2 holds for the operator M_γ and H_γ.

§III. Covering Methods and "A_1-Weights"

Conversely to what was done in section I, here I shall follow the line of thought of [5] and obtain the vector valued inequalities (A) from a weighted inequality similar to (3). The maximal operator to which this method will be applied is

$$M_R f(x) = \sup_{x \in R \in \mathcal{R}} \frac{1}{|R|} \int_R |f(y)|dy$$

where $\mathcal{R}$ is the family of all rectangles R in $\mathbb{R}^2$ parallel to some of the vectors $\vec{e}_j = (\cos 2^{-j}\pi,\ \text{sen}\ 2^{-j}\pi)$, $j = 1,2,3,\ldots$ (any other lacunary sequence of unit vectors will do just as well). The basic result for this operator was proved by A. Nagel, E.M. Stein and S. Wainger [11], establishing that M_R is bounded in $L^p(\mathbb{R}^2)$ for all $p > 1$. We shall state here two extensions of this result:

Theorem 3. *Let* $w(x)$ *be a weight in* $A_1(\mathcal{R})$, *i.e.,* $M_R w(x) \leq C\, w(x)$ *a.e. Then, for all* $1 < p < \infty$, *we have*

$$\int (M_R f(x))^p\, w(x)dx \leq C_p(w) \int |f(x)|^p\, w(x)dx$$

It is important to point out that $C_p(w)$ depends only on p and on the "A_1-constant" for w, namely: $\|w^{-1} M_R w\|_\infty$.

Theorem 4. *The operator* M_R *satisfies the strong type vector valued inequalities* (A).

Theorem 3 is what replaces here the Fefferman-Stein inequality (3). Let me first show how Theorem 4 follows from it by an argument which is quite general and, by now, well understood. The vector valued inequalities to be proved are obvious for $p = q$, and also for $q = \infty$. From here, the case $1 < p \leq q < \infty$ is obtained by interpolation. Now, if $1 < q < p < \infty$, we can write

$$\int (\sum_j [M_R f_j(x)]^q)^{p/q} dx = \sum_j \int [M_R f_j(x)]^q u(x) dx$$

for some $u \in L_+^r(\mathbb{R}^2)$ of unit norm, where $r = (\frac{p}{q})' = \frac{p}{p-q}$. If B_r denotes the norm of M_R as an operator in L^r, we define

$$w(x) = \sum_{k=o}^{\infty} (2 B_r)^{-k} (M_R)^k u(x)$$

where $(M_R)^o = Id$ and $(M_R)^k$ is the k-th iterate of the operator M_R. Then, $u(x) \leq w(x)$, $\|w\|_r \leq 2\|u\|_r$ and $w \in A_1(R)$ with A_1-constant not greater than $2B_r$. Therefore, by Theorem 3 and Hölder's inequality

$$\sum_j \int [M_R f_j(x)]^q u(x) dx \leq C_{q,r} \int \sum_j |f_j(x)|^q w(x) dx$$
$$\leq 2 C_{q,r} \|(\sum_j |f_j|^q)^{1/q}\|_p^q.$$

This proves Theorem 4. Now, the way to obtain Theorem 3 begins with the same inequality to be proved but with the operator M_R replaced by its "dual". To formulate this precisely, write

$$E_R f(x) = (\frac{1}{|R|} \int_R f) \chi_R(x)$$

for each rectangle R. Then, we have

Lemma. *Given* $w \in A_1(R)$ *and arbitrary intervals* $R_j \in R$ *and functions* $f_j \in L^1_{loc}(\mathbb{R}^2)$, *for all* $1 \leq p < \infty$, *we have*

$$\int |\sum_j E_{R_j} f_j(x)|^p w(x) dx \leq C_p(w) \int (\sum_j |f_j(x)|)^2 w(x) dx$$

Proof of the Lemma. It follows by interpolation with change of measure between the two estimates

$$(5) \qquad \int |\sum_j E_{R_j} f_j(x)|^q dx \leq C_q \int (\sum_j |f_j(x)|)^q \, dx$$

which holds for all $1 \leq q < \infty$, and

$$(6) \qquad \int |\sum_j E_{R_j} f_j(x)| w(x)^{1+\varepsilon} dx \leq C_\varepsilon \int \sum_j |f_j(x)| w(x)^{1+\varepsilon} dx$$

which holds for some $\varepsilon > 0$. Observe that (5) is equivalente (by duality) to the boundedness of $M_{\mathcal{R}}$ in $L^{q'}(\mathbb{R}^2)$, $1 < q' \leq \infty$. On the other hand, (6) is a consequence of the reverse Hölder's inequality, which holds for weights in $A_1(\mathcal{R})$ and implies that $w^{1+\varepsilon} \in A_1(\mathcal{R})$ for some $\varepsilon > 0$. In fact, assuming $f_j \geq 0$, the left hand side of (6) is equal to

$$\sum_j \int f_j(x) \, E_{R_j}(w^{1+\varepsilon})(x) dx \leq C_\varepsilon \int \sum_j f_j(x) \, w(x)^{1+\varepsilon} \, dx$$

<u>Proof of Theorem 3</u>. It is based on covering arguments similar to those of A. Córdoba and R. Fefferman (see [7]). Given $f \geq 0$, let

$$E = \{x : M_{\mathcal{R}} f(x) > \lambda\} = \bigcup R_j$$

where $R_j \in \mathcal{R}$ are rectangles such that

$$\int_{R_j} f(x) dx > \lambda \, |R_j|$$

From $\{R_j\}$ we select a subsequence $\{\tilde{R}_j\}$ exactly as in [7], i.e., we order $\{R_j\}$ so that $\ell(R_j)$ = "longest side of R_j" decreases, then take $\tilde{R}_1 = R_1$ and, once $\tilde{R}_1, \tilde{R}_2, \ldots, \tilde{R}_{k-1}$ have been selected, we call $\tilde{R}_k$ to the first rectangle in the original sequence (if any) satisfying

$$\sum_{j<k} |\tilde{R}_j \cap \tilde{R}_k| \leq \frac{1}{2} |\tilde{R}_k|$$

Two consequences of the selection method are relevant for our purposes: If f_k denotes the characteristic function of $\tilde{R}_k - (\bigcup_{j<k} \tilde{R}_j)$, then

$$(7) \qquad E_{\tilde{R}_k} f_k(x) > \frac{1}{2} \chi_{\tilde{R}_k}(x)$$

(this is obvious), and

$$(8) \qquad E \subset \{x : M(\sum_k \chi_{\tilde{R}_k})(x) > \tfrac{1}{2}\}$$

M denoting here the strong maximal operator ((8) follows by a simple geometric argument which can be seen in [7]). Since M is known to be bounded in $L^q(w)$ for all $q > 1$, we have

$$w(E) \le C_q \int |\sum_k \chi_{\tilde{R}_k}(x)|^q \, w(x)\, dx$$

On the other hand, using (7) and the previous lemma, and taking into account that $\sum_k f_k$ is the characteristic function of $\bigcup_k \tilde{R}_k$,

$$\| \sum_k \chi_{\tilde{R}_k} \|^q_{L^q(w)} \le C_q \, w(\bigcup_k \tilde{R}_k) \le$$

$$\le C_q \, \lambda^{-1} \sum_k |\tilde{R}_k|^{-1} w(\tilde{R}_k) \int_{\tilde{R}_k} f(x)dx \le \text{(because } w \in A_1(R))$$

$$\le C_q \, \lambda^{-1} \sum_k \int_{\tilde{R}_k} f(x)\, w(x)dx \le C_q \lambda^{-1} \|f\|_{L^{q'}(w)} \, \|\sum_k \chi_{R_k}\|_{L^q(w)}$$

Now, given $p > 1$, we apply the preceding inequalities with $q = p'$ to get

$$w(E) \le C_p \, \lambda^{-p} \int |f(x)|^p \, w(x)\, dx$$

which shows that M_R is of weak type (p,p) with respect to $w(x)dx$ for all $p > 1$, and the strong type result follows by interpolation.

Remarks: A different proof of Theorem 3 has been given by B. Jawerth [9] by adapting the original argument of [11], which uses the Fourier transform, to the weighted case, while our method starts with the result itself: $\|M_R f\|_p \le C_p \|f\|_p$, $1 < p \le \infty$, and uses rather general methods to extend it to $L^p(w)$, $w \in A_1(R)$. We have presented this different approach in the hope that some of the ideas involved may be useful in related problems.

It must be said that B. Jawerth has actually obtained the inequality

$$\text{(9)} \qquad \|M_R f\|_{L^p(w)} \le C_p(w) \, \|f\|_{L^p(w)}$$

for all weights $w \in A_p(R)$, $1 < p < \infty$, thus giving a complete extension of Muckenhoupt's result for the Hardy-Littlewood maximal function to the case of the operator M_R. On the other hand, starting from Theorem 3, the interpolation and duality argument used in the above Lemma gives (9) for all $w \in A_1(R) \; A_1(R)^{1-p} = \{w_0 \, w_1^{1-p} \mid w_0, w_1 \in A_1(R)\}$, which turns out to be equivalent to Jawerth's result, since factorization holds for the weights associated to $R : A_p(R) = A_1(R) \; A_1(R)^{1-p}$.

§IV. ℓ^2-Valued Inequalities for Bochner-Riesz Means

In this last section, I would like to make some almost trivial remarks concerning the result proved in [3] for the maximal Bochner--Riesz operators in R^2

$$S_*^\alpha f(x) = \sup_{0<t<\infty} |S_t^\alpha f(x)|$$

where $(S_t^\alpha f)\hat{}(\xi) = \hat{f}(\xi)(1-|\xi|^2/t^2)_+^\alpha$. Carbery's theorem states that $\|S_*^\alpha f\|_p \le C_\alpha \|f\|_p$ for all $\alpha > 0$ and $2 \le p \le 4$. The following consequences can also be obtained from the proof:

Theorem 5: (a) *For functions* $f_j \in L^p(R^2)$ *and for all* $\alpha > 0$, *the vector valued inequality*

$$\|(\sum_j (S_*^\alpha f_j)^q)^{1/q}\|_p \le C_{\alpha,p} \|(\sum_j |f_j|^q)^{1/q}\|_p$$

holds if $2 \le q \le p \le 4$.

(b) *The weighted* L^2-*estimate*

$$\int_{R^2} (S_*^\alpha f(x))^2 |x|^{-\delta} dx \le C_\alpha \int_{R^2} |f(x)|^2 |x|^{-\delta} dx$$

holds if $0 \le \delta \le 1$ *and* $\alpha > 0$.

Of course, for each fixed $\alpha > 0$, a wider range of p's, q's and δ's is allowed (by interpolation), but I have only stated the results valid for all $\alpha > 0$.

Observe that part (a) implies the vector valued inequality

$$(10) \quad \|(\sum_j |S_{t_j}^\alpha f_j|^2)^{1/2}\|_p \le C_{\alpha,p} \|(\sum_j |f_j|^2)^{1/2}\|_p \quad (4/3 \le p \le 4)$$

which was first proved in [4] (the case $4/3 \le p \le 2$ is obtained by duality). From (10), the boundedness in $L^p(R^2)$, $4/3 \le p \le 4$, of the maximal lacunary means: $\sup_{k\in Z} |S_{2^k}^\alpha f(x)|$ can be obtained by rather standard methods (see [4] again). We also point out that part (b) of the theorem is sharp for positive δ (though one should expect the best possible result: $-1 \le \delta \le 1$) and is an extension to S_*^α of results previously proved by I. Hirschman [8] for the ordinary Bochner-Riesz means S_t^α.

Proof of Theorem 5: Everything can be reduced to the inequality

$$(11) \quad \|(\sum_j (S_*^\alpha f_j)^2)^{1/2}\|_4 \le C_\alpha \|(\sum_j |f_j|^2)^{1/2}\|_4$$

Since the remaining cases stated in (a) follow by interpolation, and (b) is a consequence of (11) and [13, Theorem 3]. Now, the estimate for S_*^α given in [3] starts from the standard majorization

$$S_*^\alpha f(x) \leq C_{\alpha,\beta} \{f^*(x) + G_\beta f(x)\}$$

where f^* is the Hardy-Littlewood maximal function of f, β is taken so that $-\frac{1}{2} < \beta < \alpha - \frac{1}{2}$, and G_β is an operator of Littlewood-Paley type

$$G_\beta f(x) = \left(\int_0^\infty |T_t^\beta f(x)|^2 \frac{dt}{t}\right)^{1/2}$$

with $(T_t^\beta f)^\wedge(\xi) = \hat{f}(\xi)\,\psi\left(\frac{|\xi|}{t}\right)(1 - |\xi|^2/t^2)_+^\beta$, ψ smooth and vanishing in a neighbourhood of the origin. Then, the key point in Carbery's proof is the inequality

$$\| G_\beta f \|_4 \leq C_\beta \| f \|_4 \quad , \qquad \beta > -\frac{1}{2}$$

Since G_β can be considered as a linear operator from L^4 to L^4_H, with $H = L^2(\mathbb{R}_+, \frac{dt}{t})$ (Hilbert space), the theorem of Marcinkiewicz and Zygmund [10] proves that it has a bounded extension

$$Id_{\ell^2} \otimes G \; : L^4_{\ell^2} \longrightarrow L^4_{\ell^2(H)}$$

with the same norm. Thus, (11) Holds for G_β, and it also holds for the maximal operator: $f \longmapsto f^*$ by the Fefferman-Stein inequalities which motivated this talk. This ends the proof.

References

[1] A. Benedek, A.P. Calderón, R. Panzone: Proc. Nat. Acad. Sci. USA 48 (1962), 356-365.

[2] C.P. Calderón: Illinois J. Math. 23 (1979), 476-484.

[3] A. Carbery: Duke Math. J. 50 (1983), 409-416.

[4] A. Córdoba, B. López-Melero: Ann. Inst. Fourier 31 (1982), 147-152.

[5] C. Fefferman, E.M. Stein: Amer. J. Math. 93 (1971), 107-115.

[6] R. Fefferman: Adv. in Math. 30 (1978), 171-201.

[7] R. Fefferman: Proc. Symp. Pure Math. XXXV (1), Amer. Math. Soc., Providence R.I. (1979), 51-60.

[8] I. Hirschman: Duke Math. J. 28 (1961), 45-56.

[9] B. Jawerth: Amer. J. Math. (to appear).

[10] J. Marcinkiewicz, A. Zygmund: Fund. Math. 32 (1939), 115-121.

[11] A. Nagel, E.M. Stein, S. Wainger: Proc. Nat. Acad. Sci. USA 75 (1978), 1060-1062.

[12] J.L. Rubio de Francia: Lecture Notes in Math. 908, Springer-Verlag, Berlin and New York (1982), 86-101.

[13] ————————: Trans. Amer. Math. Soc. 275 (1983), 781-790.

[14] ————————: Amer. J. Math. 106 (1984), 533-547.

[15] ————————, F.J. Ruiz, J.L. Torrea: C.R. Acad. Sci. Paris, Sér. I, 297 (1983), 477-480.

[16] E.M. Stein, S. Wainger: Bull. Amer. Math. Soc. 84 (1978), 1239-1295.

[17] F. Zo: Studia Math. 55 (1976), 111-122.

Recent Progress in Fourier Analysis
I. Peral and J.-L. Rubio de Francia (Editors)

A FATOU THEOREM AND A MAXIMAL FUNCTION NOT INVARIANT UNDER TRANSLATION

Peter Sjögren
Chalmers, Göteborg

1. Introduction

Let $P(z,e^{i\beta})$, $z = x+iy \in U$, $\beta \in \mathbf{T} = \mathbf{R}/2\pi\,\mathbf{Z}$, be the Poisson kernel in the unit disk U. Then $P(z,e^{i\beta})^{\lambda+1/2}$ is for any λ an eigenfunction of the hyperbolic Laplacian $L_z = \frac{1}{4}(1-|z|^2)^2(\partial^2/\partial x^2 + \partial^2/\partial y^2)$ with eigenvalue $\lambda^2 - 1/4$. In the bidisk U^2,

$$P(z_1,e^{i\beta_1})^{\lambda_1+\frac{1}{2}}\,P(z_2,e^{i\beta_2})^{\lambda_2+\frac{1}{2}}$$

is therefore an eigenfunction of the bihyperbolic Laplacian $L = L_{z_1} + L_{z_2}$ with eigenvalue $\lambda_1^2 + \lambda_2^2 - \frac{1}{2}$. We fix $\lambda_1^2 + \lambda_2^2 = R^2 > 0$ and write $\lambda_1 = R\cos\theta$, $\lambda_2 = R\sin\theta$.

If f is an integrable function on the set $X = \mathbf{T}^2 \times [0,\pi/2]$, the function

$$u(z_1,z_2) = Pf(z_1,z_2) = \int_X P(z_1,e^{i\beta_1})^{R\cos\theta+\frac{1}{2}}\,P(z_2,e^{i\beta_2})^{R\sin\theta+\frac{1}{2}}\, f(\beta_1,\beta_2,\theta)\,d\beta_1 d\beta_2 d\theta$$

thus satisfies $Lu = (R^2 - \frac{1}{2})u$ in U^2. Karpelevic has shown that any positive solution of this equation can be obtained in this way, with a positive measure in X instead of f. Thus X is the natural boundary here.

A boundary point $(b_1,b_2,v) \in X$ with $0 < v < \pi/2$ can be thought of as the limit point of the geodesic $\Gamma_{b_1,b_2,v}$ defined by

$$(z_1,z_2) = (e^{ib_1}\tanh(R\,s\cos v),\ e^{ib_2}\tanh(R\,s\sin v)),\qquad s \geq 0. \tag{1.1}$$

One can thus expect the normalized function $Pf(z_1,z_2)/P1(z_1,z_2)$ to converge to $f(b_1,b_2,v)$ as (z_1,z_2) tends to $(\partial U)^2$ along or near

$\Gamma_{b_1,b_2,v}$. Here 1 is the constant function 1. This was proved for continuous f, and a.e. on X for $f \in L^p$, $p > 1$, by Linden [1]. The author proved the following in [2].

Theorem 1. *If* $f \in L^1(X)$, *then* Pf/P1 *converges to* $f(b_1,b_2,v)$ *along* $\Gamma_{b_1,b_2,v}$, *and in any tube of bounded bihyperbolic distance from* $\Gamma_{b_1,b_2,v}$, *for a.a.* $(b_1,b_2,v) \in X$.

Let

$$Mf(b_1,b_2,v) = \sup_{\Gamma_{b_1,b_2,v}} P|f|/P1, \quad (b_1,b_2,v) \in X,$$

be the relevant maximal function. In [2], it was proved that M is of weak type (1,1) in X. This implies Theorem 1. But Theorem 1 is also a consequence of the following weaker result. Let for $\delta > 0$

$$X_\delta = \mathbb{T}^2 \times \{v \in [0,\pi/2] : \cos v > \delta, \ \sin v > \delta\}.$$

Theorem 2. *For any* $\delta > 0$, *the restriction of* Mf *to* X_δ *defines a bounded operator from* $L^1(X)$ *to* $L^1_{weak}(X_\delta)$.

The aim of this note is to show how much easier it is to prove Theorem 2 rather than the full result of [2]. We refer to [2] for more details. In Sjögren [3], the analogs of Theorem 1 and 2 were proved in Riemannian symmetric spaces.

2. Proof of Theorem 2.

We fix δ, and let $C = C(R,\delta)$ denote various constants. Let z_1, z_2 be given by (1.1), and set from now on

$$t_1 = e^{-2R\, s \cos v}, \qquad t_2 = e^{-2R\, s \sin v}$$

so that $1 - |z_i|$ is close to $2t_i$.

There is a kernel K such that

$$\frac{Pf(z_1,z_2)}{P1(z_1,z_2)} = Kf(z_1,z_2) = \int_X K(z_1,z_2;\beta_1,\beta_2,\theta) f(\beta_1,\beta_2,\theta)\, d\beta_1 d\beta_2 d\theta.$$

Formula (4.1) of [2] and Harnack's inequality imply

$$K(z_1,z_2;\beta_1,\beta_2,\theta) \leq C \frac{1}{t_1}\left(1 + \frac{|\beta_1 - b_1|}{t_1}\right)^{-1-\varepsilon} \cdot \frac{1}{t_2}\left(1 + \frac{|\beta_2 - b_2|}{t_2}\right)^{-1-\varepsilon} \sqrt{s}\, e^{-\varepsilon s(\theta - v)^2}, \tag{2.1}$$

if $(b_1,b_2,v) \in X_\delta$ and $s > 1$. Here $\varepsilon = \varepsilon(R,\delta) > 0$. For $(b_1,b_2,v) \in X_\delta$, we write $Kf(z_1,z_2)$ as a sum of integrals over the sets

$$\{(\beta_1,\beta_2,\theta) \in X : 2^{q_i-1} t_i \le |\beta_i - b_i| \le 2^{q_i} t_i, \quad i=1,2, \qquad\qquad (2.2)$$
$$\text{and} \quad 2^{m-1}/\sqrt{s} \le |\theta - v| \le 2^m/\sqrt{s}\},$$

$q_1,q_2,\ m \ge 0$. Here 2^{-1} is replaced by 0. In these integrals, (2.1) allows us to replace K by

$$C\, 2^{-\varepsilon q_1}\, 2^{-\varepsilon q_2} \frac{1}{2^{q_1}t_1\, 2^{q_2}t_2} \sqrt{s}\, e^{-\varepsilon 2^{2m-2}}.$$

Then we suppress the lower bounds in (2.2), extending the integrations to the sets

$$B_s^{q_1,q_2,m} = B_s^{q_1,q_2,m}(b_1,b_2,v) =$$
$$= \{(\beta_1,\beta_2,\theta) \in X : |\beta_i - b_i| \le 2^{q_i}t_i,\ |\theta - v| \le 2^m/\sqrt{s}\}.$$

The Lebesgue measure $|B_s^{q_1,q_2,m}|$ is at most $8 \cdot 2^{q_1}t_1\, 2^{q_2}t_2\, 2^m/\sqrt{s}$, and we conclude

$$K|f|(z_1,z_2) \le C \sum_{q_1,q_2,m=0}^{\infty} 2^{-\varepsilon q_1}\, 2^{-\varepsilon q_2}\, 2^m\, e^{-\varepsilon 2^{2m-2}} \cdot$$
$$\cdot \frac{1}{|B_s^{q_1,q_2,m}|} \int_{B_s^{q_1,q_2,m}} |f|\, d\beta_1 d\beta_2 d\theta.$$

Taking sup in s, we get an estimate for Mf. By summing in weak L^1 we see that Theorem 2 will follow if we can prove that the operator

$$M^{q_1,q_2,m} f(b_1,b_2,v) = \sup_{s>1} \frac{1}{|B_s^{q_1,q_2,m}|} \int_{B_s^{q_1,q_2,m}} |f|\, d\beta_1 d\beta_2 d\theta$$

maps $L^1(X)$ into $L^1_{weak}(X)$, uniformly in q_1, q_2, m. Notice that t_1, t_2 and thus the sides of $B_s^{q_1,q_2,m}$ depend on v, so that $M^{q_1,q_2,m}$ is not invariant under translation. We shall only consider $M^{0,0,0}$, since the general case is analogous.

Let for $k=1,2,\ldots$

$$T_k f(b_1,b_2,v) = \sup_{2^{k-1}\le s\le 2^k} \frac{1}{|B_s|}\int_{B_s} |f|\, d\beta_1 d\beta_2 d\theta \quad \text{if}$$

$$(b_1,b_2,v) \in X_\delta \quad \text{and } 0 \quad \text{otherwise,}$$

where B_s means $B_s^{0,0,0}(b_1,b_2,v)$. Extending f, $T_k f$, and $M^{0,0,0}f$ by 0, we can consider them as defined in $\mathbb{T}^3 \supset X$. To these operators T_k, we shall apply a lemma from [2], which we now recall.

Let (M,μ) be a measure space, and assume that for each $k=1,2,\ldots$ there is a partition of M into an at most countable number of sets called k-pieces. The k-pieces are measurable with positive finite measure, and any $(k+1)$-piece is contained in some k-piece. The following is part of Lemma 1 of [2].

Lemma. *Let* (M,μ) *and the* k-*pieces be as just described, and let* $(T_k)_1^\infty$ *be a sequence of subadditive operators mapping functions in* $L^1 + L^\infty(\mu)$ *into nonnegative* μ-*measurable functions. Assume*

(a) the T_k *are of weak type* $(1,1)$, *uniformly in* k

(b) the restriction $T_k f|_P$ *depends only on* $f|_P$, *for any* k-*piece* P

(c) $\|T_k f\|_{L^\infty} \le C \sup \frac{1}{\mu(P)} \int_P |f|\, d\mu,$

the $\sup$ *taken over all* $(k+N)$-*pieces, where* N *is a fixed natural number. Then the operator* $f \to \sup_k T_k f$ *is of weak type* $(1,1)$.

We must thus construct k-pieces in $\mathbb{T}^3$. For $2^{k-1} \le s \le 2^k$ and $(b_1,b_2,v) \in X_\delta$, the set B_s is a box of dimensions at most

$$2e^{-R2^{k\delta}} \times 2e^{-R2^{k\delta}} \times 2/\sqrt{2^{k-1}}\,. \tag{2.3}$$

As in the proof of Lemma 2 in [2], we can then construct for each k a partition of $\mathbf{T}^3$ into congruent boxes, called k-pieces, whose side lengths are contained in the intervals

$$[10e^{-R2^{k\delta}},\ 30e^{-R2^{k\delta}}],\quad [10e^{-R2^{k\delta}},\ 30e^{-R2^{k\delta}}],\quad \text{and}$$

$$[10/\sqrt{2^{k-1}},\ 30/\sqrt{2^{k-1}}],$$

respectively. (If the right-hand endpoint of any of these intervals is $\ge 2\pi$, the corresponding sides of the k-pieces will be all of $\mathbf{T}$). This is done hierarchically, so that $(k+1)$-pieces are subsets of k-pieces. Also, 2π should be an odd multiple of each side length of a k-piece, for all k.

The operators T_k are subadditive, and to finish the proof of Theorem 2, we need only verify (a) - (c). If $x = (b_1, b_2, v)$ is in a k-piece P, $T_k f(x)$ is determined by the restriction of f to the set

$$\{x+y : |y_1| \leq e^{-R2^{k}\delta}, \; |y_2| \leq e^{-R2^{k}\delta}, \; |y_3| \leq 1/\sqrt{2^{k-1}},$$

cf (2.3); the addition is here in $\mathbb{T}^3$. This set is often but not always contained in P. To obtain (b), we replace $T_k f(x)$ by 0 in the latter case. This is permitted since we can repeat the argument below with translated k-pieces, as in the proof of Lemma 2 in [2].

Aiming at (a), we form a mean value

$$F_k^v(\beta_1, \beta_2) = \sqrt{2^{k-1}} \int_{|\theta - v| \leq 1/\sqrt{2^{k-1}}} |f(\beta_1, \beta_2, \theta)| d\theta .$$

Clearly,

$$\int_{\mathbb{T}} \| F_k^v \|_{L^1(\mathbb{T}^2)} dv \leq \| f \|_{L^1(X)} .$$

Since the third side of B_s does not vary much as s runs over $[2^{k-1}, 2^k]$, we get

$$T_k f(b_1, b_2, v) \leq C \sup_s \frac{1}{|R_s^v|} \int_{R_s^v} F_k^v(\beta_1, \beta_2) d\beta_1 \, d\beta_2 ,$$

where $R_s^v = R_s^v(b_1, b_2)$ is the rectangle

$$\{(\beta_1, \beta_2) \in \mathbb{T}^2 : |\beta_1 - b_1| \leq e^{-2Rs} \cos v , \; |\beta_2 - b_2| \leq e^{-2Rs} \sin v\},$$

For each v, we thus get a translation invariant two-dimensional maximal function operator. It is of weak type (1,1), uniformly in v, as seen by standard methods. Now (a) follows by integration in v.

To get (c), write for $2^{k-1} \leq s \leq 2^k$

$$\frac{1}{|B_s|} \int_{B_s} |f| \leq \frac{1}{|B_s|} \Sigma |P| \frac{1}{|P|} \int_P |f| ,$$

the sum taken over those (k+N)-pieces P intersecting B_s. It is enough to verify that $\Sigma |P| \leq C |B_s|$. This clearly follows if the sides of the (k+N)-pieces are no longer than the corresponding sides of B_s. But the dimensions of B_s are at least

$$e^{-2R2^k} \times e^{-2R2^k} \times 1/\sqrt{2^k} ,$$

and the dimensions of a (k+N)-piece are at most

$$Ce^{-R2^{k+N}\delta} \times Ce^{-R2^{k+N}\delta} \times C/\sqrt{2^{k+N-1}} .$$

Hence, (c) follows if we choose $N = N(R,\delta)$ suitably. This completes the proof.

References

[1] O. Linden, Fatou theorems for the eigenfunctions of the Laplace-Beltrami operator. Thesis, Yeshiva University, 1977.

[2] P. Sjögren, Fatou theorems and maximal functions for eigenfunctions of the Laplace-Beltrami operator in a bidisk. To appear in J. reine angew. Math.

[3] P. Sjögren, A Fatou theorem for eigenfunctions of the Laplace-Beltrami operator in a symmetric space. Dept. of Math., Chalmers University of Technology and the University of Goteborg, Report 1983-6.

Recent Progress in Fourier Analysis
I. Peral and J.-L. Rubio de Francia (Editors)

A COUNTER-EXAMPLE FOR THE DISC MULTIPLIER

Per Sjölin
University of Stockholm

Let (X,μ) be a measure space and set $Lf = f\log|f|$ if f is a measurable function on X.

R. Rochberg and G. Weiss have obtained the following results (see [2], Proposition (3.35) and Corollary (3.37)).

Theorem A. *Assume that* $1 \leq p_1 < p_2 \leq \infty$ *and that* T *is a bounded linear operator on* $L^{p_i}(X,\mu)$, $i = 1,2$. *If* $p_1 < p < p_2$ *then*

$$\|TLf - LTf\|_p \leq C_p \|f\|_p, \quad f \in L^p(X,\mu). \tag{1}$$

Corollary B. Suppose T is a linear operator which is bounded from L^p to L^p for a range of p that includes p_0 in its interior. Then there exists a constant C_T such that, for any function $f \in L^p$ whose modulus takes on only the values 0 or 1, we have

$$\|Tf\log|Tf|\|_{p_0} \leq C_T \{\mu(\text{supp } f)\}^{1/p_0} \tag{2}$$

The following questions are then asked in [2] : What is an interesting example of an operator T that is bounded on L^{p_0} for which (2) fails? Does (2) hold for the disc multiplier on $L^2(\mathbf{R}^2)$?.

We shall here prove that the answer to the last question is negative.

We define the disc multiplier T by setting $(Tf)^\wedge = m\hat{f}$, $f \in L^2(\mathbf{R}^2)$, where $m(x) = 1$, $|x| \leq 1$, and $m(x) = 0$, $|x| > 1$, and

$$\hat{f}(x) = \int_{\mathbf{R}^2} e^{-ix.t} f(t)\,dt, \quad x \in \mathbf{R}^2.$$

It is well-known that $m = \hat{K}$, where

$$K(x) = a\,\frac{e^{i|x|}}{|x|^{3/2}} + b\,\frac{e^{-i|x|}}{x^{3/2}} + O(|x|^{-5/2}), \quad |x| \to \infty, \tag{3}$$

for some constants a and b.

Since $K \in L^q(\mathbb{R}^2)$, $q > 4/3$, we can define Tf for $f \in L^p(\mathbb{R}^2)$, $1 \leq p < 4$, by setting

$$Tf(x) = K * f(x) = \int K(x-y)\, f(y)\, dy.$$

Hence Tf is well-defined also for $f \in L^p + L^r$, $1 \leq p < r < 4$. If $f \in L^2$ we have $Lf \in L^{2-\varepsilon} + L^{2+\varepsilon}$, $0 < \varepsilon < 1$, and hence TLf is well-defined.

We shall here prove the following theorem.

Theorem 1. *The commutator* $[T,L] = TL - LT$ *is not bounded on* $L^2(\mathbb{R}^2)$.

Proof. Assume that $[T,L]$ is bounded on $L^2(\mathbb{R}^2)$. It follows that

$$\| T(f \log|f|) - Tf \log|Tf| \|_2 \leq C \|f\|_2. \tag{4}$$

If $|f|$ takes only the values 0 and 1 then (4) implies

$$\int |Tf|^2 \log^2|Tf|\,dx \leq C \int |f|^2\, dx. \tag{5}$$

We shall prove that (5) does not hold and hence (2) fails for the disc multiplier if $p_0 = 2$.

To give a counter-example to (5) we shall use a modification of C. Fefferman's argument in [1].

We shall use the process of sprouting of triangles, which is defined in [1], p. 333.

Let $T^{(0)}$ be an equilateral triangle with height $h = \sqrt{3}/2$ whose base is the interval $[0,1]$ on the x-axis. Sprout $T^{(0)}$ from height h to height $2h$ to obtain two new triangles $T_1^{(1)}$ and $T_2^{(1)}$. Sprouting these triangles from height $2h$ to height $3h$ we get 4 triangles $T_1^{(2)},\dots,T_4^{(2)}$. We continue this process and after k steps we have 2^k triangles $T_j^{(k)}$, $j=1,2,\dots,2^k$, with base 2^{-k} and height $(k+1)h$. Now fix k and set $E = \bigcup_{j=1}^{2^k} T_j^{(k)}$. Let $|A|$ denote the Lebesgue measure of a set A in the plane. It is easy to see from similarity of triangles that

$$\Big|\bigcup_{j=1}^{2^{i+1}} T_j^{(i+1)}\Big| \leq \Big|\bigcup_{j=1}^{2^i} T_j^{(i)}\Big| + \sqrt{3}\,\frac{1}{i+2}, \quad i = 0,1,2,\dots,$$

from which it follows that $|E| \leq C \log k$ for some constant C. For each j, $j = 1,2,\dots,2^k$, we now construct two rectangles R_j

and $\tilde{R}_j$ with side lenghts k and 2^{-k} as in Fig. 1. Obviously $\sum_{j=1}^{2^k} |R_j| = k$ and hence there exist numbers η_k, $k = 1,2,\ldots,$ with

$$\lim_{k \to +\infty} \eta_k = 0 \tag{6}$$

such that

$$|E| \leq \eta_k \sum_j |R_j| \tag{7}$$

From the construction it also follows that the R_j's are pairwise disjoint and that

$$|\tilde{R}_j \cap E| \geq c\, |R_j| \tag{8}$$

for some positive constant c.

Now set

$$d = \frac{k\, 2^{2k}}{10^4}$$

and dilate the plane by the mapping $x \to dx$, $x \in \mathbf{R}^2$. Let us from now on use the notation R_j, $\tilde{R}_j$, E for the dilated sets. Obviously (7) and (8) still hold and the new rectangles have side lengths $dk = N$ and

$$d\, 2^{-k} = \frac{k\, 2^k}{20^4} = \frac{\sqrt{N}}{100},$$

where

$$N = \frac{k^2\, 2^{2k}}{10^4}.$$

Let v_j denote a unit vector parallel to the longest side of R_j (see Fig. 1). Set $f_j(x) = e^{iv_j \cdot x} \chi_{R_j}(x)$, $j = 1,2,\ldots,2^k$, so that $|f_j| = \chi_{R_j}$. The following estimate is essentially well-known.

Lemma 2. *There exists a positive constant* c *such that*

$$|Tf_j(x)| \geq c, \quad x \in \tilde{R}_j . \tag{9}$$

We postpone the proof of Lemma 2.

We set $f(x) = \sum_j \varepsilon_j f_j(x)$, where $\varepsilon_j = \pm 1$, $j = 1,2,\ldots,2^k$. Hence $|f| = \Sigma |f_j| = \Sigma\, \chi_{R_j} = \chi_{\cup R_j}$, since the R_j's are disjoint.

Choising $\{\varepsilon_j\}$ suitably we obtain

$$\int_E |Tf(x)|^2 dx = \int_E |\sum_j \varepsilon_j\, Tf_j(x)|^2 dx \geq \int_E \sum_j |Tf_j(x)|^2 dx \geq$$

$$\geq \sum_j \int_{E \cap \tilde{R}_j} |Tf_j(x)|^2 dx \geq \sum_j c\, |\tilde{R}_j \cap E| \geq c \sum_j |R_j|,$$

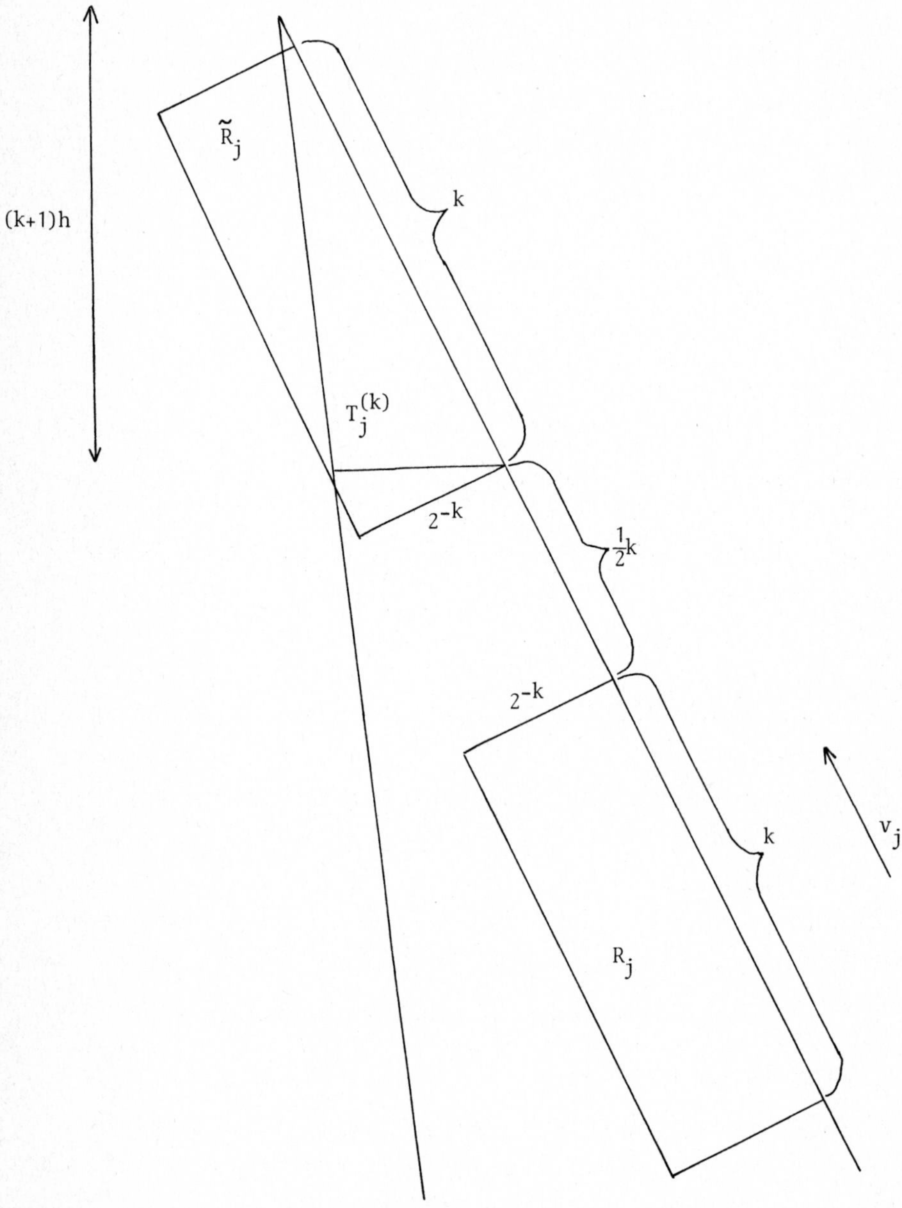

Fig. 1

where we have used (9) and (8).

Hence there exists a constant C_0 such that

$$\sum_j |R_j| \le C_0 \int_E |Tf|^2 dx. \tag{10}$$

Now set

$$A = \{x \in E;\ \ |Tf(x)| \le \lambda\}$$

and

$$B = \{x \in E;\ \ |Tf(x)| > \lambda\},$$

where $\lambda = (2C_0\eta)^{-1/2}$ and $\eta = \eta_k$. Thus $E = A \cup B$.

The inequality (10) yields

$$\sum_j |R_j| \le C_0 \int_A |Tf|^2 dx + C_0 \int_B |Tf|^2 dx.$$

On the other hand (7) implies

$$C_0 \int_A |Tf|^2 dx \le C_0\ \lambda^2 |E| \le C_0\ \lambda^2 \eta\ \sum_j |R_j| = \frac{1}{2} \sum_j |R_j|, \tag{12}$$

since $C_0 \lambda^2 \eta = 1/2$.

From (11) and (12) we conclude that

$$\frac{1}{2} \sum_j |R_j| \le C_0 \int_B |Tf|^2 dx.$$

On B we have

$$\log |Tf| \ge \log\ \lambda = \frac{1}{2} \log\ \frac{1}{2C_0\eta}$$

and it follows that

$$\frac{1}{2} \sum_j |R_j| \le C_0 \int_B |Tf|^2\ \frac{\log^2 |Tf|}{(\frac{1}{2} \log 1/2C_0\eta)^2}\ dx \le$$

$$\le C\ \frac{1}{\log^2\ 1/2C_0\eta} \int_B |Tf|^2 \log^2 |Tf| dx.$$

If (5) holds we then obtain

$$\frac{1}{2} \sum_j |R_j| \le C\ \frac{1}{\log^2\ 1/\eta} \int |f|^2 dx = C\ \frac{1}{\log^2\ 1/\eta}\ \sum_j |R_j|\ .$$

Hence

$$\log^2\ 1/\eta \le C$$

and we obtain a contradiction to (6).

We conclude that (5) and (4) do not hold and thus the proof of the theorem is complete.

For the sake of completeness we shall give a proof of Lemma 2.

Proof of Lemma 2. Since $Tf = K * f$, where K is radial, we may assume that R_j is parallel to the x-axis and $v_j = (1,0)$. We may also assume that the center of R_j is the origin. $\tilde{R}_j$ is then also parallel to the x-axis and has center $(3N/2,0)$.

We denote the three terms on the right hand side of (3) by $L(x)$, $M(x)$ and $R(x)$.

We have

$$L * f_j(x) = a \int_{R_j} e^{i|x-t|} |x-t|^{-3/2} e^{iv_j . t} dt =$$
$$= a e^{iv_j . x} \int_{R_j} e^{i(|x-t|-v_j.(x-t))} |x-t|^{-3/2} dt.$$

If θ denotes the angle between the vectors v_j and $x-t$, we obtain

$$||x-t|-v_j.(x-t)| = |x-t|(1-\cos\theta) \leq 6N \sin^2 \theta/2 \leq$$
$$\leq 6N \left(\frac{1}{100\sqrt{N}}\right)^2 \leq \frac{1}{10},$$

if $t \in R_j$ and $x \in \tilde{R}_j$.

We conclude that

$$|L * f_j(x)| \geq c \int_{R_j} N^{-3/2} dt = c, \quad x \in \tilde{R}_j, \qquad (13)$$

since $a \neq 0$.

It is also easy to see that

$$|R * f_j(x)| \leq C \int_{R_j} N^{-5/2} dt \leq C/N, \quad x \in \tilde{R}_j. \qquad (14)$$

Now fix $x \in \tilde{R}_j$. We have

$$M*f_j(x) = b \int_{R_j} e^{-|x-t|} |x-t|^{-3/2} e^{iv_j . t} dt =$$
$$= b \int_{R_j} e^{iF(t)} |x-t|^{-3/2} dt,$$

where

$$F(t) = t_1 - ((t_1 - x_1)^2 + (t_2 - x_2)^2)^{1/2}.$$

We write $F' = \partial F/\partial t_1$ and $F'' = \partial^2 F/\partial t_1^2$ and a computation shows that

$$1 \leq F' \leq 2 \tag{15}$$

and

$$|F''| \leq C/N \tag{16}$$

for $t \in R_j$.

Writing

$$I(t_2) = \int_{-N/2}^{N/2} e^{iF}\, iF' \frac{1}{iF'} |x-t|^{-3/2}\, dt_1,$$

we obtain

$$\int_{R_j} e^{iF} |x-t|^{-3/2}\, dt = \int_{-\sqrt{N}/200}^{\sqrt{N}/200} I(t_2) dt_2 .$$

An integration by parts shows that

$$I(t_2) = \left[e^{iF} \frac{1}{iF'} |x-t|^{-3/2}\right]_{-N/2}^{N/2}$$

$$-\int_{-N/2}^{N/2} e^{iF} \frac{1}{i} \left(- \frac{3}{4F'} ((t_1-x_1)^2 + (t_2-x_2)^2)^{-7/4}\, 2(t_1-x_1) - \right.$$

$$\left. - \frac{F''}{F'^2} |x-t|^{-3/2}\right) dt,$$

and it follows that

$$|I(t_2)| \leq CN^{-3/2} + C\int_{-N/2}^{N/2} N^{-5/2}\, dt_1 \leq CN^{-3/2}, \quad |t_2| \leq \sqrt{N}/200.$$

We conclude that

$$|M * f_j(x)| \leq C/N, \quad x \in \tilde{R}_j, \tag{17}$$

and a combination of (13), (14) and (17) yields

$$|K * f_j(x)| \geq c, \quad x \in \tilde{R}_j .$$

Hence the proof of the lemma is complete.

References

[1] Fefferman, C. The multiplier problem for the ball. Annals of Math. 94 (1971), 330-336.

[2] Rochberg, R., And Weiss, G. Derivatives of Analytic Families of Banach Spaces. Annals of Math. 118 (1983), 315-347.

Recent Progress in Fourier Analysis
I. Peral and J.-L. Rubio de Francia (Editors)

THREE VARIATIONS ON THE THEME OF MAXIMAL FUNCTIONS

E.M. Stein
Princeton University

Contents

It is a noteworthy fact that more than 50 years after their discovery by Hardy and Littlewood, maximal functions are still the subject of investigation. It is our purpose here to describe three recent developments in this area of research. The first deals with the question of inequalities for maximal functions and related singular integrals with bounds independent of the dimension. In this part we give complete proofs of two theorems we announced earlier. In the second and third parts we merely sketch the background and motivate two further results, whose detailed proofs will be published elsewhere. I am happy to acknowledge my indebtedness to Alexander Nagel and J.O. Strömberg for collaboration in part of the work described here. Their contributions will be cited more explicitly below.

I. ESTIMATES FOR MAXIMAL FUNCTIONS AND SINGULAR INTEGRALS IN $\mathbb{R}^n$, AS $n \to \infty$.

Here I want to present some results whose thrust is that certain fundamental estimates in harmonic analysis have formulations with bounds independent of n, as $n \to \infty$.

We begin by considering the basic maximal function given by averages with respect to centered balls.

Thus define,

$$M(f)(x) = M^{(n)}f(x) = \sup_{r>0} \frac{c_n}{r^n} \int_{|y|\leq r} |f(x-y)|dy$$

with c_n^{-1} the volume of the unit ball.

THEOREM 1.

$$\|M^{(n)}f\|_p \leq A_p\|f\|_p, \qquad 1 < p \leq \infty$$

with the bound A_p *independent of* n.

Of course it should be pointed out that the usual covering arguments of Vitali-type (used to prove weak-type (1,1) estimates) lead to bounds that grow exponentially in n, so a different approach is required. The idea of the proof (a variant of that given in [7] and [10]) leads to a strengthening of theorem 1, and can be described quite simply. It is this: Suppose B is the unit ball in $\mathbb{R}^n$, then as $n \to \infty$, most of its volume is concentrated on the surface of the sphere (in fact $(1-\varepsilon)^n$ represents the proportion of the volume in the ball of radius $1-\varepsilon$). Thus a theorem about averages over balls for large n, should be related to a corresponding result about averages over spheres. This leads to the following.

Suppose

$$M(f)(x) = M^{(n)}f(x) = \sup_{r>0} \int_{|y|=1} |f(x-ry)|d\sigma(y)$$

with $d\sigma$ the normalized uniform measure on the unit sphere of $\mathbb{R}^n$, and with $M(f)$ defined for continuous f of compact support.

THEOREM 2.

(1) $$\|M^{(n)}(f)\|_p \leq A_p\|f\|_p, \quad \textit{if} \quad 1 < p \leq \infty$$

and $n > \max(p/(p-1),2)$, *with* A_p *independent of* n.(*)

(*) This, and related results were announced in [8].

It is clear that theorem 2 implies theorem 1. In fact $M^{(n)}f(x) \leq \mathcal{M}^{(n)}(f)(x)$, which take care of those n for which $n > \max(p/(p-1),2)$. The remaining finite number of n are covered by the standard result.

We turn to the proof of theorem 2. The key idea is already contained in the version of theorem 1 (proved in [6a]) which allows the constant to grow with n. We state it as a lemma:

LEMMA 1.

$$\|\mathcal{M}^{(k)}(f)\|_p \leq A_p(k)\|f\|_p, \quad \text{if} \quad 1 < p \leq \infty \tag{2}$$

and $k > \max(p/(p-1),2)$.

Next this lemma is combined with the following observation.

LEMMA 2. *The inequality* $\|\mathcal{M}^{(k)}f\|_p \leq A\|f\|_{p_k}$ *(for all* f *which are continuous and have compact support in* $\mathbb{R}^k$*) implies the inequality* $\|\mathcal{M}^{(n)}f\|_p \leq A\|f\|_p$ *(for all* f *wich are continuous and have compact support in* $\mathbb{R}^n$*), with the same bound* A, *whenever* $n \geq k$.

This can be proved by induction. It will suffice to show that the conclusion holds with $n = k+1$, assuming it for $n = k$. To do this, let u denote any unit vector in $\mathbb{R}^{k+1}$, and consider the "great sphere" given by the hyper-plane perpendicular to $u = \{y \in \mathbb{R}^{k+1}, |y| = 1, \text{ and } u \cdot y = 0\}$. On it define $d\sigma^u(y)$ to be the normalized uniform measure, and set

$$\mathcal{M}_u^{(k)}(f)(x) = \sup_{0<r} \int_{\{|y|=1, y\cdot u=0\}} |f(x-ry)|\, d\sigma^u(y) \tag{3}$$

The inequality $\|\mathcal{M}^{(k)}(f)\|_p \leq A\|f\|_p$ easily implies $\|\mathcal{M}_u^{(k)}(f)(x)\|_p \leq A\|f\|_p$, with the same A, for each u. However

$$\mathcal{M}^{(k+1)}(f)(x) \leq \int_{|u|=1} \mathcal{M}_u^{(k)}|(f)(x)|\, d\sigma_{k+1}(u) \tag{4}$$

where $d\sigma_{k+1}$ is the normalized uniform measure on the unit sphere of $\mathbb{R}^{k+1}$. The proof of (4) is itself a straight-forward consequence of the integration formula

$$\int_{|y|=1} F(y)\, d\sigma_{k+1}(y) = \int_{|u|=1} \left(\int_{\{|y|=1, y\cdot u=0\}} F(y)\, d\sigma^u(y) \right) d\sigma_{k+1}(u) \tag{5}$$

valid for all non-negative F defined on the unit sphere of $\mathbb{R}^{k+1}$. In fact (5) can be seen by observing that the right side defines a measure on the unit sphere of $\mathbf{R}^{k+1}$, which is rotation invariant and has total integral equal to 1. This proves lemma 2 and from it and lemma 1 follows theorem 2.

There are two ideas to be retained here. The "transference" of the inequality (with the same bound) from k to n is possible only because we are dealing with the spherical maximal function, and not the usual one. Secondly, the inequalities for the spherical maximal function require the intervention of square functions and the Fourier transform. We will return to these points later.

Several problems for maximal functions, with $n \to \infty$, are left open: Does $M^{(n)}$ have weak-type $(1,1)$ bounds independent of n? What can be said when the usual balls are replaced by dilates of more general (convex) sets? Some results bearing on these question are in [10].

We now turn to the theory of singular integrals in $\mathbf{R}^n$. The most basic examples are of course the Riesz transform $R_1, R_2, \ldots, R_n$, defined by

$$(R_j(f))^\wedge(\xi) = i \frac{\xi_j}{|\xi|} \hat{f}(\xi).$$

Our goal is to give L^p bounds for these operators, which are independent of n. Of course the motivation for doing this comes from theorem 1, and the close relation that one knows exists between maximal functions and singular integrals. By analogy with the above one might hope to transfer an appropriate inequality about (variants of) the Riesz transforms from k dimensions to n dimensions. What this inequality might be is a mystery. However the second aspect of the proof of theorem 1, aluded to before, is relevant here. Thus our approach will be via appropriate results for square functions. The theorem can be formulated as follows. $R(f)$ will denote the vector $= (R_1(f), R_2(f), \ldots, R_n(f))$, and $|R(f)| = (\sum_{j=1}^{n} |R_j(f)|^2)^{1/2}$. We shall assume for simplicity here that f is real-valued, and write $|R(f)|_p$ for

$$\| R(f) \|_p$$

THEOREM 3

$$|R(f)|_p \leq A_p |f|_p \qquad 1 < p < \infty$$

with A_p *independent of* n.

As stated before the proof requires the use of correctly chosen square functions. The ones we shall use are direct generalizations of the classical square functions introduced by Littlewood and Paley when $n = 1$. (It is worthwhile to point out that the variants of the Lusin "area integral" do not seem to be appropriate here. A hint of this comes from the fact that this type of square function depends on the size of the aperture of the cone, and the corresponding factors of proportionality grow exponentially with n. For similar reason the maximal theorem 1 requires in the definition of $M^{(n)}f(x)$ balls centered at x, and not the larger class of balls containing x).

The square functions we need are

(6) $$g(f)(x) = \left(\int_0^\infty |(\nabla u)(x,t)|^2 t\,dt\right)^{1/2},$$

(7) $$g_1(f)(x) = \left(\int_0^\infty \left|\frac{\partial u}{\partial t}(x,t)\right|^2 t\,dt\right)^{1/2},$$

(8) $$g^*(f)(x) = \left(4\int_0^\infty \int_{\mathbb{R}^n} |(\nabla u)(x-y,t)|^2 P_t(y)\,dy\,t\,dt\right)^{1/2},$$

with

$$u(x,t) = \int_{\mathbb{R}^n} P_t(y)f(x-y)\,dy, \quad P_t(y) = \frac{c_n t}{(|y|^2+t^2)^{\frac{n+1}{2}}};$$

u is the Poisson integral of f, and P_t the Poisson kernel. Also

$$|(\nabla u)(x,t)|^2 = \left|\frac{\partial u}{\partial t}\right|^2 + \sum_{j=1}^n \left|\frac{\partial u}{\partial x_j}\right|^2.$$

It will be useful to extend the definition of g_1 to functions f which take their values in a finite-dimensional Hilbert space H. Let m be the dimension of this Hilbert space and write $f(x) = (f_1(x), f_2(x), \ldots, f_m(x))$, by expressing f in terms of a fixed orthonormal basis of H. It is then natural to define $g_1(f)(x)$ by $(g_1(f)(x))^2 = \sum_{j=1}^m (g_1(f_j(x)))^2$. With these definitions in mind, we can state.

<u>LEMMA 3</u>. <u>*If*</u> f <u>*is real-valued*</u>

(9) $$g_1((Rf)(x)) \leq g(f)(x).$$

This is obvious once one remarks that

$$\frac{\partial}{\partial t}(R_j(f)*P_t) = \frac{\partial}{\partial x_t}(f*P_t).$$

The two key lemmas are then the following:

LEMMA 4. *Suppose* f *is a function on* $\mathbb{R}^n$ *which takes its values in an* m *dimensional Hilbert space and is in* L^p. *Then*

(10) $$\|f\|_p \leq B_p \|g_1(f)\|_p, \qquad 1 < p < \infty.$$

with B_p *independent of* n *and* m.

LEMMA 5. *Suppose* f *is a real-valued function on* $\mathbb{R}^n$ *which is in* L^p, *then*

(11) $$\|g(f)\|_p \leq A_p\|f\|_p, \qquad 1 < p < \infty$$

with A_p *independent of* n.

To prove lemma 4 we consider first the case $m = 1$ (scalar-valued functions) and we apply the theory of symmetric diffusion semi-groups of (5), to the operators $f \to T^t(f) = f*P_t$. This semi-group satisfies the axioms (I) to (IV) in [5], p. 65. We now invoke the corollary 2 on p. 120 and the fact that the projection E_0 is zero in this case. The crucial point is to realize that the arguments leading up to that corollary make use of only the general axioms (I) to (IV) of the semi-group T^t, in which the dimension n of the underlying space never enters! Thus B_p can be taken to be independent of n, when $m = 1$. To prove the lemma for general m requires one only to review the proof just cited and to realize that this proof goes through mutatis mutandis in the (infinite dimensional) Hilbert-space-valued case, and hence the results are independent of m.

We shall see an alternate approach to lemma 4 later, but now we turn to the proof of lemma 5, and to begin with the case $1 < p \leq 2$.

In proving (11) it suffices to consider non-negative f and to make the additional limitation that f has compact support and is smooth. We have first

(12) $$\Delta(u^p) = p(p-1)u^{p-2}|\nabla u|^2$$

where $\Delta = \frac{\partial^2}{\partial t^2} + \sum_{j=1}^{n} \frac{\partial^2}{\partial x_j^2}$. Also

$$(13) \qquad \int_{\mathbf{R}^n} I(x)dx = \|f\|_p^p ,$$

if $I(x) = \int_0^\infty t\Delta(u^p)(x,t)dt$, by Green's theorem. However

$$(g(f)(x))^2 = \int_0^\infty t|\nabla(u)(x,t)|^2 dt = \frac{1}{p(p-1)} \int_0^\infty u^{2-p}(x,t)\,\Delta(u)^p(x,t)tdt \leq$$

$$\text{(by (12))} \qquad \leq 1/p(p-1)\ (\sup_{t>0} u(x,t))^{2-p}\ I(x).$$

However $\sup_{t>0} u(x,t) \leq Mf(x)$ as is well-known, and therefore

$$(14) \qquad \int (g(f))^p dx \leq (p(p-1))^{-p/2} \int_{\mathbf{R}^n} (Mf(x))^{\frac{(2-p)p}{2}} (I(x))^{p/2} dx.$$

To the right-side of (14) we apply Hölder's inequality with exponents r and r', $1/r+1/r' = 1$, where $r = 2/2-p$, and $r' = 2/p$. If we use (13) the outcome of this calculation is

$$|g(f)|_p^p \leq (p(p-1))^{-p/2} |Mf|_p{}^{p/2} |f|_p^{p/r'} ,$$

and so

$$\|g(f)\|_p \leq A_p' \|f\|_p$$

is proved for f positive, $1 < p \leq 2$, with

$$A_p' = (p(p-1))^{-1/2}\ A_p^{(p/2)(2-p)} ,$$

and A_p is the constant appearing in theorem 1.

What we have shown up to this point combined with lemma 3 suffices to prove theorem 3 for $1 < p \leq 2$. To consider the case $p \geq 2$ need only prove lemma 5 for $p \geq 2$. To do this we need the g^* function defined by (8). The first observation is

$$(15) \qquad g(f)(x) \leq g^*(f)(x).$$

In fact $|\nabla u(x,t+s)|^2 \leq \int_{\mathbf{R}^n} |\nabla u(x-y,s)|^2 P_t(y)dy$ for all $t,s > 0$. This is because for s fixed $|\nabla u(x,t+s)|^2$ is a sub-harmonic function (of (x,t)) in the half-space which is majorized by the harmonic function given by the Poisson integral, because this majorization holds on the boundary, i.e. when $t = 0$. Thus

$$|\nabla u(x,2t)|^2 \leq \int_{\mathbf{R}^n} |\nabla u(x-y,t)|^2\ P_t(y)dt ,$$

and (15) follows from the definition of g and g^* given by (6) and (8).

Since we know that $\|g(f)\|_2 \leq A_2\|f\|_2$, it suffices by (15) to show that

$$(16) \qquad \|g^*(f)\|_p \leq A_p\|f\|_p\,, \qquad 4 \leq p < \infty$$

with A_p independent of n.

As before it is enough to restrict consideration to non-negative f which are smooth and have compact support. Now with p fixed, $4 \leq p < \infty$, let q denote the exponent conjugate to $p/2$ (note $1 < q \leq 2$, and $r = p/p-2$). Let ψ be a non-negative bounded function with compact support, with $\|\psi\|_r \leq 1$, but otherwise let ψ be arbitrary. Then

$$\int (g^*(f)(x))^2\psi(x)dx = \int_0^\infty \int_{\mathbb{R}^n} 2\Delta(u^2(x,t))\psi(x,t)\ tdxdt$$

where $\psi(x,t) = \psi * P_t$ is the Poisson integral of ψ, because

$$\Delta u^2 = 2|\nabla u|^2.$$

However $\Delta(u^2)\psi = \Delta(u^2\psi) - 2(\nabla u^2).\nabla\psi$ since $\Delta\psi = 0$. Thus

$$\int g^*(f)(x)^2\psi(x)dx = 2\iint \Delta(u^2(x,t)\psi(x,t)tdxdt + E$$

where $|E| \leq 8 \iint |u(x,t)||\nabla u||\nabla\psi|tdtdx$.

Now $\sup_{t>0} |u(x,t)| \leq M(f)(x)$, and so by Schwarz's inequality

$$|E| \leq 8 \int_{\mathbb{R}^n} Mf(x)g(f)(x)g(\psi)(x)dx.$$

Since

$$2\int_0^\infty \int_{\mathbb{R}^n} \Delta((u(x,t))^2\psi(x,t))tdxdt = 2\int_{\mathbb{R}^n} (f(x))^2\psi(x)dx$$

by Green's theorem, we can conclude that

$$(17) \quad \int_{\mathbb{R}^n} (g^*(f)(x))^2\psi(x)dx \leq 2\int_{\mathbb{R}^n} (f(x))^2\psi(x)dx + {} + 8\int_{\mathbb{R}^n} M(f)g^*(f)g(\psi)dx.$$

In (17) we take the supremum over all allowable ψ. By Hölder's inequality, its converse, and theorem 1, we therefore get

$$\|g^*(f)\|_p^2 \leq 2\|f\|_p^p + 8A_p\|f\|_p\|g^*(f)\|_p\,,$$

and hence we have proved (16) with A_p replaced by the largest positive root of the equation $x^2-8A_px-2 = 0$; we have thus obtained a bound that does not exceed $8A_p+2$, and hence proved the theorem.

A final remark concerns an alternate proof of lemma 4, which does not appeal to the general theory of symmetric difusion semi-groups. We claim first that the proof we gave for lemma 5 generalizes (with only two small changes) to the case of functions which take their valued in a Hilbert space (and of course with bounds independent of the dimension of that Hilbert space). In fact the argument we gave goes through word-for-word, with the exception of two changes. The identity $\Delta(u^p) = p(p-1)u^{p-2}|\nabla u|^2$ for positive-valued functions has to be replaced by the inequality $\Delta(|u|^p) \geq p(p-1)|u|^{p-2}|\nabla u|^2$, $1 \leq p \leq 2$, where here $|.|$ stands for the norm in the Hilbert space. Also, in the case $p = 2$ we need to replace the identity $\Delta(u^2) = 2|\nabla u|^2$, by the identity $\Delta(|u|^2) = 2|\nabla u|^2$. Once the analogue of lemma 5 is proved in the Hilbert space-valued case, it of course implies that $\|g_1(f)\|_p \leq A_p\|f\|_p$, $1 < p < \infty$, which then by a well-known argument proves (10).

II. WIDER APPROACH REGIONS

In this part I should like to describe some work done jointly with A. Nagel [4]. It may be easier to understand the results obtained if they are presented as answers to a series of related questions.

Here the setting is $\mathbb{R}^n$, with n fixed, and we are given a countable collection $\mathcal{B} = \{B_m\}$, of balls B_m having center x_m, and radius r_m; $B_m = B_{r_m}(x_m)$.

<u>Question 1</u>: <u>Under what condition on</u> $\{B_m\}$ <u>does one have</u>

$$\text{(Q.1)} \qquad \lim_{m\to\infty} \frac{1}{|B_m|}\int_{B_m} f(x-y)\,dy = f(x) \quad \text{a.e.} \quad \text{for each } f \in L^1.$$

An obvious necessary condition is that

$$\text{(18)} \qquad B_m \to \{0\}, \quad \text{as} \quad m \to \infty.$$

If in addition $|x_m| \leq cr_m$, (this is the classical condition), then as is well-known (Q.1) holds. However, are there any examples where (Q.1) holds and for which $|x_m|/r_m \to \infty$? This question was raised by Walter Rudin. If we drop the condition (18) we can ask.

Question 2: Under what condition $\{B_m\}$ is the mapping

$$f \to \sup_m \frac{1}{|B_m|} \int_{B_m} |f(x-y)|\,dy \tag{Q.2}$$

of weak type (1,1)? When is it of type (p,p), $1 < p < \infty$?.

One can also raise these question in a different way. Let $u(x,y)$ denote the Poisson integral of f; so $u(x,y) = f*P_y$ with $P_y(x) = \dfrac{c_n y}{(|x|^2+y^2)^{\frac{n+1}{2}}}$. Let Ω be a given open subset of $R_+^{n+1} = \{(x,y),\ x \in R^n,\ y > 0\}$. The question is when is Ω an "approach region"; that is, in addition to the usual situation when Ω is a cone, are there other regions of interest?.

Question 3: Under what condition on Ω is the mapping

$$f(x) \to \sup_{(\bar{x},y)\in\Omega} |u(x-\bar{x},y)| \tag{Q.3}$$

of weak type (1,1)? When is it of type (p,p), $1 < p < \infty$?.

To answer the first two questions, we denote by $\hat{\mathcal{B}}$ the larger collection of balls, so that $B \in \hat{\mathcal{B}}$, if $B \supset B_m$ for some $B_m \in \mathcal{B}$. We also let A_r denote $\bigcup B$, with $B \in \hat{\mathcal{B}}$ and radius of $B = r$.

THEOREM 4. *The following are equivalent:*

i) $|A_r| \leq cr^n$

ii) The mapping (Q.2) *is of weak-type* (1,1)

iii) The mapping (Q.2) *is of type* (p,p), *for some* p, $1 < p < \infty$

The answer to Question 3 is equivalent with the above result. We define $\hat{\Omega}$ as obtained as the union of all cones (of fixed aperture) whose vertex is some point in Ω; i.e. $\hat{\Omega} = \{(x,y) \mid$ so that $\exists (x_1,y_1) \in \Omega$ with, $|x-x_1| < \alpha(y-y_1)$, where α is a fixed positive constant$\}$. We also define $\hat{\Omega}(y) \subset R^n$, by $\hat{\Omega}(y) = \{x \in R^n \mid (x,y) \in \hat{\Omega}\}$. Then the relevant condition, equivalent with the fact that the mappings (Q.3) are of weak-type (1,1), or of type (p,p), is that $|\hat{\Omega}(y)| \leq cy^n$. These results are all proved in [4] where several different methods are given.

Here I would like to discuss in more detail two examples which will help to shed light on the matter. We consider the special case of $\mathcal{B} = \{B_m\}$, where instead of the classical condition $|x_m| \leq cr_m$ relating the centers x_m and the radii, we assume

$$(19) \qquad |x_{m+1}| \leq cr_m$$

(Note that if we choose x_m, and r_m, so that $|x_m| = 2^m 2^{-m^2}$, $r_m = 2^{-m^2}$, $m \geq 1$, then $|x_m|/r_m \to \infty$ while (19) holds). One technique for proving the desired properties for the maximal operator given in (Q.2) (al least when $p = 2$) is then as follows. We denote by T_m the mapping given by $(T_m f)(x) = u(x-x_m, r_m)$, where $u(x,y)$ is the Poisson integral of f. We also define S_m, by $S_m(f) = u(x,r_m)$. Let T_m^* denote the adjoint of T_m; observe that $S_m^* = S_m$. Then we first verify the inequality

$$(20) \qquad T_k T_j^* \leq c\{T_j^* + T_k + S_j\}$$

for all integers j and k, where the inequality is meant as operators acting on positive functions.

To prove (20) observe that when $k = j$, we have actually $T_k T_j^* \leq c\ S_j$; while if $k < j$, we can show easily that $T_k T_j^* \leq c\ T_j^*$, because of (19). Similarly $T_k T_j^* \leq c\ T_k$, if $k > j$. Now let $m(x)$ denote and arbitrary integer-valued measurable function of x. We need to estimate $T = T_{m(x)}$. Of course $S = S_{m(x)}$ is bounded on L^2 by the usual maximal theorem. It follows from (20) that $TT^* \leq c(T^* + T + S)$, and hence if $f \geq 0$,

$$\int (T^*f)^2 dx = \int (TT^*f) f dx \leq c\left\{\int (T^*f) f + \int (Tf) f dx + \int (Sf) f dx\right\}.$$

Thus $\|T^*\|^2 \leq c\{\|T^*\| + \|T\| + \|S\|\}$, and hence $\|T^*\| = \|T\| \leq c'$.

Our second example allows us to raise a further question. In the above we have limited ourselves to a <u>denumerable</u> collection of balls whose radii are much smaller than the distances of their centers from the origin. What happens when instead one considers <u>continous</u> collections of such balls? For example, let $n = 1$, and take the family $\{B_{h^2}(h)\}$ of balls centered at h of radius h^2, where $0 < h \leq 1$. First of all,

$$\sup_{0<h\leq 1} \frac{1}{|B_{h^2}(h)|} \int_{B_{h^2}(h)} |f(x-y)|\,dy$$

is not bounded on any L^p, $p < \infty$. However if we set

$$\mathcal{M}(f)(x) = \sup_{0<h\leq 1} \frac{1}{h^{3/2}} \int_{B_{h^2}(h)} |f(x-y)|\,dy,$$

then while $\mathcal{M}$ may be infinite a.e. if $f \in L^p$, $p < 2$, it can be seen that $f \to \mathcal{M}(f)$ is of weak-type (2,2), because

$$\mathcal{M}(f) \leq \sup_{0<h\leq 1} \frac{1}{h^{3/2}} \left(\int_{|y|\leq 2h} |f(x-y)|^2 dy\right)^{1/2} \cdot |B_{h^2}(h)|^{1/2}$$

$$\leq \sup_{0<h\leq 1} \left\{\frac{2}{h} \int_{|y|\leq 2h} |f(x-y)|^2 dy\right\}^{1/2}.$$

However by the theory of Nagel, Rudin and Shapiro [3], which uses results of capacity theory, one has that $f \to \mathcal{M}(f)$ is actually of type (2,2). The question is then how to formulate a theorem which contains both this result and that of theorem 4.

We shall describe an answer to this question in terms of the general approach region Ω for Poisson integrals used before. Suppose, as before that we consider the region $\hat{\Omega}$ which is obtained as the union of all cones (of a fixed aperture) whose vertexes are points in Ω. We let $\hat{\Omega}(y) = \{x \in \mathbf{R}^n,\ (x,y) \in \hat{\Omega}\}$, and u denotes the Poisson integral of f.

We define $\eta(y)$ by

(21) $$\eta(y) = \inf_{t\geq y} t^n/|\hat{\Omega}(t)|.$$

With η, determined by Ω, we define $\mathcal{M}_p$ by

$$\mathcal{M}_p(f)(x) = \sup_{(\bar{x},y)\in\Omega} (\eta(y))^{1/p}\, |u(x-\bar{x},y)|.$$

<u>THEOREM 5</u>.

(a) $\|\mathcal{M}_p(f)\|_p \leq A_p\|f\|_p$, *if* $1 < p \leq \infty$

(b) $\|\mathcal{M}_p(f)\|_p \leq A_p\|f\|_{H^p}$, *if* $0 < p \leq 1$ *and* $f \in H^p$.

The proof will be found in [4].

The example described immediately above corresponds to the region

$\Omega = \{(x,y) \mid 0 < y \leq 1, \ |x|^2 < y\} \subseteq \mathbb{R}^2_+$, with $|\hat{\Omega}(t)| = 2t^{1/2}$, $0 \leq t \leq 1$, $\eta(y) = \frac{1}{2} y^{1/2}$, $0 \leq y \leq 1$.

III. GENERAL APPROXIMATIONS TO THE IDENTITY

Our final subject, which we shall only treat briefly, concerns the conditions that need be put on a non-negative integrable function Φ on $\mathbb{R}^n$, so that the family $\Phi_\varepsilon(x) = \Phi(x/\varepsilon)\varepsilon^{-n}$, gives an approximation of the identity in the sense of almost everywhere convergence. Except for an obvious normalization of Φ, the real question is what are conditions on Φ which guarantee that the mapping

$$f \longrightarrow \sup_{\varepsilon>0} |f * \Phi_\varepsilon| \tag{22}$$

is of weak-type $(1,1)$.

A very well-known sufficient condition is that Φ have a majorant which is radial, decreasing, and integrable. In that case, of course it is known that

$$\sup_{\varepsilon>0} |(f*\Phi_\varepsilon)(x)| \leq c\, M(f)(x),$$

where M is the standard maximal functions. But what about other situations, where this majorization does not hold? To gain a better understanding of this problem let us discuss four examples.

Example 1. Here we deal with $\mathbb{R}^2 = \{(x_1,x_2)\}$, and $\Phi(x) = \dfrac{1}{(1+x_1^2)(1+x_2^2)}$. This corresponds to the Poisson kernel in the product of two half-planes, and the main difficulty is due to the slow decrease of Φ at infinity along the two axes. The result in this case goes back to Marcinkiewicz and Zygmund.

Example 2. Our second example is in effect a generalization of the previous one. Here we realize $\mathbb{R}^n$ as the vector space of $m \times m$ real symmetric matrices (with $n = m\frac{(m+1)}{2}$), and Φ corresponds to the Poisson kernel of the corresponding Siegel upper half-space, $\Phi(x) = \dfrac{1}{|\det(x+iI)|^{m+1}}$; see [11, pp. 124-127]. The weak-type result was obtained in [12]. Notice that $\Phi(x)$ has different degrees of decrease at ∞; i.e. the size of $\Phi(rx)$, as $r \to \infty$, $r \in \mathbb{R}_+$, depends on which of the elementary symmetric functions of the eigenvalues of x are non-vanishing; but in any case the non-decreasing radial majorant of Φ is far from integrable at infinity.

Example 3. This example like the previous two is also suggested by symmetric spaces. Here we realize $\mathbf{R}^3$ as the strictly lower triangular matrices of the form

$$\begin{pmatrix} 1 & 0 & 0 \\ x_1 & 1 & 0 \\ x_3 & x_2 & 1 \end{pmatrix}$$

and we take $\Phi = [(1+x_1^2+x_3^2)(1+x_2^2+(x_1x_2-x_3)^2)]^{-1}$.

Because of the multiplication law of matrices, it is natural to define the dilations by $(x_1,x_2,x_3) \to (\varepsilon x_1, \varepsilon x_2, \varepsilon^2 x_3)$ (since these are then automorphisms of the group multiplication), and to redefine Φ_ε as $\Phi(x_1/\varepsilon,\ x_2/\varepsilon,\ x_3/\varepsilon^2)\varepsilon^{-4}$.

Example 4. This example is a more complicated variant of the first example and brings in the rotations of the singular directions of $\Phi(x_1,x_2) = \dfrac{1}{(1+x_1^2)(1+x_2^2)}$. For each $\theta \in \mathbf{T}$ let r_θ denote the rotation by the angle θ in the $\mathbf{R}^2$ plane. Whenever $f(x,\theta) = f(x_1,x_2,\theta) \in L^1(\mathbf{R}^2 \times \mathbf{T})$, define $M(f)$ by

$$M(f)(x) = \sup_{\varepsilon>0} \left| \int_{\mathbf{T}} \int_{\mathbf{R}^2} f(y,\theta)\Phi_\varepsilon(r_\theta(x-y))\,dy\,d\theta \right| \tag{23}$$

and the problem is to prove that the mapping $f \to M(f)$ is of weak-type $(1,1)$. This problem, in a more general context, was raised by Korányi [2] in the setting of symmetric spaces.

We now formulate a general result.

We consider $\mathbf{R}^n$ equipped with a family of dilations

$$(x_1,x_2,\ldots,x_n) \to (\varepsilon^{a_1}x_1,\ \varepsilon^{a_2}x_2,\ldots,\varepsilon^{a_n}x_n) = \varepsilon.x,$$

where a_i are fixed positive exponents, with $\varepsilon > 0$. We let Θ denote the subgroup of rotations of $\mathbf{R}^n$ that commute with these dilations. For $\theta \in \Theta$, we denote by r_θ the corresponding rotation, and write $d\theta$ for the Haar measure on Θ. We now that Φ satisfies the following conditions:

(a) $\Phi(x) = |R(x)|^{\alpha}$, where R is a rational function, and $\alpha > 0$

(b) $\int_{\mathbf{R}^n} \Phi(x)dx < \infty$

(c) $\Phi(\varepsilon.x)$ is a decreasing function for $0 < \varepsilon < \infty$.

We now write $\Phi_{\varepsilon}(x) = \Phi(\varepsilon^{-1}.x)\varepsilon^{-a}$, with $a = a_1+a_2 \ldots + a_n$.

THEOREM 6. *The mapping* $f \to M(f)$, *with*

$$M(f)(x) = \sup_{\varepsilon>0} \int_{\Theta} \int_{\mathbf{R}^n} f(y,\theta)\Phi_{\varepsilon}(r_{\theta}(x-y))dyd\theta$$

is of weak-type (1,1), *if* Φ *satisfies* (a), (b) *and* (c) *above*.

Details, and a generalization where $\mathbf{R}^n$ is replaced by a more general homogeneous group, can be found in [9]; earlier results along these lines are in, [6b], and [1, pp. 261-271].

We cannot here enter into the details of the proof. We shall only remark that the key step of the proof is to show that one has weak-type estimates with Φ replaced by the normalized characteristic function of thin rectangles of eccentricity N. The point then is that these estimates have bounds that increase only logarithmically with N.

References

[1] G.B. Folland and E.M. Stein, "Hardy spaces on homogeneous groups", Math. Notes #28, 1982, Princeton Univ. Press.

[2] A. Korányi, "A survey of harmonic functions on symmetric spaces", Proc. Sym. Pure Math. 35 (1979), I, 323-344.

[3] A. Nagel, W. Rudin, J. Shapiro, "Tangential boundary behavior of functions in Dirichlet type spaces", Ann. of Math. 116 (1982) 331-360.

[4] A. Nagel and E.M. Stein, "On certain maximal functions and approach regions", to appear in Advances in Math.

[5] E.M. Stein, "Topics in harmonic analysis", Annals of Math. Studies #63, (1970), Princeton Univ. Press.

[6] E.M. Stein, "Maximal functions",
(a) "Spherical means", Proc. Nat. Acad. Sci. 73 (1976), 2174-2175;
(b) "Poisson integrals and symmetric spaces", ibid, 2547-2549.

[7] __________, "The development of square functions in the work of A. Zygmund", Bulletin Amer. Math. Soc. 7, (1982), 359-376.

[8] __________, "Some results in harmonic analysis in $\mathbb{R}^n$, for $n\to\infty$", Bulletin Amer. Math. Soc. 9, (1983), 71-73.

[9] __________, "Boundary behavior of harmonic functions on symmetric spaces: Maximal estimates for Poisson integrals", Inventiones Math. 74 (1983), 63-83.

[10] __________, and J.O. Strömberg, "Behavior of maximal functions in $\mathbb{R}^n$ for large n", Arkiv Mat. 21 (1983), 259-269.

[11] __________, and G. Weiss, "Introduction to Fourier analysis on Euclidean spaces", Princeton Univ. Press (1971).

[12] __________, and N.J. Weiss, "On the convergence of Poisson integrals", Trans. Amer. Math. Soc. 140 (1969), 34-54.

Recent Progress in Fourier Analysis
I. Peral and J.-L. Rubio de Francia (Editors)

ESTIMATES FOR FINITE EXPANSIONS OF GEGENBAUER AND JACOBI POLYNOMIALS

Mitchell H. Taibleson*
Washington University in St. Louis
St. Louis, Missouri 63130

Estimates will be given for certain Gegenbauer polynomial and Jacobi polynomial expansions. The primary interest is on expansions of a form that are needed to extend a result of E.M. Stein [6]. Stein shows that there is a function $h \in H^1(\mathbb{R}^n)$ (resp., $h \in H^1(\mathbb{T}^n)$) such that the Bochner-Riesz means of h at the critical index $(n-1)/2$ diverge a.e. He accomplishes this as a consequence of the construction of a sequence $\{h_N\}$ that is bounded in H^1, yet the maximal operator associated with the Bochner-Riesz sum at the critical index is greater than $c \log N$ on a set of measure at least d, where c and d are positive constants.

The major elements in Stein's construction are: (1) sharp asymptotic estimates for the summability kernels, (2) number theoretic estimates that take advantage of the structure of the kernels, and (3) an intricate way of putting it all together. All of (1) and (2) and most of (3) works, almost without variation, on Σ_n. A trivial change is that on Σ_n it is convenient to use Cesàro summability.

That part of Stein's argument which can be dealt with by a wave of the hand (on $\mathbb{R}^n$) is the only part that creates special difficulty on Σ_n. Consider three families of functions: $\{f_R\}$, $\{g_R\}$, $\{\alpha_R\}$, $R \geq 1$ defined on $\mathbb{R}^n$ and all radial. Suppose that their respective Fourier transforms are C^∞, have compact support and satisfy the following conditions:

* This research was supported by NSF grant MCS 7903 122.

$$(1)\quad \hat{f}_R(\xi) = \begin{cases} 0, & 0 \le |\xi| \le R/4 \\ 1, & R/2 \le |\xi| \le 2R, \\ 0, & |\xi| \ge 4R \end{cases}$$

$$(2)\quad \hat{g}_R(\xi) = \begin{cases} 1, & 0 \le |\xi| \le R/4 \\ 0, & |\xi| \ge R/2 \end{cases}, \text{ in fact } \hat{g}_R(\xi)=1-\hat{f}_R(\xi),\ |\xi|\le 2R,$$

$$(3)\quad \hat{\alpha}_R(\xi) = \begin{cases} (1 - |\xi|^2/R^2)^{(n-1)/2}, & |\xi| \le R/2 \\ 0 & |\xi| \ge 2R \end{cases}$$

One needs to know that $\{g_R\}$ is bounded in L^1, that $\{f_R\}$ is bounded in H^1 and that the $\{\alpha_R\}$ satisfy a size condition which implies that the maximal function $\sup_{R \ge 1}|\alpha_R * g(x)|$ is dominated by a multiple of the Hardy-Littlewood maximal function. On R^n this is done by taking f_1, g_1, and α_1 as they come and then setting

$$f_R(x) = R^n f_1(Rx),\ g_R(x) = R^n g_1(Rx),\ \alpha_R(x) = R^n \alpha_1(Rx).$$

The remainder of this paper is concerned with the construction of such functions on Σ_n, the n-dimensional sphere in $\mathbb{R}^{n+1}$, as well as on other compact rank 1 symmetric spaces. When we consider Σ_n we always assume that $n \ge 2$. (See [5] where the implications of the existence of an h in H^1 with divergent means at the critical index is discussed).

For $\alpha > 0$ let $\{C_k^{(\alpha)}\}$ be the sequence of Gegenbauer (ultraspherical) polynomials. The sequence $\{C_k^{(\alpha)}\}$ is an orthogonalization of the sequence of monomials $\{t^k\}_{k=0}^{\infty}$ on $[-1,1]$ with respect to measure $(1 - t^2)^{(\alpha-(1/2))}\,dt$, normalized so $C_k^{(\alpha)}(1) = \binom{k+2\alpha-1}{k}$. It is usual to set $\cos\theta = t$, $0 \le \theta \le \pi$. The zonal spherical harmonic of degree k on Σ is the function

$$(4)\qquad (k+\alpha)\, C_k^{(\alpha)}(\cos\theta),\quad \alpha = (n-1)/2.$$

The kernel on Σ_n that induces the multiplier $\{\mu_k\}$ under "spherical convolution" is given by the distribution:

$$(5)\qquad M(\cos\theta) \sim \sum_{k=0}^{\infty} \mu_k (k+\alpha)\, C_k^{(\alpha)}(\cos\theta),\quad 0 \le \theta \le \pi$$

Good references on these matters are the paper of Bonami and Clerc [3] and the treatise of Szegö [7].

Let us define: $\Delta^0 \mu_k = \mu_k$, $\Delta^1\mu_k = \mu_k - \mu_{k+2}$, $\Delta^{s+1}\mu_k = \Delta^1(\Delta^s\mu_k)$, $s = 1,2,\ldots$ The estimate of primary interest in this paper is the conclusion of the following theorem:

Theorem. *Suppose* R *is a positive integer and* $\{\mu_k\}$ *is a sequence of real numbers with*

(6) $\mu_k = 0$ *if* $k > R$, *and*

(7) $|\Delta^\ell \mu_k| \leq A\, R^{-\ell}$, $\ell = 0,1,\ldots, [(n+3)/2]$.

If M *is defined by (5) with* $\alpha = (n-1)/2$ *there is a constant* B_n *(depending only on* n*) such that*

(a) $|M(\cos\theta)| \leq A\, B_n\, R^n$, $n \geq 2$.

(b) *If* n *is even*

$$|M(\cos\theta)| \leq \begin{cases} A\, B_n\, R^{-1/2}\theta^{-(n+1/2)}, & 0 < \theta \leq \pi/2 \\ A\, B_n\, R^{-1/2}(\pi-\theta)^{-(n+1/2)}, & \pi/2 \leq \theta < \pi \end{cases}$$

(c) *If* n *is odd*

$$|M(\cos\theta)| \leq \begin{cases} A\, B_n(1 + \log^+(R\,\theta))R^{-1}\,\theta^{-(n+1)}, & 0 < \theta \leq \pi/2 \\ A\, B_n(1+\log^+(R(\pi-\theta))R^{-1}(\pi-\theta)^{-(n+1)}, & \pi/2 \leq \theta \leq \pi \end{cases}$$

Think now of (1), (2), and (3) as describing conditions on sequences $\{\hat{f}_R(k)\}$, $\{\hat{g}_R(k)\}$, and $\{\hat{\alpha}_R(k)\}$. The simplest way to construct sequences that satisfy (6) and (7) is to start with a function ψ that is C^∞ on $[0,\infty]$, supported on $[0,1]$ and set $\mu_k = \psi(k/R)$. However, for the specific application we have in mind we need to consider a version of $\hat{\alpha}_R(k)$ for which, instead of (3), we have

$$(8) \qquad \hat{\alpha}_R(k) = \begin{cases} \binom{R-k+\frac{n-1}{2}}{R-k} \Big/ \binom{R+\frac{n-1}{2}}{R}, & 0 \leq k \leq [(R+1)/2] \\ 0, & k \geq 2R \end{cases}$$

That is, for the smaller values of k, $\hat{\alpha}_R(k)$ is the coefficient for Cesaro summability at the critical index. It is easy to show that $\{\hat{\alpha}_R(k)\}$ can be constructed to satisfy (8) and (7) uniformly in R. The estimates of the theorem show (via typical computations) that if h is integrable on Σ_n and $\alpha_R(\cos\theta) = \Sigma\hat{\alpha}_R(k)(k+\alpha)\, C_k^{(\alpha)}(\cos\theta)$

then

(9) $\sup_{R \geq 1} |(\alpha_R * h)(x)| \leq D(h^*(x) + h^*(\tilde{x}))$,

where $\tilde{x}$ is the antipode to x, h^* is the Hardy-Littlewood maximal function and D is a positive constant.

Clearly we may construct the families $\{g_R\}$ and $\{f_R\}$ to satisfy (2) and (1) as well as (7). It is an immediate consequence of the estimates of the theorem that $\{g_R\}$ is a bounded sequence in L^1 (and that $\{f_R\}$ is a bounded sequence in L^1). To complet the program we will show that $\{f_R\}$ is bounded in H^1. The point is that $\{f_R\}$ is a bounded family of H^1-molecules in the sense of [8]. The idea behind such molecules is that they have a canonical decomposition in H^1-atoms. The argument of [8] is in the setting of $\mathbb{R}^n$ and the disk in $\mathbb{C}$, but the argument for Σ_n is essentially the same. It is only necessary to allow for growth at the antipode of the center of the molecule, as well as to allow for exceptional atoms. Atomic characterizations of the H^p-spaces are discussed by Colzani [4]. Since $\hat{f}_R(0) = 0$ one sees that $\int_{\Sigma_n} f_R(u)d\mu(u) = 0$. The estimates of the theorem show that $\{f_R\}$ is a uniformly bounded sequence of $(1, \infty, 0, 1/2n)$-molecules.

The theorem follows from a proposition that is a direct consequence of an observation first made by Askey and Wainger [2] to obtain estimates for expansions of the form

(10) $\Sigma\, k^{-\beta} b(k)(k+\alpha)\, C_k^{(\alpha)}$, $\beta > 0$ and b slowly varying.

Before procedding we will gather some facts about orthogonal polynomials that we will need below.

The sequence of Jacobi polynomials $\{P_k^{(\delta,\beta)}\}$, $\delta, \beta > -\frac{1}{2}$, is an orthogonalization of the sequence of monomials $\{t^k\}_{k=0}^{\infty}$ on $[-1,1]$ with respect to the measure $(1-t)^{\delta}(1+t)^{\beta}\,dt$, normalized so $P_k^{(\delta,\beta)}(1) = \binom{k+\delta}{k}$. Up to a multiplicative constant the Gegenbauer polynomials are special cases of Jacobi polynomials:

(11) $$C_k^{(\alpha)} = \frac{\Gamma(2\alpha+k)\Gamma(\alpha+1/2)}{\Gamma(2\alpha)\Gamma(\alpha+k+1/2)} P_k^{(\alpha-1/2,\, \alpha-1/2)}$$

See [1] 22.5.20.

From Bonami and Clerc [3] p. 231, we have the estimates:

$$(12)\qquad |P_k^{(\delta,\beta)}(\cos\theta)| \leq \begin{cases} B(k+1)^{\delta} \\ B(k+1)^{-1/2}\,\theta^{-(\delta+1/2)} \end{cases}\Big\} 0 < \theta \leq \pi/2 \quad \begin{cases} B(k+1)^{-1/2}(\pi-\theta)^{-(\beta+1/2)} \\ B(k+1)^{\beta} \end{cases}\Big\} \pi/2 < \theta \leq \pi$$

where B is a positive constant that does not depend on k or θ.

The ratio of gamma functions in (11) is on the order of $(k+1)^{(\alpha-1/2)}$ as $k \to \infty$. From (11) and (12) follows:

$$(13)\qquad |(k+\alpha)\, C_k^{\alpha}(\cos\theta)| \leq \begin{cases} B(k+1)^{2\alpha}, & 0 \leq \theta \leq \pi \\ B(k+1)^{\alpha}\theta^{-\alpha}, & 0 < \theta \leq \pi/2, \\ B(k+1)^{\alpha}(\pi-\theta)^{-\alpha}, & \pi/2 \leq \theta < \pi \end{cases}$$

where B is a positive constant that does not depend on k or θ.

From Abramowitz and Stegun [1] 22.7.23 we have

$$(14)\qquad (k+\alpha)\, C_k^{(\alpha)} = \alpha(C_k^{(\alpha+1)} - C_{k-1}^{(\alpha+1)}),\quad (C_{-2}^{(\beta)} \equiv C_{-1}^{(\beta)} \equiv 0 \text{ all } \beta)$$

For convenience let us assume that $\{\mu_k\}$ has only a finite number of non-zero terms. A consequence of (14) for $s = 1,2,3,\ldots$ is

$$(15)\qquad \sum_k \mu_k(k+\alpha)\, C_k^{(\alpha)} = \alpha \sum_k \mu_k(C_k^{(\alpha+1)} - C_{k-1}^{(\alpha+1)})$$
$$= \alpha \sum_k \frac{1}{k+(\alpha+1)}\, \Delta\mu_k(k+(\alpha+1))\, C_k^{(\alpha+1)}$$
$$= (\alpha)_s \sum_k \frac{1}{k+(\alpha+s)}\, (\Delta(\ldots\Delta(\frac{1}{k+(\alpha+1)}\,\Delta\mu_k)))(k+(\alpha+s))C_k^{(\alpha+s)},$$

where $(\alpha)_s = \alpha(\alpha+1)\ \ldots\ (\alpha+s-1)$.

Proposition A. *Suppose* $\alpha > 0$, $\{C_k^{\alpha}\}$ *is the sequence of Gegenbauer polynomials and* $\{\mu_k\}$ *is a finite sequence. Then*

$$(16)\qquad |\Sigma\mu_k(k+\alpha)\, C_k^{(\alpha)}| \leq B_{\alpha} \sum_k |\mu_k|(k+1)^{2\alpha},$$

and for $s = 1,2,3,\ldots$

$$(17)\qquad |\sum_k \mu_k(k+\alpha)\, C_k^{(\alpha)}(\cos\theta)|$$
$$\leq B_{\alpha,s} \sum_k \sum_{\ell=1}^{s} \frac{1}{(k+1)^{2s-\ell}}|\Delta^{\ell}\mu_k||(k+\alpha+s)C_k^{(\alpha+s)}(\cos\theta)|$$

$$\leq B_{\alpha,s}\sum_{\ell=1}^{s}\sum_{k}|\Delta^{\ell}\mu_k|\begin{cases}(k+1)^{(2\alpha+\ell)} & , 0\leq\theta\leq\pi\\ (k+1)^{(\alpha-s+\ell)}\theta^{-(\alpha+s)} & , 0\leq\theta\leq\pi/2\\ (k+1)^{(\alpha-s+\ell)}(\pi-\theta)^{-(\alpha+s)} & , \pi/2\leq\theta<\pi\end{cases}$$

The proposition is an immediate consequence of (15) and (13).

<u>Proof of the Theorem</u>. In Proposition A we set $\alpha = (n-1)/2$, $s = [(n-1)/2] + 2$, and recall that $\mu_k = 0$ if $k > R$. Since $|\mu_k| \leq A$ we have from (16), $|M(\cos\theta)| \leq A\,B_n \sum_{k=0}^{R}(k+1)^{n-1} \leq A\,B_n R^n$.

Examine the exponent $\alpha-s+\ell$ in (17). It is equal to: $(n-1)/2 - [(n-1)/2] + \ell-2$, which is $\ell-(3/2)$ if n is even and is $(\ell-2)$ if n is odd. Since $\ell \geq 1$ we see that if n is even or n is odd and $\ell \neq 1$, then $\alpha-s+\ell > -1$. Suppose now that $0 < \theta \leq \pi/2$. We have $|\Delta^{\ell}\mu_k| \leq A\,R^{-\ell}$, $\Delta\mu_k = 0$ if $k > R$, $\ell = 1,2,3,\ldots,$ or s. If n is even, or n is odd and $\ell \neq 1$, we have the estimate:

$$B_n\sum_{k=0}^{R} A\,R^{-\ell}(k+1)^{\alpha-s+\ell}\,\theta^{-(\alpha+s)} \leq A\,B_n\,R^{-\ell}\,\theta^{-(\alpha+s)}\sum_{k=0}^{R}(k+1)^{\alpha-s+\ell}$$

$$\leq A\,B_n\,R^{-\ell}\,\theta^{-(\alpha+s)}R^{\alpha-s+\ell+1} = A\,B_n\,R^{\alpha-s+1}\,\theta^{-(\alpha+s)}$$

$$= \begin{cases}A\,B_n\,R^{-1/2}\,\theta^{-(n+1/2)}, & n \text{ even}\\ A\,B_n\,R^{-1}\,\theta^{-(n+1)}, & n \text{ odd}\end{cases}$$

For the exceptional case we have the estimate:

$$B_n\sum_{k=0}^{R} A\,R^{-1}\begin{cases}(k+1)^{n}\\ (k+1)^{-1}\,\theta^{-(n+1)}\end{cases}$$

$$\leq A\,B_n\,R^{-1}\left[\sum_{k=0}^{[1/\theta]}(k+1)^n + \sum_{k=[1/\theta]+1}^{R}(k+1)^{-1}\,\theta^{-(n+1)}\right]$$

$$\leq A\,B_n\,R^{-1}\,\theta^{-(n+1)}[1 + \log^{+}(R\theta)].$$

This establishes the theorem for $0 \leq \theta \leq \pi/2$. The case $\pi/2 \leq \theta \leq \pi$ follows from an analogous argument.

We now consider an analogue of Proposition A for Jacobi polynomials. There are several reasons why this is of interest. First, the Jacobi polynomials contain the Gegenbauer polynomials as special cases (see (11)). Second, the zonal functions for all compact symmetric spaces of rank 1 are obtained as special cases of Jacobi polynomials: Σ_n and $P_n(\mathbb{R})$, $n \geq 2$, $\{P_k^{(\frac{n}{2}-1,\,\frac{n}{2}-1)}\}$; $P_n(\mathbb{C})$, $n \geq 4$,

$\left\{P_k^{(\frac{n}{2}-1,0)}\right\}$; $P_n(\mathbb{H})$, $n \geq 8$, $\left\{P_k^{\frac{n}{2}-1,1}\right\}$; and the exceptional case, P_{16}, $\{P_k^{(7,3)}\}$. See [3] p. 257. Third, even for the special case of Σ_n, the estimates for the "lower hemisphere" can be improved significantly. In comparison with (9) we can obtain, using Proposition B, below:

$$(18) \quad |\alpha_R * h(x)| \leq \begin{cases} D(h^*(x) + R^{-1/2} h^*(\tilde{x})), & n \text{ even} \\ D(h^*(x) + R^{-1} \log R\, h^*(\tilde{x})), & n \text{ odd} \end{cases}$$

$$(19) \quad \limsup_{R \geq 1} |\alpha_R * h(x)| \leq D\, h^*(x).$$

The expansions of interest (in analogy with (5)) are of the form:

$$(20) \quad M(\cos\theta) \sim \sum_R \mu_R\, w_k^{(\delta,\beta)}\, P_k^{(\delta,\beta)}(\cos\theta)$$

where

$$(21) \quad w_k^{(\delta,\beta)} = \frac{2k+\delta+\beta+1}{2^{\delta+\beta+1}} \frac{\Gamma(k+\delta+\beta+1)}{\Gamma(\delta+1)\Gamma(k+\beta+1)}.$$

From Abramowitz and Stegun [1] 22.7.18 we have

$$(22) \quad (2k+\delta+\beta+1)\, P_k^{(\delta,\beta)} = (k+\delta+\beta+1)\, P_k^{(\delta+1,\beta)} - (k+\beta)\, P_{k-1}^{(\delta+1,\beta)},$$

where $P_{-1}^{(\delta,\beta)} \equiv 0$ for all δ,β.

From (21) and (22) we have

$$(23) \quad w_k^{(\delta,\beta)}\, P_k^{(\delta,\beta)} = $$
$$= 2(\delta+1)\left\{\frac{w_{k-1}^{(\delta+1,\beta)}}{2k+(\delta+1)+\beta+1} P_k^{(\delta+1,\beta)} - \frac{w_{k-1}^{(\delta+1,\beta)}}{2(k-1)+(\delta+1)+\beta+1} P_{k-1}^{(\delta+1,\beta)}\right\}.$$

Let $\Delta\mu_k = \mu_k - \mu_{k+1}$. For s a positive integer we have

$$(24) \quad \Sigma_k\, \mu_k\, w_k^{(\delta,\beta)}\, P_k^{(\delta,\beta)} = 2(\delta+1) \sum_k \frac{1}{2k+(\delta+1)+\beta+1} \Delta\mu_k\, w_k^{(\delta+1,\beta)} P_k^{(\delta+1,\beta)}$$
$$= 2^s (\delta+1)_s \sum_k \frac{1}{2k+(\delta+s)+\beta+1} \left(\Delta(\ldots\Delta(\frac{1}{2k+(\delta+1)+\beta+1} \Delta\mu_k)))w_k^{(\delta+s,\beta)} P_k^{(\delta+s,\beta)}\right.$$

From the observation that $w_k^{(\delta,\beta)}$ is on the order of $k^{\delta+1}$ as $k \to \infty$ and (12) we have:

$$(25)\quad |w_k^{(\delta,\beta)}\, P_k^{(\delta,\beta)}(\cos\theta)| \le \begin{cases} \left.\begin{array}{l} B(k+1)^{2\delta+1} \\ B(k+1)^{\delta+1/2}\theta^{-(\delta+1/2)} \end{array}\right\} 0<\theta\le\pi/2 \\ \left.\begin{array}{l} B(k+1)^{\delta+1/2}(\pi-\theta)^{-(\beta+1/2)} \\ B(k+1)^{\delta+\beta+1} \end{array}\right\} \pi/2\le\theta<\pi \end{cases}$$

Proposition B. *Suppose* $\delta,\beta > -\frac{1}{2}$, $\{P_k^{(\delta,\beta)}\}$ *is the sequence of Jacobi polynomials and* $\{\mu_k\}$ *is a finite sequence. Then*

$$(26)\quad |\sum_k \mu_k\, w_k^{(\delta,\beta)} P_k^{(\delta,\beta)}| \le \begin{cases} B_{\delta,\beta}\Sigma|\mu_k|(k+1)^{2\delta+1}, & 0 \le \theta \le \pi/2 \\ B_{\delta,\beta}\Sigma|\mu_k|(k+1)^{\delta+\beta+1}, & \pi/2 \le \theta \le \pi \end{cases}$$

and for $s = 1,2,3,\ldots$

$$(27)\quad |\sum_k \mu_k\, w_k^{(\delta,\beta)}\, P_k^{(\delta,\beta)}(\cos\theta)|$$
$$\le B_{\delta,\beta,s} \sum_k \sum_{\ell=1}^{s} \frac{1}{(k+1)^{2s-\ell}} |\Delta^\ell \mu_k|\,|w_k^{(\delta+s,\beta)} P_k^{(\delta+s,\beta)}(\cos\theta)|$$
$$\le B_{\delta,\beta,s} \sum_{\ell=1}^{s} \sum_k |\Delta^\ell \mu_k| \begin{cases} \left.\begin{array}{l} (k+1)^{2\delta+\ell+1} \\ (k+1)^{\delta-s+\ell+1/2}\,\theta^{-(\delta+s+1/2)} \end{array}\right. 0 < \theta \le \pi/2 \\ \left.\begin{array}{l} (k+1)^{\delta-s+\ell+1/2}(\pi-\theta)^{-(\beta+1/2)} \\ (k+1)^{\delta+\beta-s+\ell+1} \end{array}\right\} \pi/2 \le \theta < \pi \end{cases}$$

Proposition B is a direct consequence of (24) and (25).

The estimates for the kernels on Σ_n that are obtained from Proposition B with $\delta = \beta = \frac{n}{2} - 1$ are the same as those obtained from Proposition A on the upper hemisphere (i.e., $0 \le \theta \le \pi/2$). For the lower hemisphere $(\pi/2 \le \theta \le \pi)$ there is a substantial improvement. (The only difference in the argument is to use (27) for both parts of the estimate). A comparison of the estimates obtained from the two propositions follows:

From Proposition A, $\pi/2 \le \theta < \pi$

$$|M(\cos\theta)| \le \begin{cases} B_n \min\{R^n, R^{-1/2}(\pi-\theta)^{-(n+1/2)}\}, & n \text{ even} \\ B_n \min\{R^n, R^{-1}(\pi-\theta)^{-(n+1)}(1+\log^+(R(\pi-\theta))\}, & n \text{ odd} \end{cases}$$

From Proposition B, $\pi/2 \le \theta < \pi$

$$|M(\cos\theta)| \le \begin{cases} B_n \min\{R^{\frac{n}{2}-1}, R^{-\frac{1}{2}}(\pi-\theta)^{-(n-1)/2}\}, & n \text{ even} \\ B_n \min\{R^{\frac{n}{2}-1}, (\pi-\theta)^{-(n-1)/2}(1+\log^+(R(\pi-\theta))\}, & n \text{ odd} \end{cases}$$

References

[1] M. Abramowitz and I. Stegun, Handbook of Mathematical Functions, Dover, New York, 1970.

[2] R. Askey and S. Wainger, On the behaviour of special classes of ultraspherical expansions, I., Journ. d. Anal. Math. 15 (1965) 193-220.

[3] A. Bonami and J. L. Clerc, Sommes de Cesaro et multiplicateurs des developpments en harmonique sphériques, Trans. Amer. Math. Soc., 183 (1973) 223-263.

[4] L. Colzani, Hardy and Lipschitz spaces on unit sphere, Ph. D. dissertation, Washington University, St. Louis, MO, 1982.

[5] L. Colzani, M. Taibleson, and G. Weiss, Cesaro and Riesz means on the unit sphere. To appear.

[6] E. M. Stein, An H^1 function with non-summable Fourier expansion, to appear in Proc. of Cortona Conf. on harmonic analysis, 1982.

[7] G. Zsegö, Orthogonal Polynomials, Amer. Math. Soc. Colloq. Pub. No. 23, 4th ed. Providence, RI, 1975.

[8] M. Taibleson and G. Weiss, The molecular characterization of certain Hardy spaces, Asterisque 77 (1980) 67-149.

Recent Progress in Fourier Analysis
I. Peral and J.-L. Rubio de Francia (Editors)

BALLS DEFINED BY VECTOR FIELDS

Stephen Wainger[(1)]

University of Wisconsin

Recently Alexander Nagel, Eli Stein and I have written a joint paper entitled "Balls and metrics defined by vector fields I: Basic Properties" [NSW2]. My purpose here is to give an expository account of the motivation for the problems considered in that paper. Much of the motivation of the problems in that paper is from the field of several complex variables. I have learned a great deal about several complex variables from unpublised insights into several complex variables that Nagel and Stein have shared with me as well as from the beautiful talk Steve Krantz gave at the Williamstown meeting in 1978, I have incorporated many of these remarks of Krantz, Nagel, and Stein into this paper.

The problems of several complex variables with which we shall be concerned with arise in a program of Stein to understand analogues in several complex variables of certain well known phenomena in the theory of functions of one complex variable.

Let us begin then by recalling two kinds of problems (perhaps not completely distinct) that arise in the study of functions of one complex variable. The first type of problem concerns the boundary behaviour of functions holomorphic in the unit disc. Let T denote a small open isosceles triangle with base angles of say 45°. Then let T_θ be a congruent copy of T with the 90° vertex at $e^{i\theta}$, lying inside $|z| < 1$, and with axis of symmetry falling along a radius of $|z| = 1$ (see Figure I).

(1) Supported in part by an N.S.F. grant at the University of Wisconsin

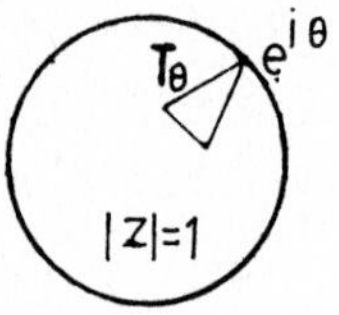

Figure I

A typical result of the first type of problem is that

1) $$\lim_{\substack{z \in T_\theta \\ z \to e^{i\theta}}} f(z) = f(e^{i\theta})$$

exists almost every where if f is in an appropriate function class. Two operators that are studied in this class of problems are the non-tangential maximal function and the area integral. The maximal function is defined by

2) $$Mf(\theta) = \sup_{z \in T_\theta} |f(z)|,$$

The area integral is given by

3) $$Sf(\theta) = \{\int_{T_\theta} |f'(z)|^2 \, dxdy\}^{1/2}.$$

Some of the main theorems in one variable assert that $f \in H^p \iff Mf \in L^p \iff Sf \in L^p, \quad 0 < p < \infty.$

The second type of problem we wish to recall is the study of natural singular integrals arising in the theory of one complex variable. "Natural" does not have a precise meaning, but surely the operator

4) $$Tf(z) = \frac{1}{2\pi i} \int_\Gamma \frac{f(\zeta)}{\zeta - Z} \, d\zeta$$

which expresses the value of a holomorphic function inside a disc bounded by a circle Γ in terms of its boundary values on Γ is a natural operator. The study of T then leads us to the singular integral transformation

5) $$Hf(\phi) = \frac{1}{2\pi} \int_0^{2\pi} \frac{f(e^{i\theta})}{e^{i\theta} - e^{i(\theta-\phi)}} \, d\theta$$

Now our understanding of both types of problems is intimately tied up with the standard notion of ball or interval on the unit circle. In fact the approach region T_θ and the balls have a simple relation. Let Pz denote the projection of z onto the unit circle, $Pz = \frac{z}{|z|}$. Let $B(\theta,\delta)$ denote the interval of points ζ on the unit circle such that $|e^{i\theta} - \zeta| < \delta$. Then, if r denotes $|z|$,

$$T_\theta = \{z \mid |z| < 1, \quad Pz \in B(\theta, C(1-r))\}$$

for a fixed positive C. So to know the balls is the same as to know the approach regions, T_θ.

Thus even to define Mf and Sf we must know the appropriate balls. Moreover as is well known a complete analysis of the singular integral Hf requires the use of balls.

We must focus on the properties of balls that make the arguments for maximal functions and singular integrals work. They are

A) $\lim_{\delta \to 0} B(\theta,\delta) = e^{i\theta}$

B) If $\delta_1 < \delta_2$

$$B(\theta,\delta_1) \subset B(\theta,\delta_2).$$

C) If $\delta_1 > \frac{\delta_2}{2}$ and $B(\theta, \delta_1) \cap B(\phi,\delta_2) \neq \phi$,

$$B(\theta, C\,\delta_1) \supset B(\phi,\delta_2)$$

for some fixed C.

D) volume $\{B(\theta,2\delta)\} \leq C$ Volume $\{B(\theta,\delta)\}$ for some fixed C.

See [S] or [CW].

It is natural then that balls having properties A, B, C, and D should play a role in the study of our problems in several complex variables. It will turn out that the balls in several complex variables will not be the usual balls. To see that the appropriate balls in several complex variables will be different from the usual balls, let us consider the following related but trivial problems suggested to me by Alex Nagel: If $f(z)$ is holomorphic and bounded in the unit disc, $|z| < 1$, how fast can $|f'(z)|$ grow as $|z| \to 1$? Moreover, we wish to study the analogous question in two complex variables.

The problem in one dimension is a standard excercise in a course in complex variables. Let us suppose for the sake of concreteness that z is the point $1-\varepsilon$. We then write down the differentiated form of Cauchy's integral theorem

$$f'(1-\varepsilon) = \frac{1}{2\pi i} \int_{\Gamma_\varepsilon} \frac{f(\zeta)}{(\zeta - (1-\varepsilon))^2}\, d\zeta$$

where

$$\Gamma_\varepsilon = \{\zeta \mid |\zeta - (1-\varepsilon)| = \tfrac{\varepsilon}{2}\}.$$

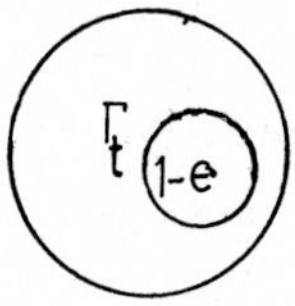

Figure II

We then get the estimate 6) $|f'(1-\varepsilon)| \leq \frac{C}{\varepsilon}$.

Of course if a much bigger disc with center at $1-\varepsilon$ fit into the unit disc a better (and false) estimate could be obtained. It is interesting to note that even though the estimate is optimal, it is possible to put an ellipse inside $|z| = 1$ containing $1-\varepsilon$ which is essentially larger than Γ_ε. See figure III.

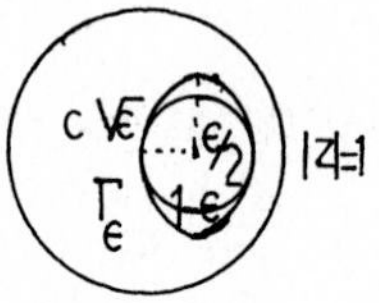

Figure III

This ellipse will be of diameter ε along the x-axis, but of length $C\sqrt{\varepsilon}$ in the y-direction. Thus we see we can obtain an improvement in an estimate like 6) only by putting in full discs.

Let us now consider the case of a function $f(z,\zeta)$ bounded and holomorphic in the unit ball $|z|^2 + |\zeta|^2 < 1$. We would like to consider the size of $\frac{\partial f}{\partial z}$ and $\frac{\partial f}{\partial \zeta}$ at a point close to the boundary. To be precise let us choose the point $z = 1-\varepsilon$, $\zeta = 0$. The same argument as in the one dimensional case shows

$$7) \qquad \left| \frac{\partial f}{\partial Z}\bigg|_{1-\varepsilon,0} \right| \leq \frac{C}{\varepsilon}$$

Let us now consider $\frac{\partial f}{\partial \zeta}$. Notice that we can put inside the unit ball a disc of radius $C\sqrt{\varepsilon}$ with center at $1-\varepsilon, 0$, namely the disc $z = 1-\varepsilon$, $|\zeta| < C\sqrt{\varepsilon}$.

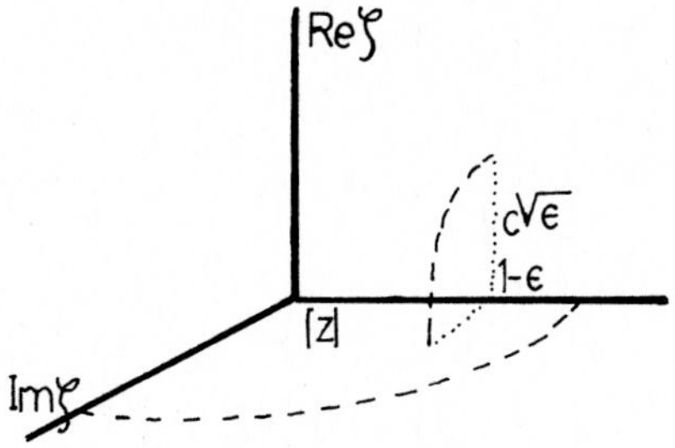

Figure IV

We may use the Cauchy integral Theorem in ζ on this large disc to calculate $\frac{\partial f}{\partial \zeta}$. We obtain

8) $$\left|\frac{\partial f}{\partial \zeta}\right| \le \frac{C}{\sqrt{\varepsilon}}.$$

In analogy with the one dimensional situation we might then expect the approach reach region at $(1,0)$ to be those pairs z, ζ with $\text{Re } z = \varepsilon$, $|\text{Im } z| < \varepsilon$, $|\zeta| < C\sqrt{\varepsilon}$, $\varepsilon > 0$ and small. Furthermore the general relation between approach regions and balls suggests that the "ball" about $(1,0)$ on $|\zeta|^2 + |z|^2 = 1$ of radius ε should have length ε in the z-direction and length $\sqrt{\varepsilon}$ in the ζ-direction.

To have a clearer picture and to go further it is convenient to think in terms of the Siegel upper half space $\text{Im } z_1 > |z_2|^2$. (Here we let z_1 and z_2 denote 2 complex variables).

The domain is a cylinder in 4-space as the defining equation of the boundary is independent of $\text{Re} z_1$. In the 3-space $\text{Re } z_1 = 0$, the domain looks like a parabaloid. In many ways this domain is more natural than the unit ball. (For example the boundary of this domain is a group, the so called Heisenberg group).

Our approach region to the origin at "height" ε are now the points $\text{Im } z_1 = \varepsilon$, $|\text{Re } z_1| < \varepsilon/2$ $|z_2| < \frac{1}{8}\sqrt{\varepsilon}$. (We could insert figures in our domain which are arbitrarily long in the direction $\text{Re } z_1$, but this is of no consequence just as was the insertion of "long ellipses" in the unit disc in one variable.

Furthermore the bull about $(0,0)$ on $\text{Im } z_1 = |z_2|^2$ of radius ε is that portion of the surface lying above

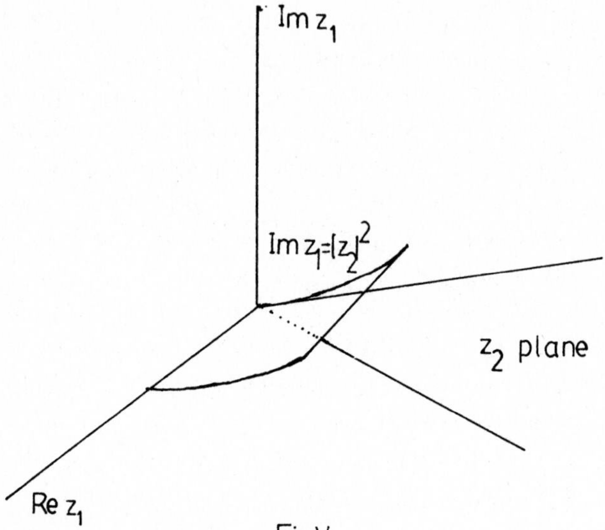

Fig V

$\operatorname{Im} z_1 > |z_2|^2 \qquad \operatorname{Re} z_1 = 0$

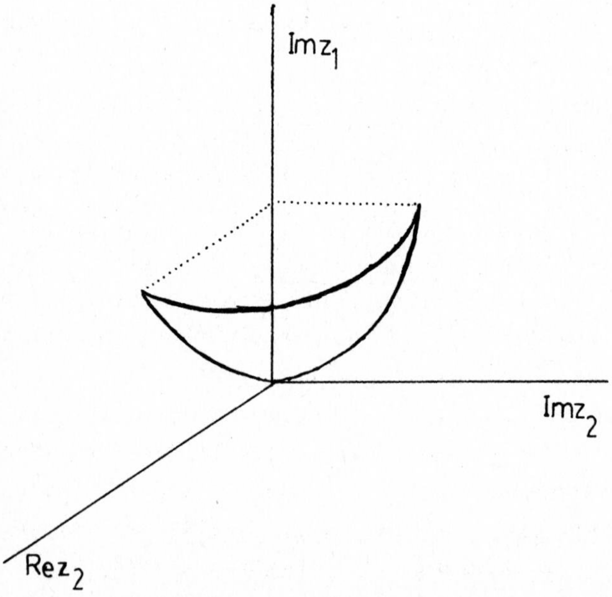

Fig VI

$$|z_2| < \sqrt{\varepsilon}$$
$$|\mathrm{Re}\, z_1| < \varepsilon$$

We must now consider how to tell in general what are the directions in which our balls should be long. Let us write $z_1 = x_1+iy_1$ and $z_2 = x_2+iy_2$. Then the tangent space at the origin to the surface $\mathrm{Im}\, z_1 = |z_2|^2$ consists of points $iy_1 + z_2$. Notice that if $y_1 = 0$, $i(iy_1 + z_2) = i(0 + z_2) = 0 + iz_2$ is again in the tangent space. While if $y_1 \neq 0$, $i[iy_1 + z_2]$ is no longer in this tangent space. This turns out to be the clue to the general situation. That is at a point on the boundary of a domain in $\mathbf{C}^2$ the real 3 dimensional tangent space can be written as a direct sum of a complex vector space, cone invariant under multiplication by i, and a complementary one.

A direction pointing in the direction of this complex vector space is called a good direction. It is in this direction that one can embed "long" complex discs in the domain. It is in this direction that we can expect the approach regions to be long, and it is in these directions that we can expect balls to be long.

It is sometimes convenient (and very important for our story) to express this concept of good directions in terms of vector fields. Suppose we have a smooth domain $\mathcal{D}$ in $\mathbf{C}^2$ and a smooth vector field L with non-vanishing real and imaginary parts defined on $\partial\mathcal{D}$ (the boundary of $\mathcal{D}$) which lies in the tangent space to $\partial\mathcal{D}$ at each point of $\partial\mathcal{D}$. If iL is also a tangent vector at each point (or equivalently if the real and imaginary parts of L are also tangent vectors) L is called a Levy vector field. Then the real and imaginary parts of L point in the good directions. If $\mathcal{D}$ is defined by $r(z_1,z_2) < 0$ (with grad $r \neq 0$ when $r = 0$) so that the boundary of $\mathcal{D}$ is defined by $r(z_1,z_2) = 0$ then

$$L = \frac{\partial r}{\partial z_1}\frac{\partial}{\partial z_2} - \frac{\partial r}{\partial z_2}\frac{\partial}{\partial z_1}$$

defines a Levy vector. Notice that both L and $\overline{L}$ annhilate r and so are tangential.

In general then we can define a family of balls on $\partial\mathcal{D}$ by taking them to have length $\sqrt{\varepsilon}$ in the good directions and length ε in the complementary directions. It turns out that these balis satisfy properties A), B), C) and D) given earlier. Also these balls turn out to be the appropriate balls for what are called strictly pseudo-convex domains. (One definition of strictly pseudo-convex domain is that for some defining function r, the matrix

$$\left(\frac{\partial^2 r}{\partial \bar{z}_j \partial z_i}\right)$$

is strictly positive definite. Another definition is that locally the domain is biholomorphically equivalent to a strictly convex domain). Note that $\operatorname{Im} z_1 > |z_2|^2$ is not strictly convex as it stands since the line $\operatorname{Im} z_1 = 0$, $z_2 = 0$ lies in the boundary. For the special case of the ball and Siegel upper half space appropriate results on the maximal function and singular integrals were obtained by Koranyi and Vagi, [KV]. The problems of type 1 were solved by Stein [S].

For general strictly pseudo convex domains. For general strictly pseudo convex domains the main results are to a large extend found in [FS], [GS], [KR], [KS], [PS], [NS], and [ST].

The question is now what happens when the domain in question has "flat" points. (Here we shall restrict attention to two complex variables). For example, in a domain like $\operatorname{Im} z_1 > |z_2|^4$, clearly one can insert discs $\operatorname{Im} z_1 = \varepsilon$ $|\operatorname{Re} z_1| < \varepsilon$, $|z_2| < C\,\varepsilon^{1/4}$. So if f is bounded and holomorphic in

$\operatorname{Im} z_1 > |z_2|^4$, $\left|\frac{\partial f}{\partial z_2}(\varepsilon,0)\right| < \frac{C}{\varepsilon^{1/4}}$. So the approach regions to the origin will be bigger than those for $\operatorname{Im} z_1 > |z_2|^2$, and the balls around the origin on the boundary of the domain will be stretched more. Of course we must decide what kind of domain we want to allow. One standard generalization of strict pseudo convexity is that of pseudo convexity-and this is the right notion for many kinds of problems. (We shall not enter into a discussion of this notion since it will in fact not concern us). However recent work of J. Kohn [K] suggests that a notion called finite type might be more appropriate for analytic estimates. We will now explain what it means for a domain $\mathcal{D}$ to be of finite type m at a point P in $\partial\mathcal{D}$. ($\mathcal{D}$ will be said to be of finite type m if it is of type m at each point of $\partial\mathcal{D}$ and m is the smallest integer for which the property holds). Let L be a Levy vector field. Denote by X and Y the real and imaginary parts of L. Then for $P \in \partial\mathcal{D}$ we say is of type m at P if X, Y, $[X,Y]$, and all commutators of X and Y of length at most m span the tangent space to $\partial\mathcal{D}$ at P and this is not true for all commutators of length j with $j < m$. (X and Y are commutators of length 1, $[X,Y]$ is of length 2, $[X,[X,Y]]$ is of length 3) etc. If $\mathcal{D}$ is of type 2, $\mathcal{D}$ is strictly pseudo convex.

Bloom and Graham [BG] have shown that a point P on $\partial\mathcal{D}$ is of type m if and only if the holomorphic image of a disc has order of contact with $\mathcal{D}$ at P of order m, and no disc has order of contact greater than m.

Our point of departure is now to find the balls appropriate for the study of problems of type 1 and type 2 in domains of finite type.

Here we face two problems: First if we consider a domain such as $\operatorname{Im} z_1 > |z_2|^{2k}$, it is pretty clear that the balls at the origin will have length ε in a short direction and $\varepsilon^{1/2k}$ in long directions. So we will have to consider balls which are of the form ε in some direction and $\varepsilon^{1/2k}$ in others. Moreover these balls will be turning. So to see if property C) for balls holds easily, Stein asked himself the following question about rectangles in $\mathbb{R}^2$. Suppose at each point we have a rectangle of size ε along a short axis and $\varepsilon^{1/2k}$ along a long axis and the axis are turning smoothly from point to point, Does the analogue of property C) hold? The answer is yes if $k = 1$, (the strictly pseudo convex case), but the answer is no if $k > 1$. So to prove property C) for our balls, we shall have to study very carefully how the balls are turning.

The second problem (which we perhaps should have called the first problem) is how to define the balls any way. One can not make the definition for example by intersecting planes with the boundary of the domain and then projecting on the boundary because the domain need not be convex (or even biholomorphically equivalent to a convex domain). We had some guide to the correct definition of the balls because in certain special cases (that is for the domains $\operatorname{Im} z_1 > |z_2|^{2k}$) natural singular integrals had been constructed by Greiner and Stein [GS2].

We were able to formulate a probable definition in terms of the defining function $r(z_1,z_2)$ or more precisely in terms of the "polarization" of r. See [NSW]. However we were unable to prove property C) in general. Largely this was because we couldn't find a good coordinate system in which to make computations. However we noticed that our definition has an equivalent formulation in terms of vector fields. In order to state this definition let us denote by X and Y the real and imaginary parts of a Levy vector field L on the boundary of a domain $\mathcal{D}$.

We let T be a vector field on the boundary of $\mathcal{D}$ tangent to the boundary of $\mathcal{D}$ so that X, Y and T form a basis for the tangent space to $\partial\mathcal{D}$ at each point of $\partial\mathcal{D}$. We then form 2^j functions λ_j^i by writing

$$\underbrace{[W_1, [W_2, [W_3, \ldots [W_{j-1}, W_j], \ldots]}_{\text{commutator of length } j}$$

$$= \lambda_j^i\, T + \text{something} \,.\, X + $$

$$+ \text{something else} \,.\, Y$$

where each W_i is either X or Y.

We then form a function

$$\Lambda(P,\delta) = \Lambda(\delta) = \sum_{j=2}^{m} \Big(\sum_{i=1}^{2^j} |\lambda_j(P)| \Big)\, \delta^j ,$$

where P is a point of the boundary of $\mathcal{D}$. Finally we write

$$(*) \qquad B(P,\delta) = \{Q \text{ in } \partial\mathcal{D} \mid Q = \exp\,(\alpha X + \beta Y + \gamma T)P$$
$$\text{with } |\alpha| < \delta, \quad |\beta| < \delta, \quad |\gamma| < \Lambda(P,\delta)\}$$

(If V is a vector field we say

$$Q = (\exp V)P$$

If there is a curve $\phi(t)$ with

$$\dot{\phi}(t) = V\phi(t), \qquad \phi(0) = P$$

and $\phi(1) = Q$).

Having found this definition of ball, we almost automatically also found a good coordinate system in which to compute-namely canonical coordinates. This means that for Q near P we could choose as coordinates of Q α, β, and γ such that

$$Q = \exp\,(\alpha\, X + \beta\, Y + \gamma\, T).$$

We were then able to prove property C for our balls. (Properties A, B, and D are obvious).

Essentially the definition (*) means that we are obtaining balls on the boundary of $\mathcal{D}$ as images under an appropriate mapping (the exponential map) of balls in the tangent space to the boundary of $\mathcal{D}$. However this mapping is not any old pushing of balls from the tangent space of $\partial\mathcal{D}$ to $\partial\mathcal{D}$, but it is twisting these balls in a very careful way. This careful twisting becomes very important as soon as the type of $\mathcal{D}$ becomes greater than 2. (That is when $\mathcal{D}$ is not strictly pseudo convex). This notion of ball enabled us to solve problems of type 1 in domains of finite type in $\mathbb{C}^2$. See [NSW]. Problems of type 2) are unsolved because no one knows (except in special cases) a good formula or approximate formula for the various singular integral kernels in the case of non-strictly pseudo convex domains.

This is however not the end of the story because of an operator introduced by Hörmander [H]. Suppose $X_1,\ldots,X_n$ are vector fields on say $\mathbb{R}^N$. Assume that $X_1,\ldots,X_n$ together with their commutators of length $\leq m$, for some fixed m, span $\mathbb{R}^N$ at every point in $\mathbb{R}^N$. Let

$$H = X_1^2 + X_2^2 + \ldots + X_n^2$$

Hörmander showed that H is hypoelliptic. (i.e, if $Hf = g$ and g is smooth so is f).

Further refined estimates including a "formula" for the fundamental solution of H were obtained by Rothschild and Stein [RS]. This suggested that kernels associated to H should have estimates in terms of an appropriate family of balls. Now we can not use a definition like (*) because there may not be N vector fields among the $X_1,\ldots,X_n$ and its commutators which form a basis for $\mathbb{R}^N$ at every point. However we were able to reformulate the definition (*) so that it would make sense in the more general setting. To facilitate this definition, let $Y_1,Y_2,\ldots,Y_g$ denote some ordering of $X_1,\ldots,X_n$ together with their commutators of length $\leq m$. We say Y_j has degree d_j if Y_j arises as a commutator of $X_1,\ldots,X_n$ of length d_j. Then our equivalent definition becomes

$$(**)\quad B(x,\delta) = \{y \mid \text{there is a curve } \phi(t) \text{ with } \phi(0) = x, \; \phi(1) = y \text{ and } \phi'(t) = \sum_{j=1}^{q} a_j(t)\, Y_j(\phi(t))$$

with

$$a_j < \delta^{\deg Y_j}\}$$

Actually with the definition (**) property C) became obvious, but it became hard to prove property D). One had to find a good coordinate system in which to compute volumes of balls. It seemed as though at each point x one should use canonical coordinates at x, but this required a choice of N vectors $Y_{i_1} \dots Y_{i_N}$ which span $\mathbb{R}^N$ at x, and the question was how to make the choice. Now suppose we take any N of the Y's say $Y_{i_1} \dots Y_{i_N}$. Let

$$B_{i_1 \dots i_N} = \{\omega \mid \omega = \exp(\alpha_1 Y_{i_1} + \dots + \alpha_N Y_{i_N})\, x \text{ with } |\alpha_j| < \delta^{\deg Y_j}\}.$$

Then

$$\operatorname{Vol} B_{i_1 \dots i_N} = \delta^{\deg Y_{i_1} + \dots + \deg Y_{i_N}} \cdot J_{i_1 \dots i_N}$$

where $J_{i_1 \dots i_N}$ is an appropriate Jacobian. Now suppose one guesses that

$$\operatorname{Vol} B(x,\delta) = \sum_{\substack{\text{all } N\text{-tuples} \\ i_1, \dots, i_N}} \operatorname{Vol} B_{i_1 \dots i_N}.$$

Then one might expect that if the distance (in terms of our balls) from x to y were aproximately δ, a good choice for $Y_{i_1} \dots Y_{i_N}$ would be a choice that maximizes

$$\delta^{\deg Y_{i_1} + \dots + \deg Y_{i_N}} \; J_{i_1 \dots i_N}$$

This is the right choice give or take a constant. We could then show by some very intricate calculations that property D) holds for our balls and that satisfactory estimates could be given to kernels associated to H. One final modification was necessary to estimate kernels associated to operators such as

$$\Delta = X_1 + X_2^2 + \dots + X_n^2$$

(notice X_1 is not squared). This operator requires balls in which the definition of $\deg Y_j$ must be slightly altered. For example X_1 would have degree 2 even though it is a commutator of length one, and $[X_1, X_2]$ would have degree three. With this final modification of our notion of ball, we were able to give satisfactory estimates for kernels associated to Δ.

We close by pointing out some recent related results in the mathematical literature. A definition of ball (or distance) equivalent to ours was introduced by Folland and Hung [FH] and a more general definition was given by C. Fefferman and Phong [FP]. Finally estimates on H similar to ours were obtained by Sanchez [SA] by different methods.

References

[BG] T. Bloom and I. Graham, "A geometric characterization of points of type m on real submanifolds of $\mathbf{C}^n$". Journal of Diff. Geom. Vol. 12 (1977), pp 171-182.

[CW] R. Coifman and G. Weiss, Analyse harmonique non-commutative sur certain espaces homogenes, Lecture Notes in Mathematics, No. 242 (Springer Verlag, New York).

[FP] C. Fefferman and D. Phong, "Subelliptic eigenvalue problems" in Conference on Harmonic Analysis in Honor of Antoni Zygmund, vol. 2, pp. 590-606, Wadsworth, 1983.

[FH] G. Folland and H. Hung, "Non-isotropic Lipschitz spaces" in Harmonic Analysis in Euclidean Spaces, Amer. Math. Soc. Part 2, pp. 391-397, Providence, 1979.

[FS] G. Folland and E. Stein, "Estimates for the $\bar{\partial}_b$ complex and analysis on the Heisenberg group", Comm. Pure Appl. Math., vol. 27 (1974), pp. 429-522.

[GS] P. Greiner and E. Stein, Estimates for the $\bar{\partial}$-Neuman Problem Mathematical Notes (Princeton Univ. Press) 1976.

[GS2] P. Greiner and E. Stein, "On the solvability of some differential operators of type $\square_b$" in Several Comples Variables, Proceedings of International Conferences Cortona, Italy, 1976-1977, pp. 106-165.

[H] L. Hörmander, "Hypoelliptic second order differential equations" Acta Math. 119 (1967), pp. 147-171.

[KS] N. Kerzman and E. Stein, "The Cauchy kernel in terms of Cauchy-Fantappie kernels", Duke Math. J. 45 (1978) pp. 197-224
and
N. Kerzman and E. Stein, "The Cauchy Kernel, the Szëgo kernel and the Riemann mapping function", Math. Ann.

[K] J. Kohn, "Boundary Behaviour of $\bar{\partial}$ on Weakly Pseudoconvex Manifolds of dimension Two", Jour. Diff. Geom. 6 (1972), pp. 523-42.

[KR] S. Krantz, "Intrisic Lipschitz Classes on Manifolds with Applications to Complex Function Theory and Estimates for the $\bar{\partial}$ and $\bar{\partial}_b$ Equations". Manuscripta Math. 24 (1978) pp. 351-78.

[KV] A. Koranyi and S. Vagi, "Singular integrals in homogeneous spaces and some problems of classical analysis", Ann. Scuola Norm. Sup. Pisa, 25 (1971), pp. 575-649.

[NS] A. Nagel and E. Stein, "Lecture notes on pseudo-differential operators", Math. notes 24, 1979, Princeton Univ. Press.

[NSW] A. Nagel, E. Stein, and S. Wainger, "Boundary behaviour of functions holomorphic in domains of finite type" Proc. Nat. Acad. Sci. Vol. 78 (1981), pp. 6596-6599.

[NSW2] A. Nagel, E. Stein, and S. Wainger, "Balls and metrics defined by vector fields I: Basic Properties" (to appear).

[PS] D. Phong and E. Stein, "Estimates for the Bergman and Szëgo projections on strongly pseudoconvex domains", Duke Math. J. 44 (1977), pp. 695-704.

Also "D. Phong and E. Stein"
"Some further classes of pseudo differential and singular-integral operators arising in boundary-value problem I composition of operators", Amer. J. Math. Vol. 104, 1982, pp. 141-172.
Also "D. Phong and E. Stein"
To appear in the Proc. of Nat. Acad. of Sci.

[RS] L. Rothschild and E. Stein, "Hypoelliptic differential operators and nilpotent groups", Acta Math. 137 (1976), pp. 247-320.

[SA] A. Sanchez, "Estimates for kernels associated to some subelliptic operators", Thesis, Princeton Univ.

[S] E. Stein, Boundary Behaviour of Holomorphic Functions of Several Complex Variables, Math. Notes Series 11 Princeton Univ. Press, 1972.

[S1] E. Stein, "Singular Integrals and estimates for the Cauchy Riemann equations", Bull. Amer. Math. Soc. 29 (1973), pp. 440-445.